R for Business Analytics

A. Ohri

R for Business Analytics

Springer

A. Ohri
Founder-Decisionstats.com
Delhi, India

ISBN 978-1-4614-4342-1 ISBN 978-1-4614-4343-8 (eBook)
DOI 10.1007/978-1-4614-4343-8
Springer New York Heidelberg Dordrecht London

Library of Congress Control Number: 2012941362

Printed on acid-free paper

Springer is part of Springer Science+Business Media (www.springer.com)

Dedicated to Dad (A. K. Ohri)
and Father (Jesus Christ)
and my 5 year old son Kush

Foreword

I basically structured the book according to the tasks that I have been doing most frequently in my decade-long career as a business analyst. The chapters are thus divided into most frequently used tasks, and I have added references to multiple sources to help the reader explore a particular subject in more depth. Again, I emphasize that this is a business analytics book, not for statistics, and my own experience as an MBA and with the literature available for MBAs in business analytics (particularly R) led me to these choices.

This book is thus organized for a business analyst rather than a statistician. It will not help you get a better grade on your graduate school thesis, but it will definitely help you get, or retain, a job in analytics. If you are a student studying R, it should help you do your homework faster.

In the current business environment, I believe that focus will shift back to the analyst rather than the software tool, and having multiple platform skills, especially in both high-end and low-cost analytical platforms, can be of some benefit to the user.

This book will focus explicitly on graphical user interfaces, tricks, tips, techniques, and shortcuts, and focus on case studies from the most commonly used tasks that a business analyst will face on a day-to-day basis. Things will be made as simple as possible but no simpler than that. Each chapter will have a case study, tutorial, or example problem. Functions and packages used in each chapter will be listed at the end to help the reader review. There might be times when some functions appear to have been repeated or stressed again; this has more to do with their analytical use and necessity. The brief interviews with creators, authors, and developers is aimed at making it easy for the business analyst to absorb aspects of R. The use cases of existing R deployments are designed to help decision makers within the analytics community to give R a chance, if they have not done so already.

The book has a pragmatic purpose and is aimed at those using or hoping to use R in a corporate business environment. Adequate references will be provided to help the reader with theoretical aspects or advanced levels and to assist the reader on his or her journey in R with other available resources.

Readers are encouraged to skip chapters that have no immediate relevance to them and go directly to those that are of maximum utility for their purposes. One issue that I faced was that the R project released almost four new versions of the software by the time I wrote the book, so please feel free to let me know about any inaccuracies or errata.

Organization of the Book

Chapter 1: Why R? Reasons for Using R in Business Analytics. In this chapter we discuss reasons for choosing R as an analytical and not just a statistical computing platform, comparisons with other analytical software, and some general costs and benefits in using R in a business environment. It lists the various reasons R should be chosen for learning by business analysts and the additional features that R has over other analytical platforms. The benefits of R are subdivided into three major categories: business analytics, data mining, and business intelligence/data visualization.

Chapter 2: R Infrastructure: Setting up Your R Analytical Infrastructure. In this chapter we discuss the practical realities in setting up an analytical environment based on R, including which hardware, operating system, additional software, requisite budgeting, and the training and software support needs. We discuss the various operating systems, hardware choices, and various providers of R-based solutions. We walk through the basics of installing R, R's library of packages, updating R, and accessing the comprehensive user help and a simple basic tutorial for starting R.

Chapter 3: R Interfaces: Ways to Use Your R Analytics Based on Your Needs. In this chapter we compare and contrast the various ways to interact with the R analytical platform. R can be used in a command line, with graphical user interfaces (GUIs), and via Web interfaces (including cloud computing interfaces). The chapter outlines both the advantages and disadvantages of a GUI-based approach. The chapter compares the features of nine kinds of GUI, including a summary sheet of comparative advantages and disadvantages. It also discusses using R from other software and from Web interfaces. Lastly, the chapter includes some useful tutorials on running R from the Amazon cloud.

Chapter 4: Manipulating Data: Obtaining Your Data Within R. This chapter talks of various ways of obtaining data within R, including the basic syntax. It deals specifically with data in databases as this is often the case for business data. The chapter shows how the user can connect to MySQL and Pentaho databases, which are the two leading open source databases used. Specific sections are devoted to using SQLite with R and to business intelligence practitioners. We take a brief look at Jaspersoft and Pentaho, the leading open source solutions within BI, and how they interact with R. While the chapter briefly mentions additional resources to handle larger datasets, it also gives a series of common analytical tasks that a business analyst is expected to perform on any data to help someone transitioning to R.

Chapter 5: Exploring Data—The Booming Business in Data Visualization. This chapter discusses exploring data in R using visual and graphical means. It talks about the basic graphs and a series of advanced graphs in R that can be easily created by an average programmer learning R in a very short time. It introduces a specialized GUI for data exploration, grapheR and Deducer, and also features ggplot2 creator Hadley Wickham 's interview. Aspects of graphs include code, easy-to-recreate examples, and information on interactive graphs. The chapter is aimed at demystifying the sometimes intimidating art of data visualization for someone who has been creating graphs mostly using spreadsheet programs.

Chapter 6: Building Regression Models. Regression models are the statistical workhorses of the business analytics industry. They are used perhaps to the point of overuse because of the inherent simplicity in communicating them to business audiences internally. We learn how to build linear and logistic regression models, study some risk models and scorecards, and discuss PMML as the way to implement models. We also have a brief case study for simplifying the process of building logistic models in R, and no theory of regression is introduced, in keeping with the business analytics focus of the book.

Chapter 7: Data Mining Using R. Data mining using R employs the Rattle GUI for simplifying and speeding the process of data modeling in R. However, it begins by introducing the concepts of information ladder and various data mining methodologies to the reader including, briefly, CRISP-DM, SEMMA, and KDD. It also features extracts from brief interviews with two authors who have written books on data mining using R. As special cases, text mining, Web mining, and the Google Prediction application interface (API) are featured.

Chapter 8: Data Segmentation. Data segmentation in this book deals mainly with cluster analysis, and we walk through the various types of clustering. Clustering is added here because of the inherent and increasing need for data reduction techniques in business environments of big data and because the size of datasets is increasing rapidly. We refer once again to the Rattle GUI but also briefly touch on other clustering GUIs in R. A small case study is presented regarding the use of Revolution R for clustering large amounts of data.

Chapter 9: Forecasting and Time Series. While business uses business intelligence and reporting for knowing the past and present of operations, their focus is to improve decision making for the future. Powerful but underutilized in many businesses, time series and forecasting are explained here with a case study and an R Commander epack GUI. I have tried to make it a practical chapter to help your business team do more forecasting utilizing the nascent data present in all organizations.

Chapter 10: Data Export and Output. Obtaining results for analytics is only part of the job. Output should be presented in a manner that convinces decision makers to take actionable decisions R provides many flexible ways to generate and embed output, and they are presented here.

Chapter 11: Optimizing R Coding. Now that you have learned how to functionally use R for business analytics, the next step is to understand how to utilize its powerful flexibility without drowning in the huge library available. This chapter discusses tips, tweaks, and tools including code editors to help you code better and faster.

Chapter 12: Additional Training Literature. This chapter is meant as a follow-up for readers interested in expanding their R knowledge and in a more structured view of the documentation environment of R.

Chapter 13: Case Studies Using R. This chapter presents coding case studies based on various business uses, including Web analytics, and is designed to help readers find a ready reference for using R for business analytics in their operational business context.

Preface

There are 115 books listed on the R project Web site, more books have been published since that list was last updated, and if you are wondering why one more is needed, well, I am here to explain why. I have been working in the field known as business analytics for nearly a decade now. You may the know the field as statistical computing, data mining, business intelligence, or, lately, data science. I generally prefer the term decision sciences, but overall the field refers to using data, statistical techniques, and specialized tools to assist decision makers in government, research, and industry with insights to maximize positive outcomes and minimize costs.

I found the field of business analytics to be both very enjoyable and lucrative from a career point of view. Selling credit cards and loans and getting more revenue from the same people was a lot of fun, at least in the early years for me. The reason I found business analytics interesting was that it combined the disciplines of investigative and diligent thinking with business insights.

I also found the field of business analytics a bit confusing. There were two kinds of people with almost opposing views on how to apply business analytics: technically minded people (like computer science and statistics majors) who aimed for robust results and business-minded people (like MBAs) who aimed for revenue and quicker results. I was caught in the middle of the occasional crossfire.

When I started learning all this—in 2001—the predominant platform was the SPSS and SAS languages. In 2007, following the birth of my first son, I discovered the R programming language almost as a necessity to keep my fledgling consulting shop open as I needed a reliable analytics platform without high annual fees. In launching an analytics startup at the age of 30, I found that I could not afford the tools that I had been using in some of the world's largest corporations (and I discovered that there were no separate discounts for small enterprises in India). So I moved to R because it was free. Or so I thought.

But R took me a lot of time to learn, and time is not free. Something was wrong – either with me, or the language, or the whole universe. In the meantime, I started writing an analytics blog called DecisionStats (http://decisionstats.com) to network with other business analysts, and with almost 20,000 views every month and nearly

100 interviews with leading analytics practitioners, I slowly obtained more practical insights into the field of business analytics.

Over a period of time, as I slowly immersed myself in R, I discovered that it, like all languages, had its own set of tricks and techniques with tradeoffs to get things done faster, and I did need not to memorize huge chunks of code or be overawed by those who could. I was aided in this by a few good books including R for SAS and SPSS users by Bob Muenchen, great blogs like those by David Smith and Tal G. Overall, I learned R at a much faster pace than I had expected to initially, thanks to Prof. John Fox who made R Commander, Prof. Graham Williams who made Rattle, and many others who have helped to make R what it is today. The R community exploded in size, acceptability, and organization, and I was no longer fighting a lonely battle to learn R on my own for analytical purposes.

In 2012, R has grown at a pace I could not have imagined in 2007. I am both humbled and blessed to know that R is the leading statistical computing language and used and completely supported by leading technology companies including Microsoft, SAS Institute, Oracle, Google, and others. *No R&D budget can compete with nearly ALL the statistics departments of the world and their professors working for free on this project.*

If you are a decision maker thinking of using R in combination with your existing analytical infrastructure, you will find the brief interviews of various partners and contributors to R to be very enlightening and helpful. They have specifically been added to increase the book's readability for a business analytics audience that may prefer reading English to code.

I would like to thank Anne Milley of JMP, Jill Dyche of Baseline Consulting, Bob Muenchen of the University of Tennessee, Karl Rexer of Rexer Analytics, and Gregory Shapiro of KD Nuggets specifically for their help in mentoring me through my analytical wanderings. I would also like to thank Marc Strauss and Susan Westendorf of Springer USA and Leo Augustine for their help.

I am humbled by the patient readers of Decisionstats and appreciate the attention that the R and broader analytics community has shown towards the author.

Delhi, India A. Ohri

Contents

1 Why R 1

1.1 Reasons for Classifying R as a Complete Analytical Environment 1

1.2 Additional Advantages of R over Other Analytical Packages 2

1.3 Differences Between R as a Statistical Language and R as an Analytical Platform 3

1.4 Costs and Benefits of Using R 3

1.4.1 Business Analytics 3

1.4.2 Data Mining 4

1.4.3 Business Dashboards and Reporting 4

1.5 Using SAS and R Together 5

1.6 Brief Interview: Using R with JMP 5

2 R Infrastructure 9

2.1 Choices in Setting up R for Business Analytics 9

2.1.1 Licensing Choices: Academic, Free, or Enterprise Version of R 9

2.1.2 Operating System Choices 10

2.1.3 Operating System Subchoice: 32- or 64-bit 11

2.1.4 Hardware Choices: Cost-Benefit Tradeoffs for Additional Hardware for R 11

2.1.5 Interface Choices: Command Line Versus GUI. Which GUI Should You Choose as the Default Startup Option? 12

2.1.6 Software Component Choice 13

2.1.7 Additional Software Choices 13

2.2 Downloading and Installing R 13

2.3 Installing R Packages 14

2.4 Starting up Tutorial in R 20

2.5 Types of Data in R 21

2.6 Brief Interview with John Fox, Creator of Rcmdr GUI for R 21

2.7 Summary of Commands Used in This Chapter 23
2.7.1 Packages 23
2.7.2 Functions 23

3 R Interfaces 25
3.1 Interfaces to the R Statistical Language 25
3.2 Basic R 26
3.3 Advantages and Limitations of Graphical User Interfaces to R 27
3.3.1 Advantages of Using GUIs for R 27
3.3.2 Limitations of Using GUIs for R 27
3.4 List of GUI 27
3.4.1 R Commander 28
3.4.2 R Commander E Plugins or Extensions 31
3.5 Summary of R GUIs 34
3.6 Using R from Other Software 34
3.6.1 RExcel: Using R from Microsoft Excel 36
3.7 Web Interfaces to R 37
3.8 Interview: Using R as a Web-Based Application 40
3.9 Cloud Computing to Use R 41
3.9.1 Benefits of R on the Cloud 42
3.9.2 Tutorial: Using Amazon EC2 and R (Linux) 42
3.9.3 Tutorial: Using Amazon EC2 and R (Windows) 44
3.9.4 Installing R on a Basic Linux Amazon EC2 Instance (Which Is Free) 46
3.9.5 Using R Studio on Amazon EC2 46
3.9.6 Running R on the Cloud Using cloudnumbers.com 47
3.10 Google and R 47
3.10.1 Google Style Guide 47
3.10.2 Using R at Google 48
3.10.3 Using Google Services and R Packages 50
3.11 Interview: Using R at Google 51
3.12 Interview: Using R Through Cloud Computing at cloudnumbers.com 53
3.13 Summary of Commands Used in This Chapter 54
3.13.1 Packages 54

4 Manipulating Data 57
4.1 Challenges of Analytical Data Processing 57
4.1.1 Data Formats 57
4.1.2 Data Quality 58
4.1.3 Project Scope 58
4.1.4 Output Results vis-à-vis Stakeholder Expectation Management 58
4.2 Methods for Reading in Smaller Dataset Sizes 59
4.2.1 CSV and Spreadsheets 59

4.2.2 Reading Data from Packages 61
4.2.3 Reading Data from Web/APIs 61
4.2.4 Missing Value Treatment in R 62
4.2.5 Using the as Operator to Change the Structure of Data 63
4.3 Some Common Analytical Tasks 64
4.3.1 Exploring a Dataset 64
4.3.2 Conditional Manipulation of a Dataset 65
4.3.3 Merging Data 69
4.3.4 Aggregating and Group Processing of a Variable 70
4.3.5 Manipulating Text in Data 72
4.4 A Simple Analysis Using R 73
4.4.1 Input 73
4.4.2 Describe Data Structure 73
4.4.3 Describe Variable Structure 74
4.4.4 Output 76
4.5 Comparison of R Graphical User Interfaces for Data Input 76
4.6 Using R with Databases and Business Intelligence Systems 77
4.6.1 RODBC 78
4.6.2 Using MySQL and R 78
4.6.3 Using PostGresSQL and R 84
4.6.4 Using SQLite and R 85
4.6.5 Using JasperDB and R 86
4.6.6 Using Pentaho and R 87
4.7 Summary of Commands Used in This Chapter 88
4.7.1 Packages 88
4.7.2 Functions 88
4.8 Citations and References 89
4.9 Additional Resources 89
4.9.1 Methods for Larger Dataset Sizes 90

5 Exploring Data 103
5.1 Business Metrics 103
5.2 Data Visualization 104
5.3 Parameters for Graphs 105
5.4 Creating Graphs in R 106
5.4.1 Basic Graphs 106
5.4.2 Summary of Basic Graphs in R 118
5.4.3 Advanced Graphs 119
5.4.4 Additional Graphs 140
5.5 Using ggplot2 for Advanced Graphics in R 151
5.6 Interactive Plots 152
5.7 GrapheR: R GUI for Simple Graphs 154
5.7.1 Advantages of GrapheR 155
5.7.2 Disadvantages of GrapheR 155

5.8 Deducer: GUI for Advanced Data Visualization 156
5.8.1 Advantages of JGR and Deducer......................... 157
5.8.2 Disadvantages of Deducer............................... 160
5.8.3 Description of Deducer................................. 161
5.9 Color Palettes 161
5.10 Interview: Hadley Wickham, Author of *ggplot2: Elegant Graphics for Data Analysis* 164
5.11 Summary of Commands Used in This Chapter.................... 165
5.11.1 Packages.. 165
5.11.2 Functions 166

6 Building Regression Models 171
6.1 Linear Regression.. 171
6.2 Logistic Regression 172
6.3 Risk Models.. 172
6.4 Scorecards... 172
6.4.1 Credit Scorecards 173
6.4.2 Fraud Models 173
6.4.3 Marketing Propensity Models........................... 173
6.5 Useful Functions in Building Regression Models in R 173
6.6 Using R Cmdr to Build a Regression Model 174
6.7 Other Packages for Regression Models............................ 188
6.7.1 ROCR for Performance Curves 188
6.7.2 rms Package... 188
6.8 PMML.. 189
6.8.1 Zementis: Amazon EC2 for Scoring.................... 189
6.9 Summary of Commands Used in This Chapter.................... 190
6.9.1 Packages... 190
6.9.2 Functions 191

7 Data Mining Using R.. 193
7.1 Definition.. 193
7.1.1 Information Ladder 194
7.1.2 KDD 194
7.1.3 CRISP-DM... 195
7.1.4 SEMMA.. 196
7.1.5 Four Phases of Data-Driven Projects 197
7.1.6 Data Mining Methods 198
7.2 Rattle: A GUI for Data Mining in R 199
7.2.1 Advantages of Rattle.................................... 199
7.2.2 Comparative Disadvantages of Using Rattle 201
7.2.3 Description of Rattle.................................... 201
7.3 Interview: Graham Williams, Author of *Data Mining with Rattle and R*.. 212

7.4 Text Mining Analytics Using R ... 214
7.4.1 Text Mining a Local Document ... 214
7.4.2 Text Mining from the Web and Cleaning Text Data ... 215
7.5 Google Prediction API ... 220
7.6 Data Privacy for Data Miners ... 222
7.7 Summary of Commands Used in This Chapter ... 222
7.7.1 Packages ... 222
7.7.2 Functions ... 223

8 Clustering and Data Segmentation ... 225
8.1 When to Use Data Segmentation and Clustering ... 225
8.2 R Support for Clustering ... 225
8.2.1 Clustering View ... 225
8.2.2 GUI-Based Method for Clustering ... 226
8.3 Using RevoScaleR for Revolution Analytics ... 226
8.4 A GUI Called Playwith ... 227
8.5 Cluster Analysis Using Rattle ... 229
8.6 Summary of Commands Used in This Chapter ... 239
8.6.1 Packages ... 239
8.6.2 Functions ... 239

9 Forecasting and Time Series Models ... 241
9.1 Introduction to Time Series ... 241
9.2 Time Series and Forecasting Methodology ... 241
9.3 Time Series Model Types ... 246
9.4 Handling Date-Time Data ... 247
9.5 Using R Commander GUI with epack Plugin ... 248
9.5.1 Syntax Generated Using R Commander GUI with epack Plugin ... 255
9.6 Summary of Commands Used in This Chapter ... 256
9.6.1 Packages ... 256
9.6.2 Functions ... 256

10 Data Export and Output ... 259
10.1 Summary of Commands Used in This Chapter ... 262
10.1.1 Packages ... 262
10.1.2 Functions ... 262

11 Optimizing R Code ... 263
11.1 Examples of Efficient Coding ... 263
11.2 Customizing R Software Startup ... 265
11.2.1 Where is the R Profile File? ... 265
11.2.2 Modify Settings ... 266
11.3 Code Editors ... 266
11.4 Advantages of Enhanced Code Editors ... 272
11.5 Interview: J.J. Allaire, Creator of R Studio ... 273

11.6 Revolution R Productivity Environment 274
11.7 Evaluating Code Efficiency 275
11.8 Using system.time to Evaluate Coding Efficiency 277
11.9 Using GUIs to Learn and Code R Faster 278
11.10 Parallel Programming 278
11.11 Using Hardware Solutions 279
11.12 Summary of Commands Used in This Chapter 279
11.12.1 Packages 279
11.12.2 Functions 280

12 Additional Training Literature 281
12.1 Cran Views 281
12.2 Reading Material 282
12.3 Other GUIs Used in R 283
12.3.1 Red-R: A Dataflow User Interface for R 283
12.3.2 RKWard 284
12.3.3 Komodo Sciviews-K 288
12.3.4 PMG (or Poor Man's GUI) 289
12.3.5 R Analytic Flow 290
12.4 Summary of Commands Used in This Chapter 290
12.4.1 Packages 290
12.4.2 Functions 290

13 Appendix 293
13.1 Web Analytics Using R 293
13.1.1 Google Analytics with R 293
13.2 Social Media Analytics Using R 295
13.2.1 Using Facebook Data with R 295
13.2.2 Using Twitter Data with R 297
13.3 RFM Analytics Using R 299
13.4 Propensity Models using R 300
13.5 Risk Models in Finance Using R 300
13.6 Pharmaceutical Analytics Using R 300
13.7 Selected Essays on Analytics by the Author 301
13.7.1 What Is Analytics? 301
13.7.2 What Are the Basic Business Domains Within Analytics? 302
13.8 Reasons a Business Analyst Should Learn R 304
13.9 Careers in Analytics 305
13.10 Summary of Commands Used in This Chapter 307
13.10.1 Packages 307
13.10.2 Functions 307

Index 309

Chapter 1
Why R

Chapter summary: In this chapter we introduce the reader to R, discuss reasons for choosing R as an analytical and not just a statistical computing platform, make comparisons with other analytical software, and present some broad costs and benefits in using R in a business environment.

R is also known as GNU S, as it is basically an open source derivative and descendant of the S language. In various forms and avatars, R has been around for almost two decades now, with an ever growing library of specialized data visualization, data analysis, and data manipulation packages. With around two million users, R has one of the largest libraries of statistical algorithms and packages.

While R was initially a statistical computing language, in 2012 you could call it a complete analytical environment.

1.1 Reasons for Classifying R as a Complete Analytical Environment

R may be classified as a complete analytical environment for the following reasons.

- Multiple platforms and interfaces to input commands: R has multiple interfaces ranging from command line to numerous specialized graphical user interfaces (GUIs) (Chap. 2) for working on desktops. For clusters, cloud computing, and remote server environments, R now has extensive packages including SNOW, RApache, RMpi, R Web, and Rserve.
- Software compatibility: Official commercial interfaces to R have been developed by numerous commercial vendors including software makers who had previously thought of R as a challenger in the analytical space (Chap. 4). Oracle, ODBC, Microsoft Excel, PostgreSQL, MySQL, SPSS, Oracle Data Miner, SAS/IML, JMP, Pentaho Kettle, and Jaspersoft BI are just a few examples of commercial

A. Ohri, *R for Business Analytics*, DOI 10.1007/978-1-4614-4343-8_1,

software that are compatible with R usage. In terms of the basic SAS language, a WPS software reseller offers a separate add-on called the Bridge to R. Revolution Analytics offers primarily analytical products licensed in the R language, but other small companies have built successful R packages and applications commercially.
- Interoperability of data: Data from various file formats as well as various databases can be used directly in R, connected via a package, or reduced to an intermediate format for importing into R (Chap. 2).
- Extensive data visualization capabilities: These include much better animation and graphing than other software (Chap. 5).
- Largest and fastest growing open source statistical library: The current number of statistical packages and the rate of growth at which new packages continue to be upgraded ensures the continuity of R as a long-term solution to analytical problems.
- A wide range of solutions from the R package library for statistical, analytical, data mining, dashboard, data visualization, and online applications make it the broadest analytical platform in the field.

1.2 Additional Advantages of R over Other Analytical Packages

So what all is extra in R? The list below shows some of the additional features in R that make it superior to other analytical software.

- R's source code is designed to ensure complete custom solutions and embedding for a particular application. Open source code has the advantage of being extensively peer-reviewed in journals and the scientific literature. This means bugs will found, information about them shared, and solutions delivered transparently.
- A wide range of training material in the form of books is available for the R analytical platform (Chap. 12).
- R offers the best data visualization tools in analytical software (apart from Tableau Software's latest version). The extensive data visualization available in R comprises a wide variety of customizable graphics as well as animation. The principal reason why third-party software initially started creating interfaces to R is because the graphical library of packages in R was more advanced and was acquiring more features by the day.
- An R license is free for academics and thus budget friendly for small and large analytical teams.
- R offers flexible programming for your data environment. This includes packages that ensure compatibility with Java, Python, and C++.
- It is easy to migrate from other analytical platforms to the R platform. It is relatively easy for a non-R platform user to migrate to the R platform, and there is no danger of vendor lock-in due to the GPL nature of the source code and the open community, the GPL can be seen at http://www.gnu.org/copyleft/gpl.html.

- The latest and broadest range of statistical algorithms are available in R. This is due to R's package structure in which it is rather easier for developers to create new packages than in any other comparable analytics platform.

1.3 Differences Between R as a Statistical Language and R as an Analytical Platform

Sometimes the distinction between statistical computing and analytics does come up. While statistics is a tool- and technique-based approach, analytics is more concerned with business objectives. Statistics are basically numbers that inform (descriptive), advise (prescriptive), or forecast (predictive). Analytics is a decision-making-assistance tool. Analytics on which no decision is to be made or is being considered can be classified as purely statistical and nonanalytical. Thus the ease with which a correct decision can be made separates a good analytical platform from a not-so-good one. The distinction is likely to be disputed by people of either background, and business analysis requires more emphasis on how practical or actionable the results are and less emphasis on the statistical metrics in a particular data analysis task. I believe one way in which business analytics differs from statistical analysis is the cost of perfect information (data costs in the real world) and the opportunity cost of delayed and distorted decision making.

1.4 Costs and Benefits of Using R

The only cost of using R is the time spent learning it. The lack of a package or application marketplace in which developers can be rewarded for creating new packages hinders the professional mainstream programmer's interest in R to the degree that several other platforms like iOS and Android and Salesforce offer better commercial opportunities to coding professionals. However, given the existing enthusiasm and engagement of the vast numbers of mostly academia-supported R developers, the number of R packages has grown exponentially over the past several years. The following list enumerates the advantages of R by business analytics, data mining, and business intelligence/data visualization as these have three different domains in the data sciences.

1.4.1 Business Analytics

R is available for free download.

1. R is one of the few analytical platforms that work on Mac OS.

2. Its results have been established in journals like the *Journal of Statistical Software*, in places such as LinkedIn and Google, and by Facebook's analytical teams.
3. It has open source code for customization as per GPL and adequate intellectual protection for developers wanting to create commercial packages.
4. It also has a flexible option for enterprise users from commercial vendors like Revolution Analytics (who support 64-bit Windows and now Linux) as well as big data processing through its RevoScaleR package.
5. It has interfaces from almost all other analytical software including SAS, SPSS, JMP, Oracle Data Mining, and RapidMiner. Exist huge library of packages is available for regression, time series, finance, and modeling.
6. High-quality data visualization packages are available for use with R.

1.4.2 Data Mining

As a computing platform, R is better suited to the needs of data mining for the following reasons.

1. R has a vast array of packages covering standard regression, decision trees, association rules, cluster analysis, machine learning, neural networks, and exotic specialized algorithms like those based on chaos models.
2. R provides flexibility in tweaking a standard algorithm by allowing one to see the source code.
3. The Rattle GUI remains the standard GUI for data miners using R. This GUI offers easy access to a wide variety of data mining techniques. It was created and developed in Australia by Prof. Graham Williams. Rattle offers a very powerful and convenient free and open source alternative to data mining software.

1.4.3 Business Dashboards and Reporting

Business dashboards and reporting are an essential piece of business intelligence and decision making systems in organizations.

1. R offers data visualization through ggplot, and GUIs such as Deducer, GrapheR, and Red-R can help even business analysts who know none or very little of the R language in creating a metrics dashboard.
2. For online dashboards R has packages like RWeb, RServe, and R Apache that, in combination with data visualization packages, offer powerful dashboard capabilities. Well-known examples of these will be shown later.
3. R can also be combined with Microsoft Excel using the R Excel package to enable R capabilities for importing within Excel. Thus an Excel user with no knowledge of R can use the GUI within the R Excel plug-in to take advantage of the powerful graphical and statistical capabilities.

4. R has extensive capabilities to interact with and pull data from databases including those by Oracle, MySQL, PostGresSQL, and Hadoop-based data. This ability to connect to databases enables R to pull data and summarize them for processing in the previsualization stage.

1.5 Using SAS and R Together

What follows is a brief collection of resources that describe how to use SAS Institute products and R: Base SAS, SAS/Stat, SAS/Graph.

- A great blog on using both SAS and R together is http://sas-and-r.blogspot.com/.
- The corresponding book *SAS and R* http://www.amazon.com/gp/product/1420070576.
- Sam Croker's paper on the use of time series analysis with Base SAS and R: http://www.nesug.org/proceedings/nesug08/sa/sa07.pdf.
- Phil Holland's paper "SAS to R to SAS," available at http://www.hollandnumerics.co.uk/pdf/SAS2R2SAS_paper.pdf, describes passing SAS data from SAS to R, using R to produce a graph, then passing that graph back to SAS for inclusion in an ODS document.
- One of the first books on R for SAS and SPSS users was by Bob Muenchen: http://www.amazon.com/SAS-SPSS-Users-Statistics-Computing/dp/0387094172.
- A free online document by Bob Muenchen for SAS users of R is at https://sites.google.com/site/r4statistics/books/free-version.
- A case study, "Experiences with using SAS and R in insurance and banking," can be found at http://files.meetup.com/1685538/R%20ans%20SAS%20in%20Banking.ppt.
- The document "Doing More than Just the Basics with SAS/Graph and R: Tips, Tricks, and Techniques" is available at http://biostat.mc.vanderbilt.edu/wiki/pub/Main/RafeDonahue/doingmore_currentversion.pdf.
- The paper "Multiple Methods in JMP® to Interact with R" can be downloaded at http://www.nesug.org/Proceedings/nesug10/po/po06.pdf.
- Official documentation on using R from within SAS/IML is available at http://support.sas.com/documentation/cdl/en/imlug/63541/HTML/default/viewer.htm#imlug_r_sect010.htm.

1.6 Brief Interview: Using R with JMP

An indicator of the long way R has come from being a niche player to a broadly accepted statistical computing platform is the SAS Institute's acceptance of R as a complementary language. What follows is a brief extract from a February 2012 interview with researcher Kelci Miclaus from the JMP division at SAS Institute that includes a case study on how adding R can help analytics organizations even more.

Ajay: How has JMP been integrating with R? What has been the feedback from customers so far? Is there a single case study you can point to where the combination of JMP and R was better than either one of them alone?

Kelci: Feedback from customers has been very positive. Some customers use JMP to foster collaboration between SAS and R modelers within their organizations. Many use JMP's interactive visualization to complement their use of R. Many SAS and JMP users use JMP's integration with R to experiment with more bleeding-edge methods not yet available in commercial software. It can be used simply to smooth the transition with regard to sending data between the two tools or to build complete custom applications that take advantage of both JMP and R.

One customer has been using JMP and R together for Bayesian analysis. He uses R to create MCMC chains and has found that JMP is a great tool for preparing data for analysis and for displaying the results of the MCMC simulation. For example, the control chart and bubble plot platforms in JMP can be used to quickly verify convergence of an algorithm. The use of both tools together can increase productivity since the results of an analysis can be achieved faster than through scripting and static graphics alone.

I, along with a few other JMP developers, have written applications that use JMP scripting to call out to R packages and perform analysis like multidimensional scaling, bootstrapping, support vector machines, and modern variable selection methods. These really show the benefit of interactive visual analysis coupled with modern statistical algorithms. We've packaged these scripts as JMP add-ins and made them freely available on our JMP User Community file exchange. Customers can download them and employ these methods as they would a regular JMP platform. We hope that our customers familiar with scripting will also begin to contribute their own add-ins so a wider audience can take advantage of these new tools (see http://www.decisionstats.com/jmp-and-r-rstats/).

Ajay: How is R a complementary fit to JMP's technical capabilities?

Kelci: R has an incredible breadth of capabilities. JMP has extensive interactive, dynamic visualization intrinsic to its largely visual analysis paradigm, in addition to a strong core of statistical platforms. Since our brains are designed to visually process pictures and animated graphics more efficiently than numbers and text, this environment is all about supporting faster discovery. Of course, JMP also has a scripting language (JSL) that allows you to incorporate SAS code and R code and to build analytical applications for others to leverage SAS, R, and other applications for users who don't code or who don't want to code. JSL is a powerful scripting language on its own.

It can be used for dialog creation, automation of JMP statistical platforms, and custom graphic scripting. In other ways, JSL is very similar to the R language. It can also be used for data and matrix manipulation and to create new analysis functions. With the scripting capabilities of JMP, you can create custom applications that provide both a user interface and an interactive visual backend to R functionality.

Alternatively, you could create a dashboard using statistical or graphical platforms in JMP to explore the data and, with the click of a button, send a portion of the data to R for further analysis.

Another JMP feature that complements R is the add-in architecture, which is similar to how R packages work. If you've written a cool script or analysis workflow, you can package it into a JMP add-in file and send it to your colleagues so they can easily use it.

Ajay: What is the official view on R at your organization? Do you think it is a threat or a complementary product or statistical platform that coexists with your offerings?

Kelci: Most definitely, we view R as complementary. R contributors provide a tremendous service to practitioners, allowing them to try a wide variety of methods in the pursuit of more insight and better results. The R community as a whole provides a valued role to the greater analytical community by focusing attention on newer methods that hold the most promise in so many application areas. Data analysts should be encouraged to use the tools available to them in order to drive discovery, and JMP can help with that by providing an analytic hub that supports both SAS and R integration.

Ajay: **Since you do use R, are there any plans to give back something to the R community in terms of your involvement and participation (say at useR events) or sponsoring contests?**

Kelci: We are certainly open to participating in useR groups. At Predictive Analytics World in New York last October, they didn't have a local useR group, but they did have a Predictive Analytics meet-up group comprised of many R users. We were happy to sponsor this. Some of us within the JMP division have joined local R user groups, myself included. Given that some local R user groups have entertained topics like Excel and R, Python and R, databases and R, we would be happy to participate more fully here. I also hope to attend the useR annual meeting later this year to gain more insight on how we can continue to provide tools to help both the JMP and R communities with their work. We are also exploring options to sponsor contests and would invite participants to use their favorite tools, languages, etc. in pursuit of the best model. Statistics is about learning from data, and this is how we make the world a better place.

Citations and References

- R Development Core Team (2010). R: A language and environment for statistical computing. R Foundation for Statistical Computing, Vienna, Austria. ISBN 3-900051-07-0. http://www.R-project.org
- SAS/IML and JMP are analytical software applications that are © 2011 SAS Institute, SAS Campus Drive, Cary, NC 27513, USA

In the next chapter we will discuss setting up the basic R infrastructure.

Chapter 2
R Infrastructure

Chapter summary: In this chapter we discuss the practical realities in setting up an analytical environment based on R, including hardware, software, budgeting, and training needs. We will also walk through the basics of installing R, R's library of packages, updating R, and accessing the comprehensive user help.

Congratulations if you decided to install R! As choices go, this is the best one in open source statistical software that you could make for at least the next decade.

2.1 Choices in Setting up R for Business Analytics

Some options await you now before you set up your new analytical environment:

2.1.1 Licensing Choices: Academic, Free, or Enterprise Version of R

You can choose between two kinds of R installations. One is free and open source and is available at http://r-project.org; the other is commercial and offered by many vendors including Revolution Analytics. However, there are other commercial vendors too.

Commercial Vendors of R Language Products:

- Revolution Analytics: http://www.revolutionanalytics.com/
- XL Solutions: http://www.experience-rplus.com/
- Information Builder: http://www.informationbuilders.com/products/webfocus/PredictiveModeling.html
- Blue Reference (Inference for R): http://inferenceforr.com/default.aspx
- R for RExcel: http://www.statconn.com/

A. Ohri, *R for Business Analytics*, DOI 10.1007/978-1-4614-4343-8_2,

Enterprise R from Revolution Analytics has a complete R Development environment for Windows including the use of code snippets to make programming faster. Revolution is also expected to make a GUI available by 2012. Revolution Analytics claims several enhancements for its version of R including the use of optimized libraries for faster performance and the RevoScaleR package that uses the xdf format to handle large datasets.

2.1.2 Operating System Choices

Which operating system should the business analyst choose, Unix, Windows, or Mac OS? Often the choice is dictated by the information technology group in the business. However, we compare some of the advantages and disadvantages of each.

1. Microsoft Windows: This remains the most widely used operating system on the planet. If you are experienced in Windows-based computing and are active on analytical projects, it would not make sense for you to move to other operating systems unless there are significant cost savings and minimal business disruption as a result of the transition. In addition, compatibility issues are minimal for Microsoft Windows, and extensive help documentation is available. However, there may be some R packages that would not function well under Windows; in that case, a multiple operating system is your next option.
2. MacOS and iOS: The reasons for choosing MacOS remain its considerable appeal in esthetically designed software and performance in art or graphics related work, but MacOS is not a standard operating system for enterprise systems or statistical computing. However, open source R claims to be quite optimized and can be used for existing Mac users.
3. Linux: This is the operating system of choice for many R users due to the fact that it has the same open source credentials and so is a much better fit for all R packages. In addition, it is customizable for large-scale data analytics. The most popular versions of Linux are Ubuntu/Debian, Red Hat Enterprise Linux, OpenSUSE, CentOS, and Linux Mint.

   (a) Ubuntu Linux is recommended for people making the transition to Linux for the first time. Ubuntu Linux had a marketing agreement with Revolution Analytics for an earlier version of Ubuntu, and many R packages can be installed in a straightforward way. Ubuntu/Debian packages are also available.
   (b) Red Hat Enterprise Linux is officially supported by Revolution Analytics for its enterprise module.

4. Multiple operating systems

Virtualization versus dual boot: if you are using more than two operating systems on your PC. You can also choose between having VMware Player from VMware (http://www.vmware.com/products/player/), if you want a virtual partition on your computer that is dedicated to R-based computing, and having a choice of operating system at startup. In addition, you can dual boot your computer with a USB installer from Ubuntu's Netbook remix (http://www.ubuntu.com/desktop/get-ubuntu/windows-installer).

A software program called wubi helps with the dual installation of Linux and Windows.

2.1.3 Operating System Subchoice: 32- or 64-bit

Given a choice between a 32-bit versus 64-bit version of an operating system like Linux Ubuntu, keep in mind that the 64-bit version would speed up processing by an approximate factor of 2. However, you need to check whether your current hardware can support 64-bit operating systems; if so, you may want to ask your information technology manager to upgrade at least some of the operating systems in your analytics work environment to 64-bit versions. Smaller hardware like netbooks do not support 64-bit Linux, whereas Windows Home Edition computers may have 32-bit version installed on it. There are cost differences due to both hardware and software. One more advantage for 64-bit computing is the support from Revolution Analytics for its version of R Enterprise.

2.1.4 Hardware Choices: Cost-Benefit Tradeoffs for Additional Hardware for R

At the time of writing of this book, the dominant computing paradigm is workstation computing followed by server–client computing. However, with the introduction of cloud computing, netbooks, and tablet PCs, hardware choices are much more flexible in 2011 than just a couple years ago.

Hardware costs represent a significant expense for an analytics environment and are also remarkably depreciated over a short period of time. Thus, it is advisable to examine your legacy hardware and your future analytical computing needs and decide accordingly regarding the various hardware options available for R.

Unlike other analytical software that can charge by the number of processors, or servers, which can be more expensive than workstations, or grid computing, which can be very costly as well if it is even available, R is well suited for all kinds of hardware environments with flexible costs.

Given the fact that R is memory intensive (it limits the size of data analyzed to the RAM size of the machine unless special formats or chunking is used), the speed at which R can process data depends on the size of the datasets used and the number of users analyzing a dataset concurrently. Thus the defining issue is not R but the size of the data being analyzed and the frequency, repeatability, and level of detail of analysis required.

2.1.4.1 Choices Between Local, Cluster, and Cloud Computing

- Local computing: This denotes when the software is installed locally. For big data, the data to be analyzed are stored in the form of databases. The server version—Revolution Analytics has differential pricing for server–client versions (as is true for all analytical software pricing), but for the open source version it is free, as it is for server or workstation versions. The issue of number of servers versus workstations is best determined by the size of the data. R processes data in RAM, so it needs more RAM than other software of its class.

Cloud computing is defined as the delivery of data, processing, and systems via remote computers. It is similar to server–client computing, but the remote server (also called the cloud) has flexible computing in terms of number of processors, memory, and data storage. Cloud computing in the form of a public cloud enables people to do analytical tasks on massive datasets without investing in permanent hardware or software as most public clouds are priced on pay per usage. The biggest cloud computing provider is Amazon, and many other vendors provide services on top of it. Google also does data storage in the form of clouds (Google Storage) and uses machine learning in the form of an API (Google Prediction API).

1. *Amazon*: We will describe how to set up an R session on an Amazon EC2 machine.
2. *Google*: We will describe how to use Google Cloud Storage as well as Google Prediction API using packages.
3. *Cluster-grid computing/parallel processing*: To build a cluster, you need the RMpi and SNOW packages, plus other packages that help with parallel processing. This will be covered in general detail but detailed instructions for building a big cluster will not be provided as that is more suitable for a high-performance computing environment.

2.1.5 Interface Choices: Command Line Versus GUI. Which GUI Should You Choose as the Default Startup Option?

R can be used in various ways depending on the level of customization. The main GUIs suitable for business analyst audiences are as follows:

1. R Commander
2. Rattle
3. Deducer and JGR
4. GrapheR

5. RKWard
6. Red-R
7. Others including Sciviews-K

The interfaces to R will be covered in detail in Chap. 3, where a detailed description will also be given of how to access R from other mainstream analytical software applications like Oracle Data Miner, JMP, SAS/IML, KNIME, and Microsoft Excel. In addition to the standard desktop GUI, there are Web interfaces that use R and command line for default coding.

2.1.6 Software Component Choice

Which R packages should you install? There are almost 3,000 packages, some of them are complementary, others depend on each other, and almost all are free.

Throughout this book we will describe specialized packages that are best suited for creating the results of certain analytical tasks. In the R Programming language, multiple approaches, code, functions, and packages can be used to achieve the same result. The objective of this book is to focus on analysis rather than the language, and accordingly we will indicate the easiest approach to accomplishing the given business analysis task and mention other options and the advantages and disadvantages of using multiple options and approaches.

2.1.7 Additional Software Choices

What other applications do you need to achieve maximum accuracy, robustness, and speed of computing and how do you make use of existing legacy software and hardware to achieve the best complementary results with R?

Once we have covered the basics, we will describe, in Chap. 11, additional tips, tricks, and tweaks to help you optimize your R code. These include setting up benchmarks to measure and improve code efficiency and using syntax editors and integrated development environments.

2.2 Downloading and Installing R

To download and install the open source version of R, visit R's home page at http://www.r-project.org/.

You will be directed to the CRAN mirror closest to your location. CRAN, which stands for Comprehensive R Archive Network, is a set of online mirror servers that enable you to download R and its various packages. The global network thus ensures a fast, dedicated, reliable network for downloading and accessing software. In this manner, CRAN guarantees the highest likelihood of availability of R as it is very difficult to bring down servers of the entire CRAN, but an isolated server might get overwhelmed due to traffic (especially at new product release times). It consists of 79 sites in 34 regions. R can be downloaded from http://cran.r-project.org/mirrors.html.

For Windows-R, installers exist in the form of downloadable binaries. Download the Windows.exe file and install the program. In addition, read the Frequently Asked Questions.

On Linux (Ubuntu): To install the complete R system, open a terminal window and use *sudo apt-get update sudo apt-get install r-bas*e.

Debian packages for R are a bit dated, but this is the easiest way to install. The other way is to modify your source file with a CRAN mirror before running apt-get. Documentation for this is on the Web site given previously.

Mac OS has a separate, downloadable installer. You will need to refer to the main R Web site http://www.r-project.org.

The Australian CRAN Mirror can be accessed at http://cran.ms.unimelb.edu.au/bin/windows/base/README.R-2.15.1 and FAQs at http://cran.ms.unimelb.edu.au/bin/windows/base/rw-FAQ.html#Introduction.

<CRAN MIRROR>/bin/windows/base/release.htm is the generic link for Windows releases. The latest version was 2.14.1 in January 2012, but this will change every 6 months.

2.3 Installing R Packages

Unlike other traditional software applications that come in bundles, R comes in the form of one installer and a large number of small packages. There are an estimated 3,000 packages in R—so if you have a specific analytical need, chances are someone has created a package already for it. To launch R, simply click the icon that was created (for Windows users) or type R (at command terminal for Linux users).

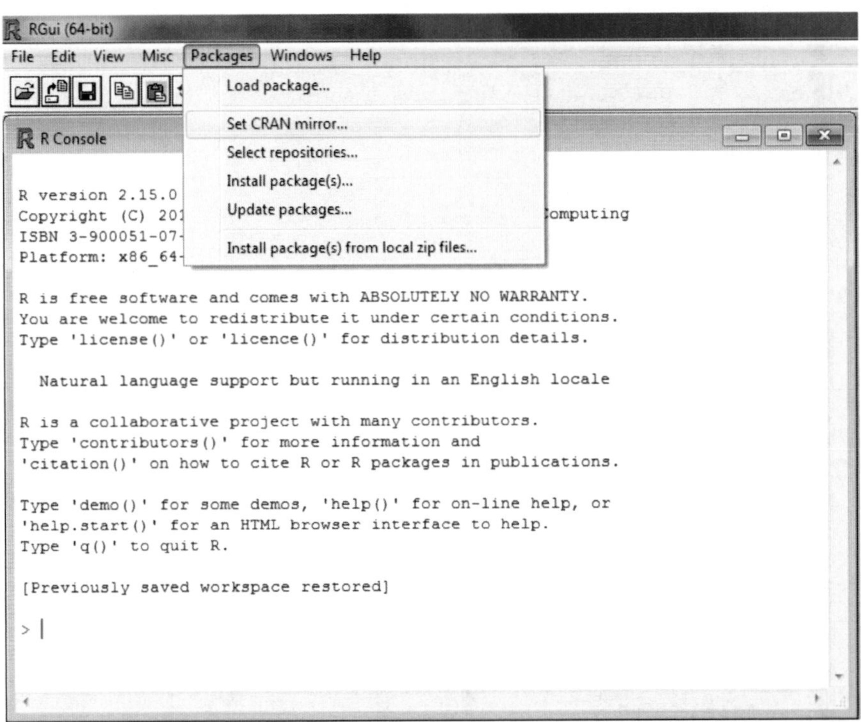

Type *install.packages()* or select Install packages from the menu. You will then be asked to choose a CRAN mirror (or nearest location for download). Click on the nearest CRAN mirror.

Click on the package name and on OK to install that package as well as packages that are required for it to operate (dependencies).

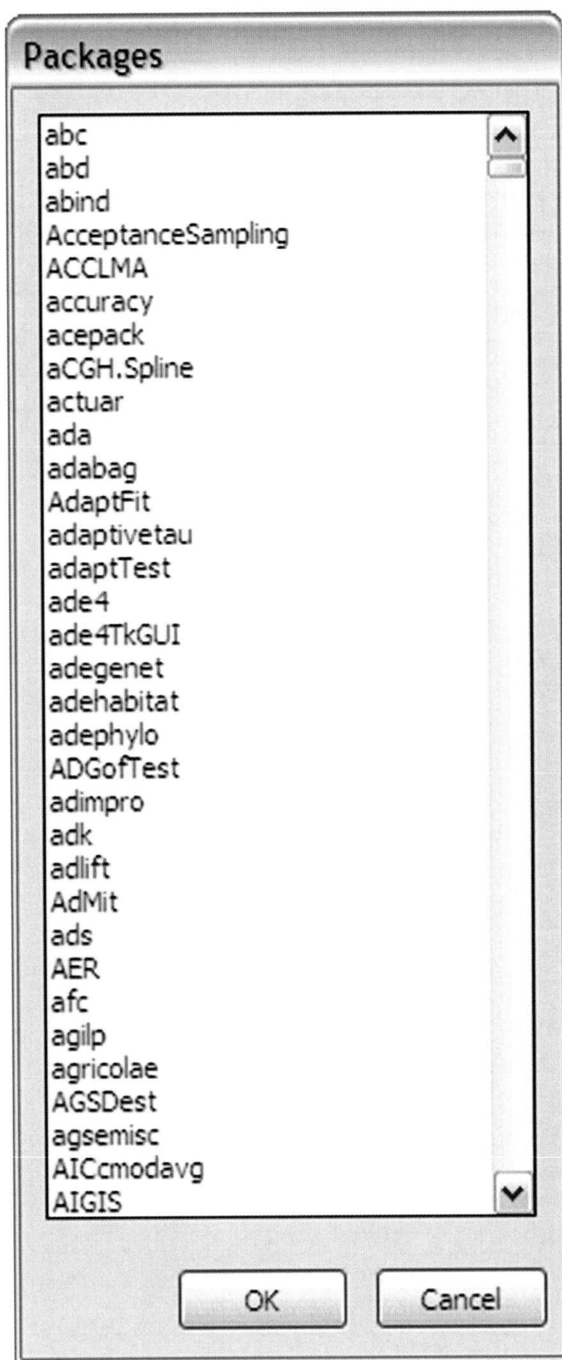
Packages
abc
abd
abind
AcceptanceSampling
ACCLMA
accuracy
acepack
aCGH.Spline
actuar
ada
adabag
AdaptFit
adaptivetau
adaptTest
ade4
ade4TkGUI
adegenet
adehabitat
adephylo
ADGofTest
adimpro
adk
adlift
AdMit
ads
AER
afc
agilp
agricolae
AGSDest
agsemisc
AICcmodavg
AIGIS
OK
Cancel

Once a package is installed, type library(package-name) to check if it is working. In this example, we are trying to see if the GUI R Commander has installed.
>*library(RCmdr)*

(a) Internet: To install a package from the Internet, you can use the following code as an example for modifying the installation to your needs:
install.packages("bigmemory", repos="http://R-Forge.R-project.org")
(b) Local file: To set a local source code file (a tar.bz file in Linux or a .zip file in Windows) set repos = null in
install.packages(pkgs, lib, repos = getOption("repos"), contriburl = contrib.url(repos, type)).

Package Troubleshooting and Dependencies

(a) Troubleshooting installations: For purposes of troubleshooting, read the package documentation as well as the official online R documentation at http://cran.r-project.org/doc/manuals/R-admin.html#Installing-packages.
Entering ? at the command prompt will give you help query results:

```
Type 'contributors()' for more information and
'citation()' on how to cite R or R packages in publications.

Type 'demo()' for some demos, 'help()' for on-line help, or
'help.start()' for an HTML browser interface to help.
Type 'q()' to quit R.

> ?timeseries
No documentation for 'timeseries' in specified packages and libraries:
you could try '??timeseries'
> ??timeseries
> |
```

Entering a double question mark gives you more comprehensive help:

```
Help files with alias or concept or title matching 'timeseries' using
fuzzy matching:


aplpack::slider.split.plot.ts
                         interactive splitting of time series
aplpack::slider.zoom.plot.ts
                         interactive zooming of time series
boot::tsboot             Bootstrapping of Time Series
car::Hartnagel           Canadian Crime-Rates Time Series
coda::as.ts.mcmc         Coerce mcmc object to time series
datasets::austres        Quarterly Time Series of the Number of
                         Australian Residents
dynlm::dynlm             Dynamic Linear Models and Time Series
                         Regression
Ecdat::Index.Time.Series
                         Time Series
Ecdat::Macrodat          Macroeconomic Time Series for the United States
fBasics::boxPlot         Time Series Box Plots
fBasics::seriesPlot      Financial Time Series Plots
fBasics::basicStats      Basic Time Series Statistics
forecast::Arima          Fit ARIMA model to univariate time series
forecast::best.arima     Fit best ARIMA model to univariate time series
forecast::forecast       Forecasting time series
forecast::forecast.StructTS
                         Forecasting using Structural Time Series models
forecast::na.interp      Interpolate missing values in a time series
forecast::simulate.ets
```

(b) Special notes:

1. GTK+ -GTK+ is a special requirement of some GUI packages like Rattle. To use GTK+ or the package RGTK, you first need to install the dependencies of GTK+ like, for example, Cairo, Pango, ATK, libglade. From the GTK Web site (http://www.gtk.org/download-windows-64bit.html) you will need the GLib, Cairo, Pango, ATK, gdk-pixbuf, and GTK+ developer packages to build software against GTK+. To run GTK+ programs, you will also need the gettext-runtime, fontconfig, freetype, expat, libpng, and zlib packages. It is ideal for the user to avoid troubleshooting problems by getting a Linux-based system for GTK-based packages; you can do this either by creating a dual-boot machine or using VMware Player to switch between operating systems.
2. ODBC: To connect to specific databases, you need to create an ODBC connection from the driver to the type of database. This will be covered in detail later.
3. In R we use the " # " symbol to comment out sentences. Commenting of code is done to make it more readable.

(c) Updating packages: You can use the following command to update all packages:
>update.packages()

2.4 Starting up Tutorial in R

- Get working directory
  - *getwd()*
- Set working directory
  - *setwd("C:/Users/KUs/Desktop")*
- List all objects
  - *ls()*
- List objects with certain words within name (i.e.fun)
  - *ls(pattern="fun")*
- Remove object "ajay"
  - *rm(ajay)*
- Remove all objects
  - *rm(list=ls())*
- Comment code: Use " # " in front of the part to be commented out.
  - *#comments*
- List what is in an object called ajay. Note: simply typing the name of the object is sufficient to print out the contents, so you need to be careful when dealing with huge amounts of data.
  - *> ajay*
  - *[1] 1 3 4 50*
  - List second value within an object
  - *> ajay[2]* (note bracket type)
  - *[1] 3 >*
  - List second to fourth values within an object (note colon ":")
  - *> ajay[2:4]*
  - *[1] 3 4 50*
  - Look at the class of an object
  - *class(ajay)*
  - *[1] "numeric" #Classes can be numeric, logical, character, data frame, linear model (lm), time series (ts), etc.*
  - Use *as(object, value)* to coerce an object into a particular class
  - *ajay=as.data.frame(ajay) #turns the list into a data frame. This will change the dimensions and length of the object*
  - Find the dimensions of an object called "ajay"
  - *> dim(ajay)*
  - *[1] 4 1*

  - Find the length of an object called "ajay"
  - > *length(ajay)*
  - *[1] 1*

- Suppose we change the object from data frame back into a list; we now have different dimensions
  - > *ajay=as.list(ajay)*
  - > *dim(ajay)*
  - *NULL*
  - > *length(ajay)*
  - *[1] 4*

- The scope of classes can be further investigated for building custom analytical solutions, but object-oriented programming is beyond the scope of this book as it is aimed at business analytics users. Interested readers may consult a brief tutorial in classes and methods at http://www.biostat.jhsph.edu/~rpeng/biostat776/classes-methods.pdf and the documentation at http://cran.r-project.org/doc/manuals/R-lang.html#Objects.

2.5 Types of Data in R

Unlike SAS and SPSS, which predominantly use the dataset/data frame structure for data, R has tremendous flexibility in reading data. Data can be read and stored as a list, matrix, or data.frame. This has great advantages for the power programmer experienced in object-oriented programming, but for the average business analyst the multiple ways of doing things in R can lead to some confusion and even more prolonged agony in the famous "learning curve" of R.

The various types of data in R are vectors, lists, arrays, matrixes, and data frames. For the purposes of this book, most data types will be data frames.

Data frames are rectangular formats of data with column names as variables and unique rows as records.

For other types of data please see http://www.statmethods.net/input/datatypes.html and http://www.cyclismo.org/tutorial/R/types.html.

2.6 Brief Interview with John Fox, Creator of Rcmdr GUI for R

What follows is a brief extract from a September 2009 interview with Prof. John Fox, creator of R Commander, one of R's most commonly used GUIs.

Ajay: What prompted you to create R Commander? How would you describe R Commander as a tool, say, for a user of other languages and who want to learn R but are afraid of the syntax?

John: I originally programmed the R Commander so that I could use R to teach introductory statistics courses to sociology undergraduates. I previously taught this course with Minitab or SPSS, which were programs that I never used for my own work. I waited for someone to come up with a simple, portable, easily installed point-and-click interface to R, but nothing appeared on the horizon, and so I decided to give it a try myself.

I suppose that the R Commander can ease users into writing commands, inasmuch as the commands are displayed, but I suspect that most users don't look at them. I think that serious prospective users of R should be encouraged to use the command-line interface along with a script editor of some sort.

I wouldn't exaggerate the difficulty of learning R: I came to R—actually S then—after having programmed in perhaps a dozen other languages, most recently at that point Lisp, and found the S language particularly easy to pick up.

Ajay: I particularly like the R Cmdr plugins. Can anyone expand the R Commander's capabilities with a customized package plugin?

John: That's the basic idea, though the plugin author has to be able to program in R and must learn a little Tcl/Tk.

Ajay: What are the best ways to use the R Commander as a teaching tool (I noticed the help is a bit outdated).

John: Is the help outdated? My intention is that the R Commander should be largely self-explanatory. Most people know how to use point-and-click interfaces.

In the basic courses for which it is principally designed, my goals are to teach the essential ideas of statistical reasoning and some skills in data analysis. In this kind of course, statistical software should facilitate the basic goals of the course. As I said, for serious data analysis, I believe that it's a good idea to encourage use of the command-line interface.

Ajay: Do people on the R core team recognize the importance of GUIs? How does the rest of the R community feel? What kind of feedback have you gotten from users?

John: I feel that the R Commander GUI has been generally positively received, both by members of the R core team who have said something about it to me and by others in the R community. Of course, a nice feature of the R package system is that people can simply ignore packages in which they have no interest. I noticed recently that a paper I wrote several years ago for the *Journal of Statistical Software* on the Rcmdr package has been downloaded nearly 35,000 times. Because I wouldn't expect many students using the Rcmdr package in a course to read that paper, I expect that the package is being used fairly widely. [*Update: As of February 2012 it has been downloaded 81,477 times.*]

For more details on John's work see http://socserv.mcmaster.ca/jfox/

2.7 Summary of Commands Used in This Chapter

2.7.1 Packages

R Commander
Rattle
Deducer and JGR
GrapheR
RKWard
Red-R
Others including Sciviews-K
Snow
Rmpi
bigmemory

2.7.2 Functions

- install.packages(FUN): Installs the package named FUN if available.
- update.packages(): Updates all packages on local system.
- library(FUN): Loads the package FUN from local machine to R system.
- ?FUN: Searches for help on keyword FUN.
- ??FUN: Searches for comprehensive help on keyword FUN.
- sudo apt-get install FUN: Installs a software called FUN in Linux-based systems.

Citations and References

- Fox, J. The R Commander: A basic-statistics graphical user interface to R. J. Stat. Softw. **14**(9), 1–42 (2005). http://www.jstatsoft.org/v14/i09/paper
- *R Blogger http://www.r-bloggers.com/* aggregates blogs from almost 140 R blogs
- R email help lists: http://www.r-project.org/mail.html.
- Full interview with John Fox: http://www.decisionstats.com/interview-professor-john-fox-creator-r-commander/

Chapter 3
R Interfaces

Chapter summary: In this chapter we discuss the various ways to interface R and to use R analytics based on one's needs. We will cover how to minimize the time spent learning to perform tasks in R by using a GUI instead of the command line. In addition, we will learn how to interface to R from other software as well as use it from an Amazon cloud computing environment. We will also discuss the relative merits and demerits of various R interfaces.

3.1 Interfaces to the R Statistical Language

Command line: The default method is to use the command prompt by the installed software on download from http://r-project.org. For Windows users there is a simple GUI that has an option for packages (loading package, installing package, setting CRAN mirror for downloading packages), Misc (useful for listing all objects loaded in a workspace as well as clearing objects to free up memory), and a help menu.

GUI (Click and point): Besides the command prompt, many GUIs enable the analyst to use click-and-point methods to analyze data without getting into the details of learning complex and, at times, overwhelming R syntax. R GUIs are very popular both as modes of instruction in academia as well as in actual usage as they cut down considerably on the time needed to adapt to the language. As with all command-line and GUI software, for advanced tweaks and techniques, the command prompt will come in handy as well.

Web interfaces

Cloud computing

A. Ohri, *R for Business Analytics*, DOI 10.1007/978-1-4614-4343-8_3,

3.2 Basic R

Some R programmers use the second [<-] or third [->] assignment operator, but this book will mostly use the simpler first [=] method.

There are multiple ways to refer to data objects in R, as the following example will show.

```
> ajay=c(1,2,4,8,16,32) # Assign the object named "ajay" to the list of numbers
> print(ajay) # Print the object named "ajay"
[1] 1 2 4 8 16 32
> ajay # Note: if you type just the name of the object, it will print out its
contents as well
[1] 1 2 4 8 16 32
> ajay[1:3] # This refers to the first three members of the object "ajay"
[1] 1 2 4
> ajay[3] # This refers to the third member of the object "ajay"
[1] 4
> ajay[1] # This refers to the first member of the object "ajay"
[1] 1
ajay[-(1:2)] # This includes everything EXCEPT the first two elements
[1] 4 8 16 32
> ajay=as.data.frame(ajay) # Convert a data object from a list to a data frame
> length(ajay) # This gives the length of an object
[1] 1
> dim(ajay) # This gives the dimensions of an object
[1] 6 1
> nrow(ajay)# This gives the number of rows of an object
[1] 6
> ncol(ajay) # This gives the number of columns of an object
[1] 1
> class(ajay) # This gives the class of an object
[1] "data.frame"
> head(ajay,2) # This gives the first two (or specified) records of an object
ajay
1 1
2 2
> tail(ajay,2) # This gives the last two (or specified) records of an object
ajay
5 16
6 32
> str(ajay) # This gives the structure of an object
'data.frame': 6 obs. of 1 variable:
$ ajay: num 1 2 4 8 16 32
> names(ajay) # This gives the variables names of an object
[1] "ajay"
```

3.3 Advantages and Limitations of Graphical User Interfaces to R

Traditionally, the command line has been the programming interface of choice for R developers and users in academia. For business analytics users who need to execute tasks faster, learn tasks faster in R, and do a wide majority of tasks repeatedly, it makes sense to at least be aware of GUIs.

3.3.1 Advantages of Using GUIs for R

- Faster learning for new programmers or people transitioning to doing business analytics within R.
- Easier creation advanced models or graphics.
- Repeatability of analysis for doing the same task over and over.
- Syntax is auto-generated in many GUIs.

3.3.2 Limitations of Using GUIs for R

- Risk of junk analysis, especially by nonexperienced users, by clicking menus in GUI; prevented in many GUIs by disabling (or graying) the incorrect options.
- Lack of capability to create custom functions using a GUI.
- Limited scope and exposure in learning R syntax.
- Advanced techniques and flexibility of data handling possible only in command line.

3.4 List of GUI

What follows is a brief list of the notable GUIs.

- R Commander: most commonly used GUI in R; extendible by plugins and will also be used in this book for time series forecasting.
- Rattle: most commonly used GUI in this book for data mining (Chap. 7).
- Grapher: for simple data visualization (Chap. 5).
- Deducer (used with JGR): for advanced data visualization (Chap. 5).

Other GUIs are covered more extensively in Chap. 12.

- Other GUIs used in R
  - Sciviews-K
  - RKWard
  - PMG

- Work-flow-based GUIs
  - Red-R
  - R Analytic Flow

The R package gWidgets can be used for building toolkit-independent, interactive GUIs. Most GUIs in R are GTK or Tcl/Tk based.

3.4.1 R Commander

- R Commander was primarily created by Professor John Fox of McMaster University to cover the content of a basic statistics course. However, it is extensible, and many other packages can be added to it in menu form—in the form R Commander Plugins. It is one of the most widely used R GUIs and has a script window so you can write R code in combination with the menus. As you point and click a particular menu item, the corresponding R code is automatically generated in the log window and executed. It can be found on CRAN at http://cran.r-project.org/web/packages/Rcmdr/index.html, and you can read about it in more detail at http://socserv.mcmaster.ca/jfox/Misc/Rcmdr/.

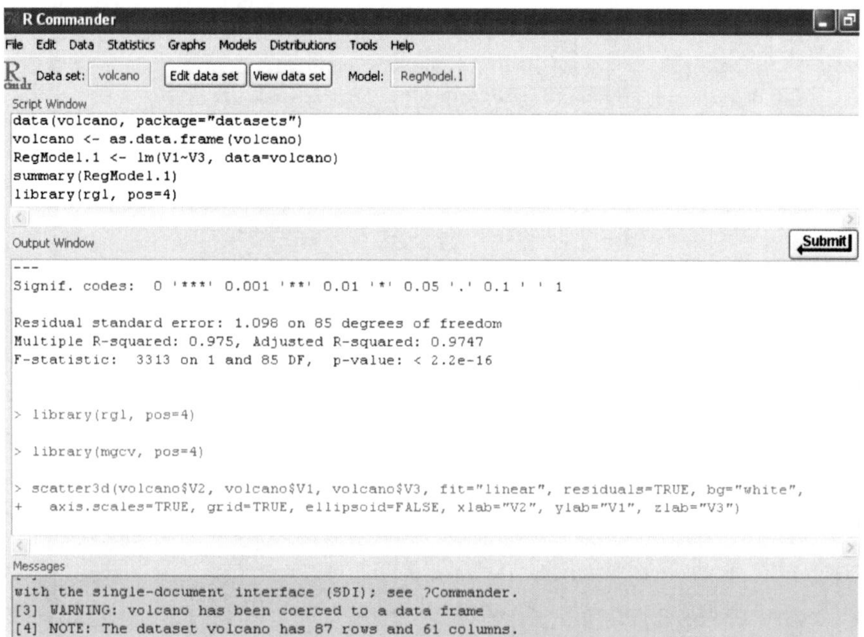

3.4.1.1 Advantages of R Commander

1. Useful for beginners in R language for creating basic graphs and analysis and building models.
2. Has script window, output window, and log window (called messages) in same screen that helps user as code is auto-generated upon clicking menus and can be customized easily (e.g., to change labels and options in graphs); graphical output shown in window separate from output window.
3. Extensible for other R packages like qcc (for quality control), teaching demos (for training), survival analysis, and design of experiments.
4. Easy-to-understand interface even for first-time users of R or statistical software.
5. Irrelevant menu items automatically grayed out (e.g., given only two variables, if you try to build a 3D scatterplot graph, then that menu would simply not be available and is grayed out).

Creating 3D Graph Using R Commander

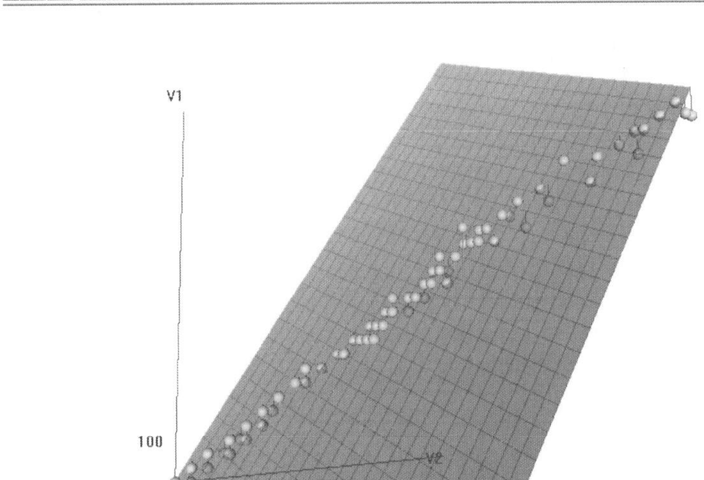

3.4.1.2 Comparative Disadvantages of R Commander

1. R Commander is basically aimed at a statistical audience (originally students in statistics), and thus the terms and menus are accordingly labeled. Hence it is more of a statistical GUI rather than an analytics GUI.

2. R Commander has limited ability to evaluate models from a business analyst's perspective, even though it has extensive statistical tests for model evaluation in a model submenu.
3. R Commander is not suited for data mining or advanced graphical visualization.

3.4.1.3 Description of R Commander

File menu: for loading and saving script files, output, and workspace, also for changing the present working directory and for exiting R.

Edit menu: for editing scripts and code in the script window.

Data menu: for creating new datasets, inputting or importing data, and manipulating data through variables. Data import can be from text, comma-delimited values, clipboard, SPSS datasets, Stata, Minitab, Excel, dbase, Access files, or from Web sites. Data manipulation includes deleting rows of data as well as manipulating variables. This menu has the option to merge two datasets by row or column.

Statistics menu: for descriptive statistics, hypothesis tests, factor analysis, clustering and creating models; there is a separate menu for evaluating models thus created.

Graphs menu: for creating various kinds of graphs including boxplots, histograms, lines, pie charts, and x-y plots. The first option is color palette, which can be used for customizing colors. Adjust colors based on publication or presentation needs. A notable option is 3D graphs for evaluating three variables at a time, a very useful and impressive feature that exposes the user to advanced graphs in R at just a few clicks; spruce up presentations using these graphs.

Also consider scatterplot matrix graphs for graphical display of variables. Graphical display in R surpasses that of any other statistical software, both in terms of appeal and ease of creation; GUIs can be used to create graphs that further help users to get the most of data insights using R at minimal effort.

Models menu: somewhat of a labeling peculiarity of R Commander as this menu is only for evaluating models that created using the statistics menu-model submenu. It includes options for graphical interpretation of model results, residuals, leverage, confidence intervals, and adding back residuals to the dataset.

Distributions menu: for cumulative probabilities, probability density, graphs of distributions, quantiles, and features for standard distributions; can be used in lieu of standard statistical tables for distributions. It has 13 standard statistical continuous distributions and five discrete distributions.

Tools menu: for loading other packages and R Commander plugins (which are then added to the interface menu after the R Commander GUI is restarted); also contains options submenu for fine-tuning (like opting to send output to R Menu).

Help menu: standard documentation and help menu. Essential reading is the short 25-page manual in it called "Getting Started with the R Commander".

3.4.2 R Commander E Plugins or Extensions

R Commander has almost 30 extensions in the form of e-plugins. Some of the more prominent ones are featured below. Plugins are simply packages that can be downloaded using install.packages(), but they can be loaded only once RCommander has been invoked through the load plugin menu option.

Rcmdr R Commander
RcmdrPlugin.DoE R Commander plugin for (industrial) design of experiments
RcmdrPlugin.EHESsampling Tools for sampling in European Health Examination Surveys (EHES)
RcmdrPlugin.Export Graphically export output to LaTeX or HTML
RcmdrPlugin.FactoMineR GUI for FactoMineR
RcmdrPlugin.HH Rcmdr support for HH package
RcmdrPlugin.IPSUR IPSUR plugin for R Commander
RcmdrPlugin.MAc Meta-analysis with correlations (MAc) Rcmdr plugin
RcmdrPlugin.MAd Meta-analysis with mean differences (MAd) Rcmdr plug-in
RcmdrPlugin.PT Some discrete exponential dispersion models: Poisson-Tweedie
RcmdrPlugin.SLC SLC Rcmdr plugin
RcmdrPlugin.SensoMineR GUI for SensoMineR
RcmdrPlugin.SurvivalT Rcmdr survival plugin
RcmdrPlugin.TeachingDemos Rcmdr teaching demos plugin
RcmdrPlugin.TextMining Rcommander plugin for "tm" package
RcmdrPlugin.doex Rcmdr plugin for Stat 4309 course
RcmdrPlugin.epack Rcmdr plugin for time series
RcmdrPlugin.orloca orloca Rcmdr plugin
RcmdrPlugin.qcc Rcmdr qcc plugin
RcmdrPlugin.qual Rcmdr plugin for quality control course
RcmdrPlugin.sos Efficiently search R help pages
RcmdrPlugin.steepness Steepness Rcmdr plugin
RcmdrPlugin.survival R Commander plugin for survival package

The following screenshot shows the complete and updated plugin list:

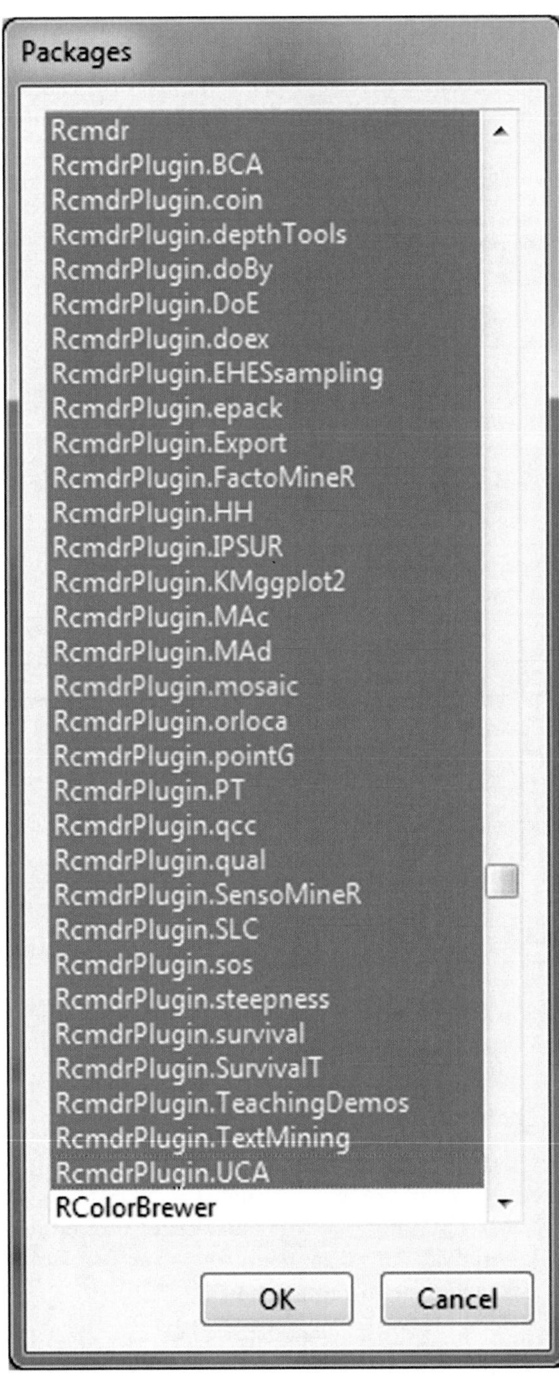
Packages
Rcmdr
RcmdrPlugin.BCA
RcmdrPlugin.coin
RcmdrPlugin.depthTools
RcmdrPlugin.doBy
RcmdrPlugin.DoE
RcmdrPlugin.doex
RcmdrPlugin.EHESsampling
RcmdrPlugin.epack
RcmdrPlugin.Export
RcmdrPlugin.FactoMineR
RcmdrPlugin.HH
RcmdrPlugin.IPSUR
RcmdrPlugin.KMggplot2
RcmdrPlugin.MAc
RcmdrPlugin.MAd
RcmdrPlugin.mosaic
RcmdrPlugin.orloca
RcmdrPlugin.pointG
RcmdrPlugin.PT
RcmdrPlugin.qcc
RcmdrPlugin.qual
RcmdrPlugin.SensoMineR
RcmdrPlugin.SLC
RcmdrPlugin.sos
RcmdrPlugin.steepness
RcmdrPlugin.survival
RcmdrPlugin.SurvivalT
RcmdrPlugin.TeachingDemos
RcmdrPlugin.TextMining
RcmdrPlugin.UCA
RColorBrewer
OK
Cancel

3.5 Summary of R GUIs

The following list gives the advantages and disadvantages of various R GUIs.

This book recommends that R Commander and its extensible plugins be used for various business analytics uses but that for data mining the Rattle GUI be used (Chap. 7) and for data visualization the Deducer GUI (Chap. 5) and its plugins be used.

For more details on other GUIs summarized here, see Chap. 12, Sect. 12.3.

Name of GUI	*Advantages*
Sciviews-K	Flexible GUI, can be used to create other GUIs
RKWard	Comprehensive GUI with lots of detail
Red-R	Workflow style
R Commander	Suitable for basic statistics, plotting, time series; has some 20 extensions
R Analytic Flow	Has Japanese interface and nice design workflow style
Rattle	Most suitable for data mining
PMG	Simple interface
JGR/Deducer	Most suitable for data visualization
Grapher	Simple graphing interface

Name of GUI	*Disadvantages*
Sciviews-K	Not widely publicized for business use
RKWard	Not ready for Windows installer
Red-R	No Linux installer
R Commander	Cluttered design interface
R Analytic Flow	Not widely publicized for business use
Rattle	Not suitable for data visualization
PMG	Not suitable for data mining or complex analysis
JGR-Deducer	Not suitable for data mining
Grapher	Not suitable for complex graphs, analysis, data mining

3.6 Using R from Other Software

Interfaces to R from other software exist as well. These include software from SAS Institute, IBM SPSS, RapidMiner, KNIME, and Oracle. The additional links below complement the links provided in Chap. 1.

A brief list of these interfaces is given below.

- SAS/IML interface to R: Read about the SAS Institute's SAS/IML Studio interface to R at http://www.sas.com/technologies/analytics/statistics/iml/index.html.

- WPS/Bridge to R: Connect to WPS, a SAS language clone at much lower cost, using Bridge to R.
- RapidMiner Extension to R: View integration with RapidMiner's extension to R at http://www.youtube.com/watch?v=utKJzXc1Cow.
- The IBM SPSS plugin for R: SPSS software has R integration in the form of a plugin. This was one of the earliest third-party software applications offering interaction with R; read more at http://www.spss.com/software/statistics/developer/.
- KNIME: Konstanz Information Miner also has R integration. View it at http://www.knime.org/downloads/extensions.
- Oracle Data Miner: Oracle has a data mining offering for its very popular database software, which is integrated with the R language. The R Interface to Oracle Data Mining (R–ODM) allows R users to access the power of Oracle Data Mining's in-database functions using the familiar R syntax. Visit http://www.oracle.com/technetwork/database/options/odm/odm-r-integration-089013.html.
- JMP: JMP offers integration with R. Read http://blogs.sas.com/content/jmp/2010/04/12/jmp-into-r/
- SAP: SAP offers integration with R through its SAP RHANA package and uses R internally for its business objects predictive analytics. See http://www.slideshare.net/JitenderAswani/na-6693-r-and-sap-hana-dkom-jitenderaswanijensdoeprmund.
- Teradata: The TeradataR package allows R users to interact with a Teradata database. See the package at http://downloads.teradata.com/download/applications/teradata-r and read more about it at http://developer.teradata.com/applications/articles/in-database-analytics-with-teradata-r.
- Oracle also offers its own commercial version of R called Enterprise R.

3.6.1 *RExcel: Using R from Microsoft Excel*

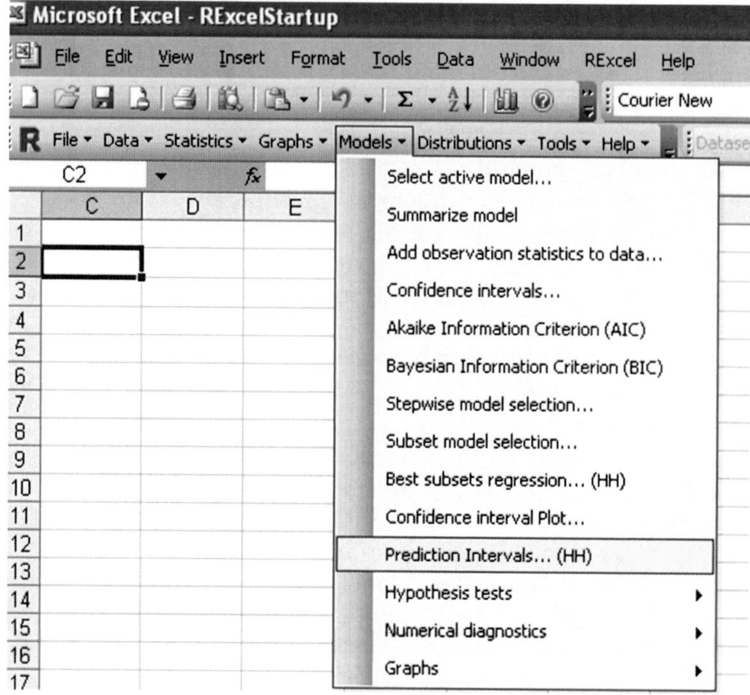

Microsoft Excel is the most widely used spreadsheet program for data manipulation, entry, and graphics. Yet as dataset sizes have increased, Excel's statistical capabilities have lagged, though its design has moved ahead in various product versions. RExcel basically works by adding an .xla plugin to Excel just like other plugins. It does so by connecting to R through R packages. Basically, it provides the functionality of R functions and capabilities for the most widely distributed spreadsheet program. All data summaries, reports, and analysis end up in a spreadsheet; RExcel makes R very useful for people who do know R. In addition, it adds (optional) menus of R Commander as menus in Excel spreadsheets.

Advantages: Enables R and Excel to communicate, thus tying an advanced statistical tool to the most widely used business analytics tool.

Disadvantages: No major disadvantages to business users. For a data statistical user, Microsoft Excel is limited to 100,000 rows, so R data need to be summarized or reduced. It is not available on MacOS or Linux.

R's graphical capabilities are very useful, but to new users, interactive graphics in Excel may be easier than, say, using ggplot or GGobi.

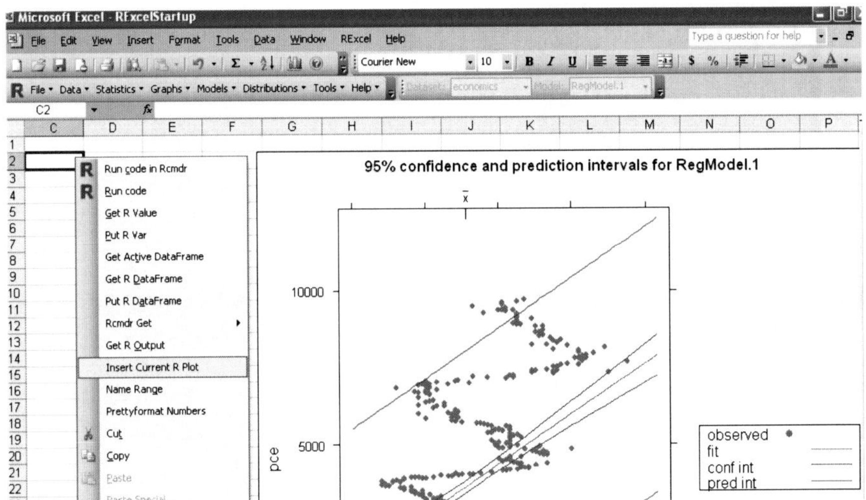

Read more on this at http://rcom.univie.ac.at/ or the complete Springer text http://www.springer.com/statistics/computanional+statistics/book/978-1-4419-0051-7.

3.7 Web Interfaces to R

If Web-enabled business intelligence (BI) is what you need, to show both summarized and multiple slices of data, then this section will greatly facilitate use of R to reduce BI dashboard costs.

What follows is a list of Web interfaces to R.

For a Web interface usually only a browser and Internet connection to upload data are required. Web interfaces can be customized to create internal corporate BI solutions and dashboards.

- R-Node: http://squirelove.net/r-node/doku.php
- Rweb: see JSS paper (http://www.jstatsoft.org/v04/i01/) for RWeb. This was one of the first Web interfaces to R; available at http://www.math.montana.edu/Rweb/
- Rcgi: http://www.ms.uky.edu/~statweb/testR3.html
- Rphp (realized in PHP and MySQL): http://dssm.unipa.it/R-php/
- RWui (Web interface used to design custom forms and other Web interfaces): http://sysbio.mrc-bsu.cam.ac.uk/Rwui
- R.Rsp: An RSP document is a text-based document containing an R-embedded template of the final document: http://cran.r-project.org/web/packages/R.rsp/index.html
- RServe: TCP/IP server that allows other programs to use R facilities: http://www.rforge.net/Rserve/. Typical uses of Rserve include interactive applications using R for model construction (see KLIMT project) or Web servlets performing online

computations. See paper at http://www.ci.tuwien.ac.at/Conferences/DSC-2003/Proceedings/Urbanek.pdf

  - JRI is a Java/R interface that lets you run R inside Java applications as a single thread: http://www.rforge.net/JRI/

- RPad: http://rpad.googlecode.com/svn-history/r76/Rpad_homepage/index.html and https://code.google.com/p/rpad/ Screenshot:

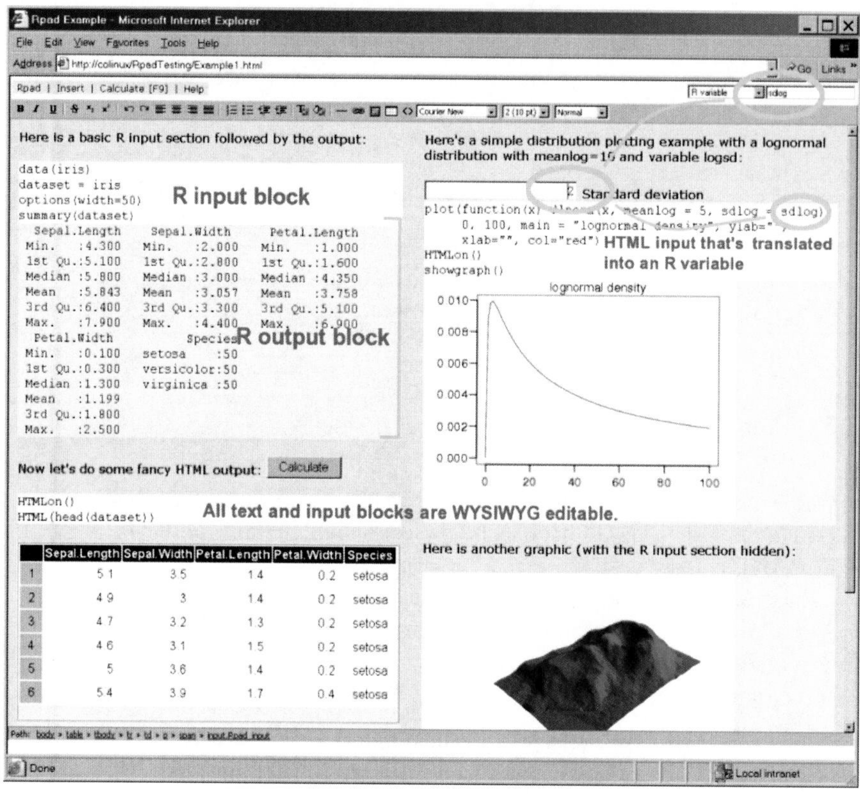

- Concerto: http://www.psychometrics.cam.ac.uk/page/300/concerto-testing-platform.htm and http://code.google.com/p/concerto-platform/. Concerto is a Web-based, adaptive testing platform for creating and running rich, dynamic tests. It combines the flexibility of HTML presentation with the computing power of the R language and the safety and performance of the MySQL database. It is free for commercial and academic use and is open source. The project originated at the Psychometrics Centre of Cambridge University. A screenshot-based easy-to-understand tutorial is available at http://code.google.com/p/concerto-platform/downloads/detail?name=screenshots_download.doc

- RApache: a project supporting Web application development using the R statistical language and environment and the Apache Web server: http://biostat.mc.vanderbilt.edu/rapache/
  - Some very good applications of RApache at UCLA:
    - http://rweb.stat.ucla.edu/stockplot
    - http://rweb.stat.ucla.edu/ggplot2
    - http://rweb.stat.ucla.edu/lme4
    - http://rweb.stat.ucla.edu/irttool/
  - and one more:
    - http://vps.stefvanbuuren.nl/puberty/
  - visualizing baseball at University of Vanderbilt: http://data.vanderbilt.edu/rapache/bbplot
  - visualizing baseball pitches: http://labs.dataspora.com/gameday/
- Rook-Rook (Web server interface and software package for R): http://cran.r-project.org/web/packages/Rook/Rook.pdf
- Biocep-R: http://biocep-distrib.r-forge.r-project.org and Elastic R for cloud-based R services: http://www.elasticr.net/
- This Web application for making graphs is also based on R https://app.prettygraph.com/
- RevoDeployR for a Web-services-based commercial approach by Revolution Analytics: http://www.revolutionanalytics.com/products/pdf/RevoDeployR.pdf
  - Key technical features of RevoDeployR:
  - Collection of Web services implemented as a RESTful API
  - JavaScript and Java client libraries, allowing users to easily build custom Web applications on top of R
  - Management console for securely administrating servers, scripts, and users
  - XML and JSON format for data exchange
  - Built-in security model for authenticated or anonymous invocation of R scripts
  - Repository for storing R objects and R script execution artifacts
- OpenCPU: OpenCPU provides a full RESTful RPC interface to R. It provides a convenient API to publish and call R functions: http://cran.r-project.org/web/packages/opencpu.demo/index.html and http://opencpu.org/examples/
  -

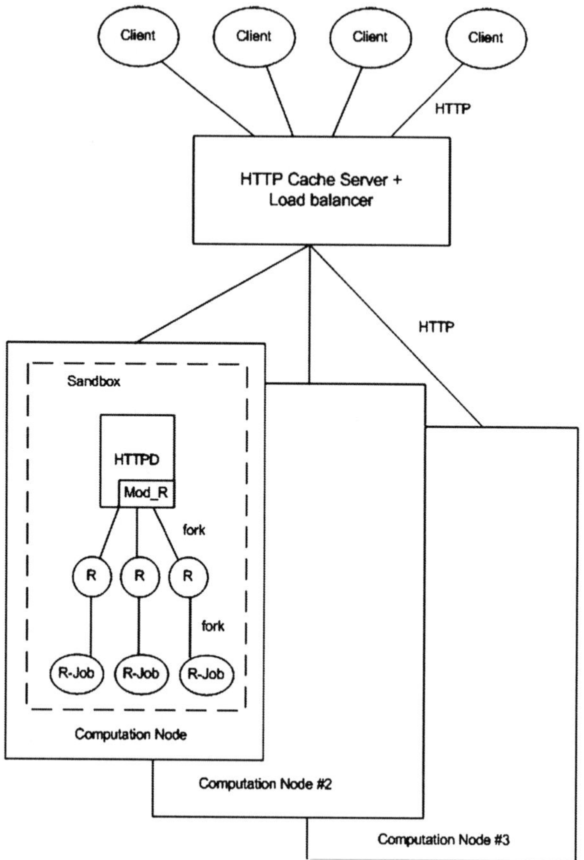

- CloudStat (Web-based platform that uses the R language): http://www.cloudstat.org/index.php?do=/about/

3.8 Interview: Using R as a Web-Based Application

Michal Kosinski heads the team that created Concerto –, a Web-based application using R. Kosinski is Director of Operations for the Psychometrics Centre at the University of Cambridge, UK.

Ajay: Why did you choose R as the background for Concerto? What other languages and platforms did you consider? Apart from Concerto, how else do you utilize R in your center, department, and university?

Michal: R was a natural choice as it is open source and free and nicely integrates with a server environment. Also, we believe that it is becoming a universal statistical and data processing language in science. We put increasing emphasis on teaching R to our students, and we hope that it will replace SPSS/PASW as a default statistical tool for social scientists.

Ajay: What can Concerto do besides conduct a computer adaptive test?

Michal: We did not plan it initially, but Concerto turned out to be extremely flexible. In a nutshell, it is a Web interface to R engine with a built-in MySQL database and easy-to-use developer panel. It can be installed on both Windows and Unix systems and used over a network or locally. Effectively, it can be used to build any kind of Web application that requires a powerful and quickly deployable statistical engine. For instance, I envision an easy-to-use Web site (that could look a bit like SPSS) allowing students to analyze their data using just a Web browser (learning the underlying R code simultaneously). Also, the authors of R libraries (or anyone else) could use Concerto to build user-friendly Web interfaces according to their methods. Finally, Concerto can be conveniently used to build simple nonadaptive tests and questionnaires. At first, it might seem slightly less intuitive than popular questionnaire services (such us my favorite, Survey Monkey), but it has virtually unlimited flexibility when it comes to item format, test flow, feedback options, etc. Also, it's free.

3.9 Cloud Computing to Use R

The combination of cloud computing and the Internet offers a new kind of interaction for scientists and analysts. It allows customized hardware on demand for peak computing loads as well as a secure environment for sensitive data. Cloud computing obviates the need for each department of each university to maintain its own cluster computer; thus it is of great use to the scientific community, especially for computing-resource-intensive users like the typical R user.

But what is cloud computing? As per the National Institute of Standards and Technology definition, cloud computing is a model for enabling convenient, on-demand network access to a shared pool of configurable computing resources (e.g., networks, servers, storage, applications, and services) that can be rapidly provisioned and released with minimal management effort or service provider interaction. A tool GUI for connecting to an R-based cloud is available at http://wwwdev.ebi.ac.uk/Tools/rcloud/.

A basic, and probably the first, tutorial for running R on Amazon's Elastic Compute Cloud (EC2) was created by Robert Grossman. The Amazon cloud is generally considered the most popular standard for public cloud computing infrastructure.

The RAmazonS3 package provides the basic infrastructure within R for communicating with the S3 Amazon storage server and is available at http://www.omegahat.org/RAmazonS3/.

3.9.1 *Benefits of R on the Cloud*

Running R on a cloud (like Amazon EC2) has the following benefits:

1. Elastic memory and number of processors for heavy computation
2. Affordable micro instances for smaller datasets (2 cents per hour for Unix to 3 cents per hour)
3. An easy-to-use interface console for managing datasets as well as processes

3.9.1.1 Benefits of Using Windows-based R on the Cloud

Running R on an Amazon EC2 on a Windows instance has the following additional benefits:

1. Remote desktop makes operation of R very easy. This can compensate for the slightly increased cost of using Windows-based instances compared to Linux-based instances.
2. 64-bit R can be used with the greatest ease.
3. You can also use your evaluation of Revolution R Enterprise (free to academic institutions and very inexpensive for enterprise software for corporations). Revolution R also has a version with Red Hat Enterprise Linux (RHEL).

You can thus combine R GUIs (like Rattle, Rcmdr, or Deducer, depending on your need for statistical analysis, data mining, or graphical analysis) with a 64-bit OS and Revolution's RevoScaler package on an Amazon EC2 to manage extremely large datasets in a very easy-to-use analytics solution.

3.9.2 *Tutorial: Using Amazon EC2 and R (Linux)*

Creating a new interface and using R packages for Ubuntu.

This is another method to use R on an Amazon EC2 machine, renting by the hour hardware and computing resources that are scalable to massive levels; the software is free. The following procedure shows how to connect to Amazon EC2 and run R on an Ubuntu Linux system.

1. Log onto Amazon Console http://aws.amazon.com/ec2/. You will need your Amazon ID to login (you can use the same one you use to buy books). After you login click the upper tab to get into Amazon EC2.
2. Choose the right AMI: In the left margin, click AMI Images. Now search for the image (I chose Ubuntu images; Linux images are cheaper) and the latest Ubuntu Lucid in the search. Choose whether you want 32-bit or 64-bit images (64-bit images will allow for faster processing of data). Click on the launch

instance in the upper tab (near the search feature). A popup will appear that shows the five-step process to launch your computing.

3. Choose the right compute instance: there are various compute instances; and they are all at different multiples of prices or compute units and differ in terms of RAM and number of processors. After choosing the desired compute instance, click on Continue.
4. Instance details: do not choose cloudburst monitoring if you are on a budget as it carries an extra charge. For critical production it would be advisable to choose cloudburst monitoring once you have become comfortable with handling cloud computing.
5. Add tag details: if you are running many instances, you need to create your own tags to help you manage them. This is advisable if you are going to run many instances.
6. Create a key pair: a key pair is an added layer of encryption. Click on "Create new pair" and name it (the name will come in handy in subsequent steps).
7. After clicking and downloading the key pair, enter security groups. A security group is just a set of instructions to help keep your data transfer secure. Enable access to your cloud instance to certain IP addresses (if you are going to connect from a fixed IP address and to certain ports on your computer). It is necessary in a security group to enable SSH using Port 22. The last step is to review details and click Launch.
8. In the left pane, click on Instances (you were in Images > AMI earlier). It will take some 3–5 min to launch an instance. You can see status as pending till then. A pending instance is indicated by a yellow light.
9. Once the instance is running, it is indicated by a green light. Click on the checkbox, and on the upper tab go to Instance Management. Click on Connect. You will see a popup with instructions like these:
10. Open the SSH client of your choice (e.g., PuTTY, terminal).
11. Locate your private key: *nameofkeypair.pem.*
12. Use *chmod* to make sure your key file is not publicly viewable; otherwise ssh will not work: *chmod 400 nameofkeypair.pem.*
13. Connect to your instance using the instance's public DNS [ec2-75-101-182-203.compute-1.amazonaws.com].
14. Enter the following command line: ssh -i decisionstats2.pem root@ec2-75-101-182-203.compute-1.amazonaws.com,
15. If you are using Ubuntu Linux on your desktop/laptop, you will need to change the above line to ubuntu@... from root@... (or to Amazon ec2 user).
16. Put the following in the command line *ssh -i yourkeypairname.pem* -X *ubuntu@ec2-75-101-182-203.compute-1. amazonaws.com*
17. Note: The X11 package should be installed for Linux users; Windows users will use Remote Desktop.
18. Choose to install any custom packages [like
19. install.packages('doSNOW')]
20. Work in R using the command line.

21. Install R Commander (if you want to work with a GUI) on a remote machine (running Ubuntu Linux) using the command *sudo apt-get install r-cran-rcmdr*.
22. *IMPORTANT*: When you are finished, close your instance. (click Instances in left pane and check the checkbox of the instance you are running; on the upper tab, select Instance Actions and click Terminate).
23. For a screenshot-based tutorial, refer to the Web site http://decisionstats.com/2010/09/25/running-r-on-amazon-ec2/
24. To save your work, create a new image in the top margin.

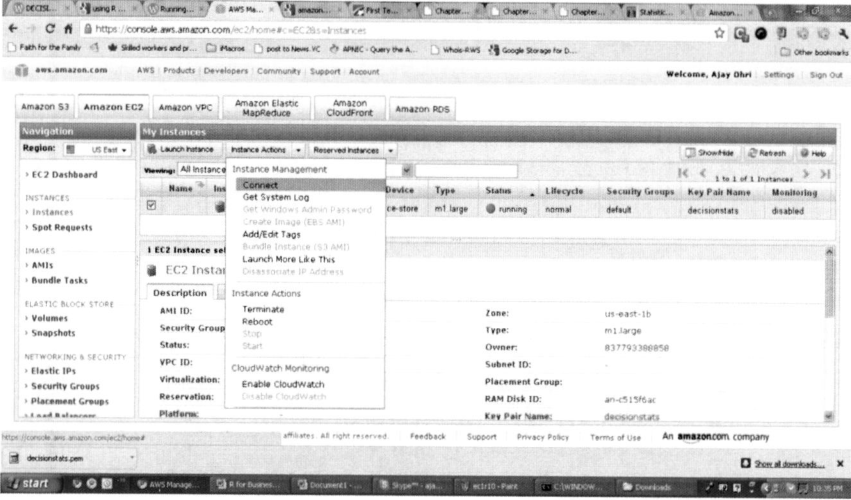

3.9.3 Tutorial: Using Amazon EC2 and R (Windows)

Note the instructions for using R on Amazon EC2 with Windows are nearly the same as those for using Linux except for using Remote Desktop, which makes it much easier for many business users. Note Windows instances cost slightly more on Amazon EC2.

1. Logon to https://console.aws.amazon.com/ec2/home.
2. Launch Windows instance.
3. Choose AMI.
4. Left pane: AMI.
5. Top Windows: – select Windows 64 AMI.
6. Note: if you select SQL Server, it will cost extra.
7. Go through the following steps and launch instance.
8. Selecting EC2 compute, depending on the number of cores, memory needs, and budget.

9. Create a key pair (a .pem file, which is basically an encrypted password) and download it. For tags, etc. just click on and pass (or read and create some tags to help you remember, and organize multiple instances). In configure firewall, remember to enable access to RDP (Remote Desktop) and HTTP. You can choose to enable the entire Internet or your own IP address/es for logging in Review and launch instance.
10. Go to instance (leftmost margin) and see status (yellow for pending). Click on Instance Actions > Connect on top bar to see the following:
11. Download the .RDP shortcut file and click on Instance Actions > Request Admin Password.
12. Wait 15 min as Microsoft creates a password for you. Click Again on Instance Actions > Request Admin Password.
13. Open the key pair file (or .pem file created earlier) using Notepad and copy-paste the private key (which will look like gibberish) and click Decrypt.
14. Retrieve password for logging on. Note the new password generated; this is your Remote Desktop password.
15. Click on the .rdp file (or Shortcut file created earlier). It will connect to your Windows instance.
16. Enter the newly generated password in Remote Desktop.
17. Login.
18. This looks like a new, clean machine with just Windows OS installed on it.
19. Install Chrome (or any other browser) if you do not use Internet Explorer.
20. Install Acrobat Reader (for documentation), Revolution R Enterprise~ 490 MB (it will automatically ask to install the .NET framework-4 files) or R.
21. Install packages (I recommend installing R Commander, Rattle, and Deducer). Apart from the fact that these GUIs are quite complimentary, they will also install almost all main packages that you need for analysis (as their dependencies). Revolution R installs parallel programming packages by default.
22. If you want to save your files to work on them later, you can make a snapshot (go to Amazon console > ec2 in the left pane > ABS > Snapshot you will see an attached memory (green light). Click on Create Snapshot to save your files to work on them later. If you want to use my Windows snapshot you can work on it; just when you start your Amazon EC2, you can click on snapshots and enter details (see snapshot name below) to make a copy or work on it to explore either 64-bit R or multicore cloud computing or just try out Revolution R's new packages for academic purposes.

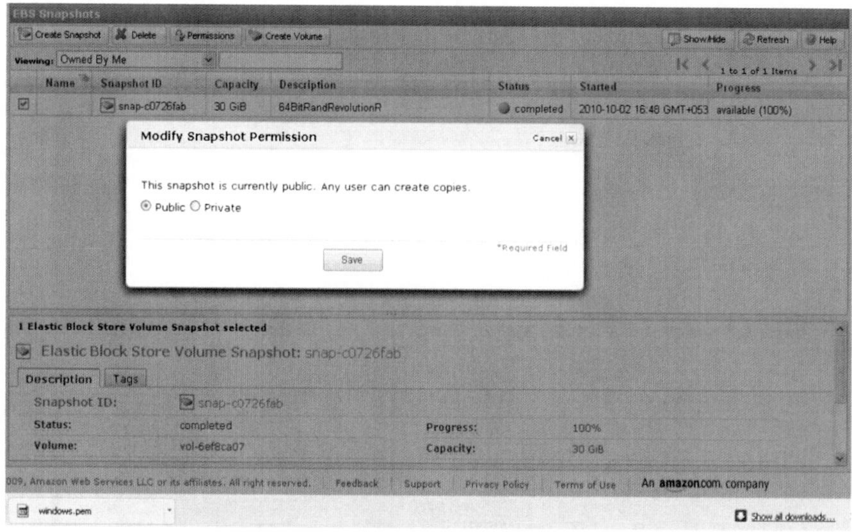

3.9.4 *Installing R on a Basic Linux Amazon EC2 Instance (Which Is Free)*

You can search for tutorials like this one: http://www.r-bloggers.com/installing-r-on-amazon-linux/. I would recommend using an existing AMI with R configured; that will cut down on learning time.

3.9.5 *Using R Studio on Amazon EC2*

There are many tutorials on using R Studio on Amazon EC2. Some of them are listed here.

- RStudio and Amazon Web Services EC2: http://community.mis.temple.edu/stevenljohnson/2011/03/12/rstudio-and-amazon-web-services-ec2/
- Securely using R and RStudio on Amazon's EC2: http://toreopsahl.com/2011/10/17/securely-using-r-and-rstudio-on-amazons-ec2/
- RStudio Server Amazon Machine Image (AMI) http://www.louisaslett.com/RStudio_AMI/

3.9.6 *Running R on the Cloud Using cloudnumbers.com*

Cloudnumbers.com allows you to work with R directly on the cloud, without getting into the business of configuring instances, by presenting a simple interface that makes it easy to transfer files and configure R packages.

- Tutorial on using Cloudnumbers.com: http://www.decisionstats.com/cloud-computing-with-rstats-and-cloudnumbers-com/

3.10 Google and R

One of the biggest users, developers, and customers of R is the Google Corporation. It has contributed to the development of R by creating a Google style guide for writing R code, helping with financing, and sponsoring R packages and conferences. These are listed below. We also look at how Google uses R internally and which R packages use Google APIs and services.

3.10.1 *Google Style Guide*

The Google R Style Guide is a list of instructions by coders who wish to work in R. It was created due to the large number of R coders at Google to standardize the documentation and help each other read the code. It aims at standardizing some elements of the production coding in R. It is available at http://google-styleguide.googlecode.com/svn/trunk/google-r-style.html.

What follows is a partial summary listing of Google's R Style Guide.

- Summary: R Style Rules

File names: end in .R
Identifiers: variable.name, FunctionName, kConstantName
Line length: maximum 80 characters
Indentation: two spaces, no tabs
Curly braces: first on same line, last on own line
Assignment: use < −, not = (I always use = because it does in one keystroke what < − does in three keystrokes)
Semicolons: do not use
General layout and ordering commenting guidelines: all comments begin with # followed by a space; inline comments need two spaces before the #

- Summary: R Language Rules

1. a. Attach: avoid
   b. Functions: errors should be raised using stop()
   c. Objects and methods: avoid S4 objects and methods when possible; never mix S3 and S4

This part is especially useful:

- Commenting Guidelines

Comment your code. Entire commented lines should begin with # and one space. Short comments can be placed after code preceded by two spaces, #, and then one space.

```
# Create a histogram of frequency of campaigns by percentage of budget spent:
hist(df$pctSpent, breaks = "scott", # method for choosing number of buckets
main = "Histogram: fraction of budget spent by campaignid",
xlab = "Fraction of budget spent",
ylab = "Frequency (count of campaignids)")
```

3.10.2 Using R at Google

Analysts and engineers at Google use a lot of R because they need a flexible analytical platform that can be customized. Any arguments that R would not work for larger datasets can be put to rest based on the fact that Google uses it for its own internal analysis for BIGDATA.

- The following image shows an example of the use of R at Google for massively parallel computational infrastructure for statistical computing using the MapReduce paradigm for R.

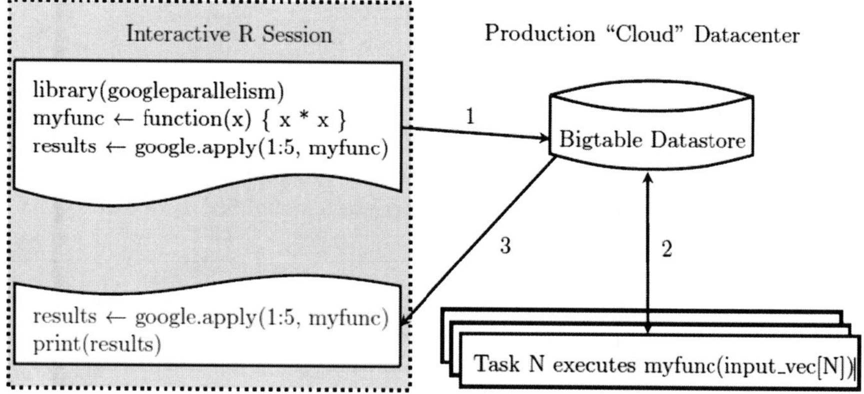

The complete research paper is available at http://research.google.com/pubs/pub37483.html.

- Visit http://google-opensource.blogspot.in/search/label/r for Google's work in R, which includes frequent daily use of R. Google is also a supporter of the R language, sponsors many student projects in R in Google's Summer of Code contest, and sponsors meetups, user groups, and conferences associated with R.
- The **int64** package was sponsored by Google to work around the lack of 64-bit integers in R for several years at Google. The package is available at http://cran.r-project.org/web/packages/int64/index.html It was created by Romain Francois and sponsored by the Google Open Source Programs Office. The following things can be done by int64.
  - Perform arithmetic operations between 64-bit operands or between int64 objects and integer or numeric types in R.
  - Read and write CSV files including 64-bit values by specifying int64 as a colClasses argument to read.csv and write.csv (with int64 version 1.1).
  - Load and save 64-bit types with the built-in serialization methods of R.
  - Compute summary statistics of int64 vectors, such as max, min, range, sum, and the other standard R functions.
- The **RProtoBuf** package makes it easy for analysts and engineers at Google to use R with Protocol Buffers read from Bigtable or other data stores. Protocol Buffers are a way of encoding structured data in an efficient yet extensible format. Google uses Protocol Buffers for almost all of its internal RPC protocols and file formats (see http://cran.r-project.org/web/packages/RProtoBuf/index.html and http://code.google.com/p/protobuf/).
- **Deducer GUI plot builder**, an example of Google giving back to the R Community, was funded by Google.
- Google Tech Talks on Engineering Data Analysis are for the many internal Google teams using ggplot2 software for visualization. Please see the Youtube video lectures at http://www.youtube.com/watch?v=TaxJwC_MP9Q and http://www.youtube.com/watch?v=iSXNfZESR5I.
- An automated R tool http://www.statistik.uni-dortmund.de/useR-2008/slides/Meyer+Lysen.pdfR (with DBI, RMySQL, and R2HTML) enabled Google to leverage statistical insights that are not accessible through standard database tools in order to identify raters that are having difficulties and communicate the results to colleagues in a production environment.
- Search pause studies were done making extensive use of R. Such studies talk about how the number of organic clicks changes when search ads are present and when search ad campaigns are turned off; see http://research.google.com/pubs/archive/37161.pdf.
- R is frequently used for analyzing Google's datacenter infrastructure. For example, this paper cites R specifically in "Failure Trends in a Large Disk Drive Population": http://research.google.com/archive/disk_failures.pdf.
- There is a Google Research paper "Availability in Globally Distributed Storage Systems", in which most of the analysis/plots are in R: http://research.google.com/pubs/pub36737.html.

- A Google employee helped create the R package **Stratasphere**, which deals with stratified spherical approximate conditional modeling. This is joint work with William Heavlin of Google and Ph.D. student Danie Percival during an internship at Google in summer 2011 (http://cran.r-project.org/web/packages/stratasphere/).
- A blog article on R use at Google can be downloaded from http://blog.revolutionanalytics.com/2011/08/google-r-effective-ads.html.

3.10.3 Using Google Services and R Packages

Most Google-related R packages are in data visualization, geographic maps, and big data computing (Map-reduce). The following R packages and projects use Google API and services for analysis.

googleVis is an R package providing an interface between R and the Google Visualization API. The functions of the package allow users to visualize data with the Google Visualization API without uploading their data to Google: http://code.google.com/p/google-motion-charts-with-r/ and http://cran.r-project.org/web/packages/googleVis/index.html.

plotGoogleMaps: Plot HTML output with Google Maps API and your own data. This helps plot SP data as HTML map mashup over Google Maps. This package provides an interactive plot device for handling geographic data for Web browsers. It is optimized for the Google Chrome browser and designed for the automatic creation of Web maps as a combination of user data and Google Maps layers: http://cran.r-project.org/web/packages/plotGoogleMaps/

RgoogleMaps: Overlays on Google map tiles in R. This package serves two purposes: (i) provide a comfortable R interface to query the Google server for static maps and (ii) use the map as a background image to overlay plots within R. This requires proper coordinate scaling: http://cran.r-project.org/web/packages/RgoogleMaps/index.html

Maptools: These also allow you to annotate maps in Google Earth: http://cran.r-project.org/web/packages/maptools/. The function kmlOverlay is used to create and write a KML file for a PNG image overlay for Google Earth.

RGoogleAnalytics: This project provides access to Google Analytics data natively from the R Statistical Computing programming language. You can use this library to retrieve an R data frame with Google Analytics data. You can then perform advanced statistical analysis such as time series analysis and regressions: http://code.google.com/p/r-google-analytics/.

RCurl to read Google documents: You can also use Rcurl and the following modified code to read a Google Docs document. Note that the key parameter changes for each document.

```
require(RCurl)
download.file(url="http://curl.haxx.se/ca/cacert.pem", destfile="cacert.pem")
url="https://docs.google.com/spreadsheet/pub?key=
0AtYMMvghK2ytcldUcWNNZTltcXdIZUZ2MWU0R1NfeWc&output=csv"
b <- getURL(url,cainfo="cacert.pem")
write.table(b,quote = FALSE, sep = ",",file="test.csv")
```

Google Docs Through Google Visualization API Query Language: Users more proficient in SQL can use the Google Visualization API query language to write SQL queries to Google Docs and use them as online databases. The query string can be added to the data source URL using the tq parameter and simply used by the preceding code.

Google query languages are defined on a spreadsheet in the following way:

See http://code.google.com/apis/chart/interactive/docs/querylanguage.html.

Translate: This R package can be used for Google Translate to translate text into multiple languages.

Bindings for the Google Translate API v2 can be found at http://cran.r-project.org/web/packages/translate/.

RWeather: This package provides programmatic access to Google Weather (and other weather) APIs:

http://cran.r-project.org/web/packages/RWeather/.

Google PredictionAPI package: Google has developed this R package, googlepredictionapi, for R users to easily access the Google Prediction API including storage objects in Google Storage. It is available at https://code.google.com/p/google-prediction-api-r-client/.

r-google-prediction-api-v12 (R package implementing Google Prediction API v1.2): This is a modification and extension of the R client library for the Google Prediction API (Markko Ko, Google Inc., 2010) for Google Prediction API v1.2 and is available at http://code.google.com/p/r-google-prediction-api-v12/.

Fusion Table R package: Andrei Lopatenko, a Google engineer, wrote an R interface to Fusion Tables to enable use of Fusion Tables as the data store for R. See http://andrei.lopatenko.com/rstat/fusion-tables.R and http://www.google.com/fusiontables/Home/.

gooJSON: A suite of helper functions for obtaining data from Google Maps API JSON objects. It calls Google Maps API and returns results as an R data frame: http://cran.r-project.org/web/packages/gooJSON/index.html.

Note a few of the preceding packages are sensitive to changes in authentication or in the Google APIs and sometimes stop working when such changes occur.

3.11 Interview: Using R at Google

What follows is an interview with Murray Stokely, software engineer at Google.

Ajay: How do engineers and analysts use R at work at Google? Name some R packages that are particularly widely used and their application use cases. How many R users work with Google (approximately)?

Murray: There are hundreds of regular R users at Google across a variety of functions, and there are dozens of open job requisitions requiring R skills. Some of the job titles for people that use R regularly include Quantitative Analyst, Statistician, Financial Analyst, Network Analyst, Network Capacity Planner, Business Strategist, User Experience Researcher, Operations Decisions Support, Site Reliability Engineer, and, of course, Software Engineer [google.com/jobs; search for "R"].

Among the R code that is checked into our main company source code management system, the lattice and ggplot2 plotting packages are the two most commonly used. These packages are used by many teams for interactive data analysis, for generating plots for reports, presentations, and papers, and for Web dashboards on our corporate network.

The Dremel package is the most popular of several dozen internal packages written to provide interactive access to our distributed datastores and computation infrastructure from within R. Our data analysts use this to perform queries and return R data.frames computed over thousands of CPU cores and petabytes of data [http://research.google.com/pubs/pub36632.html].

Ajay: Apart from R, what other alternatives in statistical computing do you use at Google (Python, SAS, SPSS)? What do you think the future of R will be in enterprises?

Murray: Python is one of the more popular languages used at Google. It is difficult to fairly compare a general scripting language like Python with a statistical computing environment like R. It is rare that the two languages are considered for the same types of tasks. Often the two languages will be used together, where a Web dashboard or data extraction pipeline written in Python will interface with R for those parts requiring higher-level statistical functions not built into any of the Python frameworks. For example, we've written an open source Web server that implements the PGI protocol and makes it easy to integrate Python and R in this way [http://code.google.com/p/polyweb/].

At Google we work with really big data, and it is not uncommon for a new statistician on his or her first week on the job to come across a problem that may require splitting up a dataset and running a few thousand copies of R simultaneously in one of our datacenters.

The fact that R is open source makes it very easy for us to integrate the tool with our internal data storage tools and cloud computing platforms, in a way that would not be possible with a commercial tool.

Inside Engineering at least, SAS and SPSS are seldom, if ever, used at Google. It is possible they are used in other parts of the company.

Ajay: Name some things that you could only do in R, and no other analytics language can help you there.

Murray: One recurring use case for R I have had at Google is in time series forecasting. None of the Python frameworks currently has the wealth of forecasting models available in R. Of course, something like linear regression is one line of code, but so too is fitting an ARIMA model with seasonality, calculating the partial autocorrelation function, or filtering outliers from data. These things can be done in other languages, but the wealth of add-on packages provided by CRAN is not available in other environments.

3.12 Interview: Using R Through Cloud Computing at cloudnumbers.com

An extract from an interview from Markus Schmidberger, Senior Community Manager for cloudnumbers.com

Ajay: What advantages does cloudnumbers.com give me over a plain vanilla Amazon EC2 instance?

Markus: If you have the time and knowledge, you can set up a computer cluster with Amazon EC2 similar to cloudnumbers.com, too. At cloudnumbers.com you get everything preinstalled and preconfigured. Everything works out of the box, and you won't need any more administration or have to deal with security keys. Furthermore, we provide a very powerful and user-friendly Web interface for starting, working, monitoring, … the computer cluster in the cloud. Security is our top priority. We meet high security standards by providing secure encryption for data transmission and storage. In the default configuration, you will not get this level of security at Amazon Web services.

Ajay: I see you have chosen R Studio as an R development platform for cloud computing as well. Can you describe what other choices you looked into and why cloudnumbers.com chose RStudio ahead of others?

Markus: If you follow the R community and the R-specific user conferences, you will detect a very strong RStudio hype. RStudio is a great development environment for R and combines many useful tools in one GUI. Due to the identical client- and Web-based environment, we decided to integrate RStudio into cloudnumbers.com as one sample programming environment. As requested by our clients, we can install any other environment, too. Based on the computer cluster infrastructure of our platform, you can connect Eclipse (StatER plugin) or TinnR with cloudnumbers.com, too.

Ajay: What are the exciting new features that you plan to roll out at cloudnumbers.com?

Markus: Cloudnumbers.com now provides a stable and efficient service for high-performance computing in the cloud. Our main features are the preinstalled computer clusters in the cloud with preinstalled applications. Currently, we provide our customers with R, openFoam, Python, Fortran, and C/C++. Due to our very wide target audience, we have to install several additional applications.

In coming weeks we are going to release BLAST, Freemat, Perl, and several more applications, especially for the life sciences sector. Besides the applications, cloudnumbers.com provides an HPC service. Today, HPC is more than just computer clusters with preinstalled applications. In the long run, we plan to improve our platform for big data analyses and GPU computing.

3.13 Summary of Commands Used in This Chapter

List of packages, datasets, and functions referred in this chapter

3.13.1 Packages

The following R GUIs were referenced; these also have multiple dependencies that are used by these packages:

1. R Commander and e-Plugins
2. Rattle
3. JGR and Deducer
4. RKWard
5. Red R
6. GrapheR
7. Komodo Sci-Views K
8. PMG
9. R Analytic Flow

The following Web Interfaces for R were mentioned in this chapter:

1. R Web http://www.math.montana.edu/Rweb/
2. R-Node http://squirelove.net/r-node/doku.php
3. Rcgi http://www.ms.uky.edu/~statweb/testR3.html
4. R-php http://dssm.unipa.it/R-php/
5. RWui http://sysbio.mrc-bsu.cam.ac.uk/Rwui/
6. Rpad http://code.google.com/p/rpad/
7. Concerto http://www.psychometrics.cam.ac.uk/page/300/concerto-testing-platform.htm
8. RApache and its applications http://biostat.mc.vanderbilt.edu/rapache/
9. Rook http://cran.r-project.org/web/packages/Rook/Rook.pdf
10. OpenCPU.demo http://cran.r-project.org/web/packages/opencpu.demo/index.html
11. RevoDeployR http://www.youtube.com/watch?v=fZtXv2ul8Ew

Citations and References

- Cloudnumbers.com An emerging cloud computing company in the field of high-performance computing (HPC) for scientific applications http://cloudnumbers.com/about-us
- BioConductor AMI for Amazon EC2: http://bioconductor.org/help/bioconductor-cloud-ami/
- A list of Web interfaces: http://cran.r-project.org/doc/FAQ/R-FAQ.html#R-Web-Interfaces
- "Large-Scale Parallel Statistical Forecasting Computations in R", Murray Stokely, Farzan Rohani, Eric Tassone, JSM Proceedings, Section on Physical and Engineering Sciences, 2011 http://research.google.com/pubs/pub37483.html
- Gesmann, M. and de Castillo, D. Using the Google visualisation API with R. R J. **3**(2), 40–44 (2011)
- The NIST Definition of Cloud Computing Authors: Peter Mell and Tim Grance Version 15, 10-7-09 National Institute of Standards and Technology, Information Technology Laboratory: [http://csrc.nist.gov/groups/SNS/cloud-computing/cloud-def-v15.doc)
- Using Google Spreadsheets as a Database with the Google Visualisation API Query Language: http://blog.ouseful.info/2009/05/18/using-google-spreadsheets-as-a-databace-with-the-google-visualisation-api-query-language/

You can read more on other analytics and business intelligence vendors that currently support or plan to support R here:

- SAP Hana with R [p. 59 https://www.experiencesaphana.com/docs/DOC-1138]
- IBM Netezza [http://thinking.netezza.com/blog/embrace-open-source-analytics]
- IBM SPSS [http://www-01.ibm.com/software/analytics/spss/products/statistics/developer/]
- Teradata R [http://developer.teradata.com/applications/articles/in-database-analytics-with-teradata-r]
- Microsoft HPC with R
  ∗[http://channel9.msdn.com/Shows/The+HPC+Show/High-performance-Analytics-with-REvolution-R-and-Microsoft-HPC-Server]
  ∗[http://www.microsoft.com/hpc/en/us/solutions/hpc-case-studies-life-sciences.aspx]
  ∗[http://blog.revolutionanalytics.com/high-performance-computing/]
- Oracle R [http://www.oracle.com/us/corporate/features/features-oracle-r-enterprise-498732.html]
- JMP with R [http://www.jmp.com/applications/analytical_apps/]
- SAS IML with R [http://support.sas.com/rnd/app/studio/Rinterface2.html]

In the next chapter, we will focus on importing data within the R analytical environment. The use of at least some GUI is presumed throughout the book for basic analytical tasks.

Chapter 4
Manipulating Data

R has different types of data storage such as lists, arrays, and data frames. This can be confusing for some analysts with a pure background in handling rectangular datasets like data (with rows for records and variables for columns). The first and often the toughest or most time-consuming task in an analytical environment for a new project is getting the data loaded into the analytical software. This chapter discusses the techniques for reading in data from various formats. The two main methods of inputting data are through the command line and a GUI, and different packages for bigger datasets (>1 GB) are discussed. In addition, obtaining data from various types of databases is specifically mentioned. Analyzing data can have many challenges associated with it. In the case of business analytics data, these challenges or constraints can have a marked effect on the quality and timeliness of the analysis as well as the expected versus actual payoff from the analytical results.

4.1 Challenges of Analytical Data Processing

Most of the challenges in a business analytics project relate to data issues in the earlier stages of the project. It is at this point that technical knowledge and analytical ability must combine a data quality mindset with some creativity. Analysis is easier; it is the data issues that are sometimes more difficult to deal with in a business project. Some challenges in processing data for business analytics are discussed.

4.1.1 Data Formats

There is a need to read in complete data without losing any part (or metadata) or adding superfluous details (which increase the project scope). Data formats account for a large chunk of data input issues. The technical constraints of data formats are relatively easy to navigate within the R language thanks to ODBC and well-documented and easily searched syntax and language.

A. Ohri, *R for Business Analytics*, DOI 10.1007/978-1-4614-4343-8_4,

4.1.2 *Data Quality*

Perfect data exist in a perfect world. The price of perfect information is one business stakeholders will mostly never budget or wait for. To deliver inferences and results based on summaries of data that have missing, invalid, or outlier data embedded within it makes the role of an analyst just as important as the tool that is chosen to remove outliers, replace missing values, or treat invalid data. Garbage in, garbage out is the term used to indicate how bad quality in data input will lead to inferior quality results.

4.1.3 *Project Scope*

Some questions that an analyst needs to ask before the project gets under way are as follows:

- How much data are required from external sources?
- How much data are required in analytical detail versus high level summary?
- What are the timelines for delivery and refreshing of data analysis? What checks will be in place (statistical as well as business)?
- How easy is it to load and implement the new analysis in the existing information technology infrastructure?

These are some of the outer parameters that can limit your analytical project scope, your analytical tool choice, and your processing methodology. The costs of additional data augmentation (should you pay to have additional credit bureau data to be appended?), time for storing and processing the data (every column needed for analysis can lead to the addition of as many rows as a whole dataset, which can be a time-consuming problem if you are considering an extra 100 variables with a few million rows), but above all that of business relevance and quality guidelines will ensure basic data input and massaging are considerable parts of the entire analytical project timeline.

4.1.4 *Output Results vis-à-vis Stakeholder Expectation Management*

Stakeholders like to see results, not constraints, hypotheses, assumptions, p-values, or chi-square values. Output results need to be streamlined to a decision management process to justify the investment of human time and effort in an analytical project; choices, training, and navigating analytical tool complexities and constraints are subsets of this. Marketing wants more sales, so they need a campaign to target certain customers via specific channels with specified collateral. To justify their business judgment, business analytics needs to validate, cross validate, and, sometimes, invalidate this business decision making with clear transparent methods and processes.

Given a dataset, the basic analytical steps that an analyst will go through are as follows:

- Data input
- Data processing
- Descriptive statistics
- Data visualization
- Model or report creation
- Output
- Presentation

4.2 Methods for Reading in Smaller Dataset Sizes

Some of the most common methods for reading in data are given below.

Detailed information on data import and export is available at http://cran.r-project.org/doc/manuals/R-data.html.

Hint: When using the command line in R, pressing the up arrow button give the last syntax that you wrote. This is useful if you are not used to writing code in a command line environment and have to troubleshoot or rerun a particular batch of code many times.

Note: R is case sensitive, unlike other analytic languages. Thus "anonymous" is not the same as "Anonymous" when referring to the name of a variable, package, dataset, or function.

4.2.1 CSV and Spreadsheets

The following command will read in a CSV file format. The CSV format is ideally suited for smaller datasets, as comma-delimited values. Note here the path and filename are given by *"file:///home/k1/Downloads/dataloss.csv"* and the input dataset is named inputdata. Note that the file path is based on Linux systems.

> inputdata <-read.csv("file:///home/k1/Downloads/dataloss.csv", na.strings = c (".", "NA", "", "?"), strip.white=TRUE, encoding="UTF-8")

The Read.table command reads in datasets in table format.

Importing data is performed by the command

Dataset <− read.table("C:/Users/Owner/Desktop/AMSsurvey.txt", header= TRUE, + sep = ",", na.strings="NA", dec=".", strip.white=TRUE)

This can be done by either writing the whole command by you or just by clicking "Import Data" or "Export Data" in a GUI.

Note: Exporting data is performed by the command
write.table(AMSsurvey, "C:/Users/Owner/Desktop/AMSsurvey.txt", sep=",",
+ col.names=TRUE, row.names=TRUE, quote=TRUE, na="NA")

The path of import and export is given as above; what is noteworthy is the kind of delimiters that are specified (here comma delimited is given by the option sep=","). Headers and variable names are also given by the options header=TRUE in the import (or read.table) and by col.names=TRUE in the export (or write.table).

4.2.1.1 Excel Data

A lot of business data reside in Microsoft Excel files. To read in Excel files you can use the R packages gdata or XLconnect. There are specific functions for reading in a particular worksheet by name, number, or even specfic areas of worksheets.

See http://cran.r-project.org/web/packages/XLConnect/XLConnect.pdf and http://cran.r-project.org/web/packages/gdata/gdata.pdf.

>library(gdata)

> read.xls("C:\\Users\\KUs\\Desktop\\test.xls") # Note we are using two slashes "\"for the Windows file path name.

We can also use one forward slash instead of two backward slashes.

Almost everything in R can be done in more than one way.

> read.xls("C:/Users/KUs/Desktop/test.xls")

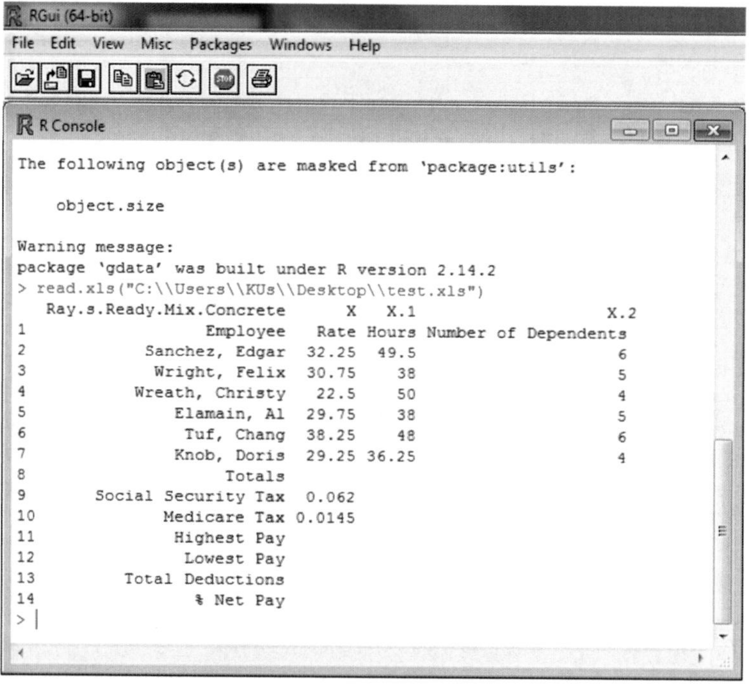

or
> *read.xls("C:\\Users\\KUs\\Desktop\\test.xls",sheet=1) #Reading Excel by worksheet number*
> *read.xls("C:\\Users\\KUs\\Desktop\\test.xls",sheet="Sheet1") #Reading Excel by worksheet name*

4.2.2 Reading Data from Packages

For the purpose of learning R or creating training tutorials you need datasets. The command data() will list all the datasets in loaded packages. Reading in data is now just a function of loading the package and using the data (datasetname) command to load the dataset. Many R packages have datasets for training purposes. One of the packages is datasets, the second is agricultural data from the package agridat at http://cran.r-project.org/web/packages/agridat/index.html, and the other is HistData for important datasets that became milestones in data visualization. You can read more on HistData at http://cran.r-project.org/web/packages/HistData/HistData.pdf.

4.2.3 Reading Data from Web/APIs

You can read in data directly from the Web using custom packages as well as by using APIs.

There are multiple packages in R to read data straight from online datasets. They are as follows:

- **Google Prediction API package**—http://code.google.com/p/r-google-prediction-api-v12/. Upload your data to Google Storage and then train them using this package for the Google Prediction API.
- **RCurl** from http://www.omegahat.org/RCurl/ allows us to download files from Web servers, post forms, use HTTPS (the secure HTTP), use persistent connections, upload files, use binary content, handle redirects and password authentication, etc. The primary top-level entry points are getURL(), getURLContent(), getForm(), and postForm().
- The basic package for Web data reading is **HttpRequest** http://cran.r-project.org/web/packages/httpRequest/. It uses the HTTP Request protocol and implements the GET, POST, and multipart POST request.
- The **Infochimps** package provides functions to access all of the APIs currently available at infochimps.com. For more information see http://api.infochimps.com/. Infochimps.com has 14,000 datasets.
- **WDI** You can access all World Bank data using the R package at http://cran.r-project.org/web/packages/WDI/index.html.
- **Quantmod** allows you to download financial data from Yahoo Finance http://cran.r-project.org/web/packages/quantmod/index.html. Also see http://www.quantmod.com/.

- **Rdatamarket** retrieves data from DataMarket.com, either as time series in zoo form (dmseries) or as long-form data frames (dmlist). Also see https://github.com/DataMarket/rdatamarket. http://datamarket.com/ has 100 million time series from the most important data providers, such as the UN, World Bank, and Eurostat.

  XML package Most packages in this category end up dependent on the XML package that is used to read and create XML (and HTML) documents (including DTDs), both local and accessible via HTTP or FTP http://cran.r-project.org/web/packages/XML/index.html.
- The **RBloomberg** package can access Bloomberg data (but requires a Bloomberg installation on a Windows PC).
- Other packages for reading online data are the **scrapeR** tools: http://cran.r-project.org/web/packages/scrapeR/index.html.
- Many people find **RJSON** useful for data interchange: http://cran.r-project.org/web/packages/rjson/index.html. Converts R objects into JSON objects and vice versa: http://www.json.org/JSON. (JavaScript Object *Notation*) is a lightweight data-interchange format. It is easy for humans to read and write and easy for machines to parse and generate.

An example of scraping a Web page is given here.

```
Downloading in R
con <- url("http://www.nytimes.com", "r")
x <- readLines(con) # We can do this using the function readLines to read the
# html or we can use the function getURL in the R Package RCurl
url=("http://www.nytimes.com")
library(RCurl)
ans <- getURL(url)
ans<-as.data.frame(ans)
```

4.2.4 *Missing Value Treatment in R*

Business data often have missing values and need to be cleaned or rectified for analysis. The R language stores missing values using NA. Other analytical software may store missing values as ". " This leads to how to find missing values in data. NA is neither numerals nor characters but simply the sign of a missing value. The function is.na returns a TRUE value if there are missing values. The function na.fail returns an error if there are missing values, and the function na.omit returns an object with missing values excluded. By default, R will not give any values for mean or standard deviation if there exists a missing value in the data.

We generate 10^8 random values with normal distribution, mean 10, and standard deviation 5 using the rnorm function. Note how we use the system.time() operator to find the time taken in seconds to do this particular operation. The function is.na returns a TRUE value if there are missing values.

The function na.fail returns an error if there are missing values in a data object, and the function na.omit returns an object with missing values excluded.

```
> ajay=c(rnorm(1e8,10,10),NA)
> system.time(is.na(ajay))
user system elapsed
0.36 0.04 0.41
> na.fail(ajay) Error in na.fail.default(ajay) : missing values in object
> mean(ajay)
[1] NA
> mean(ajay,na.rm=TRUE)
[1] 10 >sd(ajay) [1] NA
> sd(ajay,na.rm=TRUE)
[1] 5.00041
> mean(ajay)
[1] NA
> mean(na.omit(ajay) + )
[1] 10
```

Those who often work with missing values or plan to do so with R are enouraged to read the concise tutorial at http://www.ats.ucla.edu/stat/r/faq/missing.htm. For a list of software for missing value imputation including functions in R see http://www.math.smith.edu/~nhorton/muchado.pdf.

4.2.5 Using the as Operator to Change the Structure of Data

To a beginner in R, the multiple and flexible ways in which data can be stored is sometimes confusing and often problematic in the way several functions need only certain types of data objects. You can convert an R data object from one class to another by just using the "as.name" operator.

To convert a list named ajay into a data frame, use *ajay=as.data.frame(ajay)*, to convert it into a matrix use *ajay=as.matrix(ajay)*, and to convert it into a list use *ajay=as.list(ajay)*. This is important as for many business analytics purposes we will be using data.frame, and many packages/functions require other data structures to read in data.

We show an example of reading data from a csv file and converting the factor levels into numeric and character variables:

```
> DecisionStats <- read.table("C:/Users/KUs/Desktop/ga3.csv", header=TRUE,
+ sep=",", na.strings="NA", dec=".", strip.white=TRUE)
> str(DecisionStats)
'data.frame': 292 obs. of 2 variables:
$ X6.1.2011: Factor with 292 levels "1/1/2012","1/10/2012",..: 182 193 195
196 197 198 199 200 172 173 ...
$ X0: Factor with 220 levels "0","1,043","1,134",..: 1 36 217 11 69 67 67 57 74
18 ...
> names(DecisionStats)=c("Date","Visits")
> DecisionStats$Visits=as.numeric(DecisionStats$Visits)
> DecisionStats$Date=as.character(DecisionStats$Date)
> str(DecisionStats)
'data.frame': 292 obs. of 2 variables:
$ Date : chr "6/2/2011" "6/3/2011" "6/4/2011" "6/5/2011" ...
$ Visits: num 1 36 217 11 69 67 67 57 74 18 ...
```

4.3 Some Common Analytical Tasks

What follows is a list of common analytical tasks. You may want to review your own analytics practice and see if any of these are relevant. The Rattle GUI offers a wide variety of convenient data input formats. R Commander also offers a convenient way to get your data in without bothering about syntax. Other GUIs were covered in Chaps. 3 and 12.

4.3.1 Exploring a Dataset

For the purpose of business analytics using R, we will exclusively refer to data frames, though other methods and forms of data can be read by the R programming language. This section is an elaboration of the startup tutorial mentioned in Chap. 2, Sect. 2.4. Additional graphical options for data exploration or data visualization options are given in Chap. 5, Sect. 5.4.

What is in a dataset? Variables.

names(dataset) gives us variable names.

What is the distribution of numeric variables?

summary(dataset) gives distributions (quartile) for numeric variables and frequency counts for categorical variables.

This will give us both minimum and maximum values for numeric variables that may be needed to treat missing values and outliers.

In R Commander this can be done by Statistics > Summaries > Active Data Set; this will first give you a warning on the number of variables it is going to process.

What is the frequency distribution of categorical variables?

table(dataset$variable_name) gives the counts for a given categorical variable.

We have used dataset$variable_name for selecting only a particular variable for an operation/function.

ls() lists all active objects in the current workspace and is useful if you want to drop some objects to free up computing resources.

To remove an object FUN, use *rm* (FUN).

To remove all objects, use *rm(list=ls())*.

> *str(dataset)* # gives the complete structure of a dataset.

head(dataset) and *tail(dataset)*: head(dataset,n1) gives the first n1 rows of a dataset, while tail(dataset,n2) gives the last n2 rows of a dataset where n1,n2 are the number of observations that you want to print out.

class(dataset) and *dim(dataset)* gives the class and dimensions of the data object.

4.3.2 Conditional Manipulation of a Dataset

If you want to refer only to certain specific rows or columns of a dataset, use the square brackets to refer to specific parts of the data object.

For example:

```
> str(DecisionStats)
'data.frame': 292 obs. of 2 variables:
$ Date : chr "6/2/2011" "6/3/2011" "6/4/2011" "6/5/2011" ...
$ Visits: num 1 36 217 11 69 67 67 57 74 18 ...
> DecisionStats2=DecisionStats[1,] # This gives us Row 1
> str(DecisionStats2)
'data.frame': 1 obs. of 2 variables: $ Date : chr "6/2/2011" $ Visits: num 1
> DecisionStats3=DecisionStats[,1] # This gives us Column 1
> str(DecisionStats3)
chr [1:292] "6/2/2011" "6/3/2011" "6/4/2011" "6/5/2011" "6/6/2011" "6/7/2011"
"6/8/2011" "6/9/2011" "6/10/2011"
```

4.3.2.1 Attaching a Dataset from a Package

data(Titanic, package="datasets")

4.3.2.2 Variable Selection

- Keeping only some variables

Using a subset we can keep only the variables we want:

Sitka89 <- subset(Sitka89, select=c(size,Time,treat))

Will keep only the variables we have selected (size,Time,treat).

- Dropping some variables

Harman23.cor$cov.arm.span <- NULL This deletes the variable named cov.arm.span in the dataset Harman23.cor

- Renaming variables (using the package gregmisc)

library(gregmisc)
rename.vars(dataset, from="OldName", to="NewName")
data <- data.frame(x=1:10,y=1:10,z=1:10)
names(data)
data <- rename.vars(data, c("x","y","z"), c("first","second","third"))
names(data)

- Keeping records based on character condition

Titanic.sub1<-subset(Titanic,Sex=="Male")
Note the double equals sign (essential for both characters and numerals; otherwise it might give bad results)

- Keeping records based on date/time condition

subset(DF, as.Date(Date) >= '2009-09-02' & as.Date(Date) <= '2009-09-04')

- Current Date and Time

To obtain age or difference in time from a date/time object with the current date and time, we need functions to obtain the date and time of the system. The functions Sys.Date and Sys.time give the current date without and with time. You can also use the function date().

```
> Sys.time()
[1] "2012-04-06 18:37:28 IST"
> Sys.Date()
[1] "2012-04-06"
> date()
[1] "Fri Apr 06 18:39:19 2012"
```

- Converting date/time formats into other formats

If the variable dob is (01/04/1977), then the following will convert into a date object:
z=strptime(dob,"%d/%m/%Y")
and if the same date is 01Apr1977, then
z=strptime(dob,"%d%b%Y")

- Difference in date/time values and current time

The difftime function helps in creating differences in two date/time variables:
difftime(time1, time2, units='secs')
or
difftime(time1, time2, tz = "", units = c("auto", "secs", "mins", "hours", "days", "weeks"))

For the current system date/time values you can use
Sys.time()
Sys.Date()
These values can be put in the difftime function shown previously to calculate age or time elapsed.

- Keeping records based on numerical condition

Titanic.sub1<-subset(Titanic,Freq >37)

4.3.2.3 Sorting Data

> Sorting a data.frame object based on multiple columns.
> To do this, you could use the function sort_df in the package reshape:
> also

- Sorting a data frame in ascending order by a variable

AggregatedData<- sort(AggregatedData, by=~ Package)

- Sorting a data frame in descending order by a variable

AggregatedData<- sort(AggregatedData, by=~ -Installed)

- Transposing a dataset

t(datasetname)

- Transforming a dataset structure around a single variable

convert data frame

Row	Subject	Item	Score
1	Subject1	ItemA	1
2	Subject1	ItemB	0
3	Subject1	ItemC	1
4	Subject2	ItemA	0
5	Subject2	ItemB	1

to

Row	Subject	ItemA	ItemB	ItemC
1	Subject1	1	0	1
2	Subject2	0	1	0

Using the *Reshape2* package we can use the melt and acast functions:
library("reshape2")
tDat.m<- melt(tDat)
tDatCast<- acast(tDat.m,Subject~Item)

If we choose not to use the Reshape package, we can use the default reshape method in R. This requires more processing time for bigger datasets.

df.wide <- reshape(df, idvar="Subject", timevar="Item", direction="wide")

- Type in data

Using the *scan()* function we can type in data in a list:

- Using Diff for lags and the CumSum function for cumulative sums

We can use the diff function to calculate the difference between two successive values of a variable.

Diff(Dataset$X)

The cumsum function helps to give the cumulative sum

cumsum(Dataset$X)

> x=rnorm(10,20) # This gives 10 randomly distributed numbers with mean 20

> x

[1] 20.76078 19.21374 18.28483 20.18920 21.65696 19.54178 18.90592 20.67585

[9] 20.02222 18.99311

> diff(x)

[1] -1.5470415 -0.9289122 1.9043664 1.4677589 -2.1151783 -0.6358585 1.7699296

[8] -0.6536232 -1.0291181 >

cumsum(x)

[1] 20.76078 39.97453 58.25936 78.44855 100.10551 119.64728 138.55320

[8] 159.22905 179.25128 198.24438

> diff(x,2) # The diff function can be used as diff(x, lag = 1, differences = 1, . . .), where "differences" is the order of differencing.

[1] -2.4759536 0.9754542 3.3721252 -0.6474195 -2.7510368 1.1340711 1.1163064

[8] -1.6827413

4.3.3 Merging Data

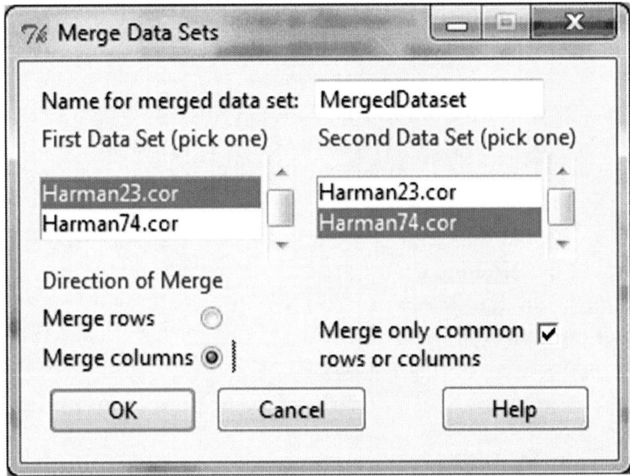

Merge using R Commander GUI

Let us compare the merge of the R Commander GUI with that of the Deducer GUI:

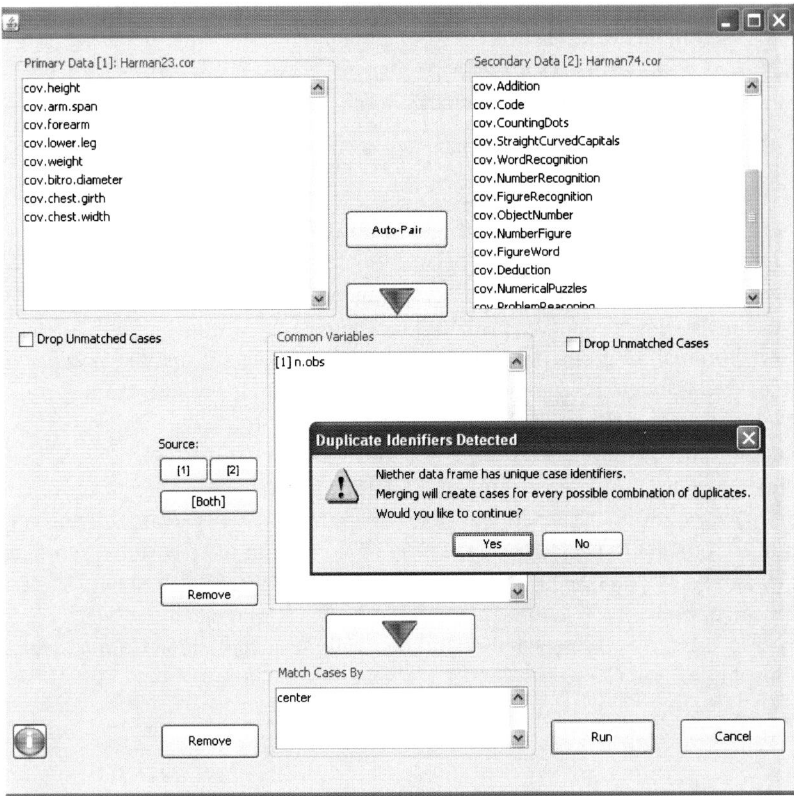

As you can see, the Deducer GUI makes it much simpler to merge datasets.

The simplest syntax for a merge statement is

totalDataframeZ <- merge(dataframeX,dataframeY,by=c("AccountId", "Region"))

Conditional joins:

Inner join: merge(df1, df2) will work for these examples because R automatically joins the frames by common variable names, but you would most likely want to specify merge(df1, df2, by="CustomerId") to make sure that you were matching on only the fields you desired.

You can also use the by.x and by.y parameters if the matching variables have different names in the different data frames.

Outer join: merge(df1, df2, all=TRUE)

Left outer: merge(df1, df2, all.x=TRUE)

Right outer: merge(df1, df2, all.y=TRUE)

4.3.4 Aggregating and Group Processing of a Variable

We can use multiple methods for aggregating by group processing of variables. Two functions we explore here are aggregate and Tapply.

Referring to the R online manual at http://stat.ethz.ch/R-manual/R-patched/library/stats/html/aggregate.html:

Compute the averages for the variables in 'state.x77', grouped
according to the region (Northeast, South, North Central, West) that
each state belongs to
aggregate(state.x77, list(Region = state.region), mean)

Using TApply

tapply(Summary Variable, Group Variable, Function)

Source:

http://www.ats.ucla.edu/stat/r/library/advanced_function_r.htm#tapply.

We can also use specialized packages for data manipulation.

For additional by-group processing, see the doBy package and the Plyr package for data manipulation. The doBy package is available at http://cran.r-project.org/web/packages/doBy/index.html. Available at http://cran.r-project.org/web/packages/plyr/index.html, Plyr is a set of tools that solves a common set of problems: you need to break a large problem down into manageable pieces, operate on each piece, and then put all the pieces back together. For example, you might want to fit a model to each spatial location or time point in your study, summarize data by panels, or collapse high-dimensional arrays into simpler summary statistics.

We can also use the R Commander GUI with the R plugin "doBy". That enables the user to use the doBy package within the GUI. To install the R Commander and load its plugins, see Chap. 3.

There are four options for using "doBy":

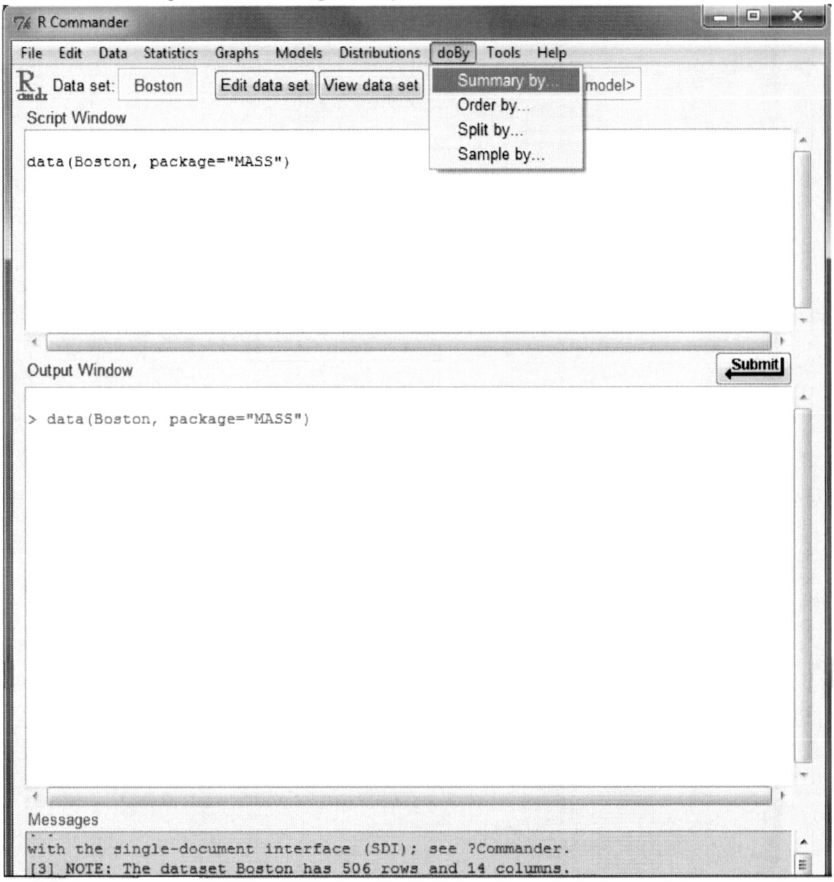

R Commander
File Edit Data Statistics Graphs Models Distributions doBy Tools Help
Data set: Boston
Edit data set
View data set
Summary by...
Order by...
Split by...
Sample by...
model>
Script Window
data(Boston, package="MASS")
Output Window
Submit
> data(Boston, package="MASS")
Messages
with the single-document interface (SDI); see ?Commander.
[3] NOTE: The dataset Boston has 506 rows and 14 columns.

Use the Summary by options (the package itself computes all numerical variables by default):

Summary by...
Name for resulting data frame (leave blank to just print):
Variables to compute statistics on
(leave blank for all numerical)
Group by
(leave blank for no grouping)
age
black
chas
crim
age
black
chas
crim
Statistic
Mean
Median
Mode
Variance
Other (specify)
OK
Cancel
Help

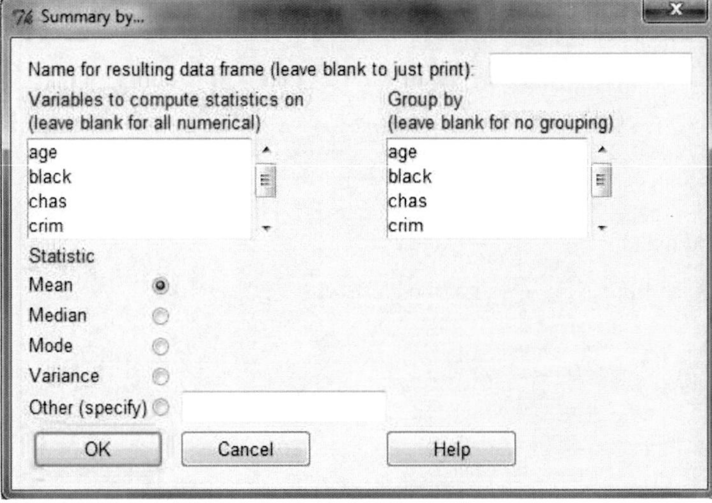

4.3.5 Manipulating Text in Data

Let us consider a data object test with three text strings:

```
> test
[1] "ajay" "AjaY O" "Vijay O"
```

- Substring: We use the substr function to take a small part of a text string. Here, Substring takes the form substr(Name of Object, Start Place of Substring, End Place of Substring).
  - We also use strsplit(x,split), which splits data objects according to the substring split:

```
> test[1]
[1] "ajay"
> substr(test[1],0,2)
[1] "aj"
> substr(test[1],1,2)
[1] "aj"
> substr(test[1],2,2)
[1] "j"
> substr(test[1],1,3)
[1] "aja"
> strsplit(test[2:3]," ")
[[1]]
[1] "AjaY" "O"
[[2]]
[1] "Vijay" "O"
```

- Concatenate: We use the paste function for concatenating two text strings and add in the separator parameter using sep= for anything else we want to use in between.

```
> paste(test[1],test[2])
[1] "ajay AjaY O"
> paste(test[1],test[2],sep=' ')
[1] "ajayAjaY O"
> paste(test[1],test[2],sep='/')
[1] "ajay/AjaY O"
```

- Search and match

grep(pattern,x) searches for matches to pattern within object x:

```
> test
[1] "ajay" "AjaY O" "Vijay O"
> grep('jay',test)
[1] 1 3
```

```
> grep('jaY',test)
[1] 2 > grep('jaY',test)
[1] 2
> grep('ja',test)
[1] 1 2 3
```

4.4 A Simple Analysis Using R

Assume you are given a dataset and you just want to perform a simple analysis on it. The R code for this is given below.

Let the dataset name be Ajay

4.4.1 Input

```
Ajay <- read.table("C://Users//KUSHU//Desktop//A.csv", header=TRUE, sep=",",
+ na.strings="NA", dec=".", strip.white=TRUE)
```

Note here the path of the input data is C:/Users/KUSHU/Desktop/A.csv

- We assume header=TRUE, which means that variable names are in the first row.
- Sep="," refers to a separator between two consecutive data elements (which is a comma here since we are reading data from a comma-delimited value).
- dec="." means we use "." to separate decimal points.
- strip.white=TRUE (for treating blank spaces)

This may look very intimidating to a new R user.

Instead, you can just use the R Commander GUI as follows:

library(Rcmdr)

and then simply click your way into the menu.

The code is automatically generated, thereby helping you learn.

4.4.2 Describe Data Structure

Now that we have inputted the data, we need to see the data quality:

- We obtain just the variable names using simply the command names:
  - *names(Ajay)*
- And we obtain the data structure using simply the command str:
  - *Str(Ajay)*
- The first five observations in a dataset can be given from
  - *head(Ajay,5)*

- The last ten observations in a dataset can be given from
  - *tail(Ajay,10)*

4.4.3 Describe Variable Structure

- We can obtain summary statistics on an entire dataset using
  - *summary(Ajay)*
- But if we only want to refer to one variable, say Surnames, and save time, we refer it to
  - *summary(Ajay$Surnames)*
  - To find out the mean and standard deviation of a particular numeric variable, use mean() and sd().
  - *mean(faithful$waiting)*
  - *sd(faithful$waiting)*
- To find the correlation between variables, it is useful to use the cor function:
  - *cor(dataset$var1,dataset$var2)*
- To find numerical distributions, we can use the command "table"
  - *table(faithful$waiting)*
- Similarly, we can plot the dataset using the simple command *plot*, histograms using *hist*, and boxplots using the *boxplot* function.
  - *plot(Ajay)*
  - *hist(Ajay)*
  - *boxplot(Ajay)*

Note: We plot the various ways to graphically represent data. The first graph in the top left is a scatterplot, the second graph in the top row is a boxplot, and the third and fourth are both histograms of particular variables, but we add a rug plot to the fourth histogram (bottom right); the fifth graph is a plot of correlations of the variables of the dataset, and the sixth and final graph is a bar plot of a variable.

```
> par(mfrow=c(2,3))
> plot(faithful)
> boxplot(faithful)
> hist(faithful$eruptions)
> hist(faithful$waiting)
> rug(faithful$waiting)
> cor(faithful) eruptions waiting
eruptions 1.0000000 0.9008112
waiting 0.9008112 1.0000000
> plot(cor(faithful))
```

+) **#The plus sign is used when we write code that spills into multiple lines.**
If we forget to close the brackets, the plus sign will remind us!
In addition, each statement is automatically submitted when we press enter. So there is no need to type "; " as a separator between various coding lines or to type and then click submit, unlike in other analytic languages you may have used.
>barplot(faithful$waiting)

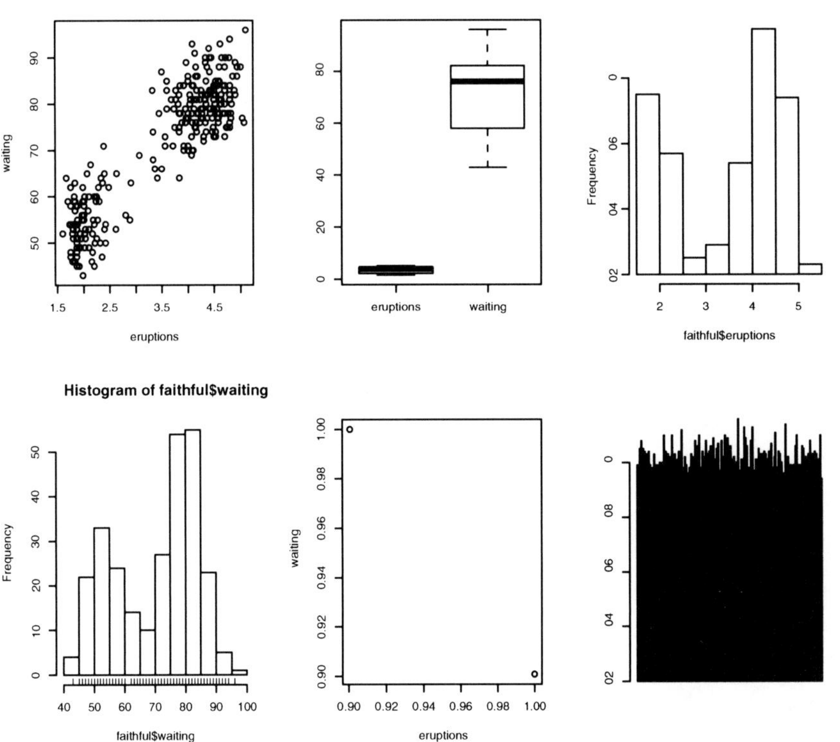

- Chapter 5 will discuss graphical analysis in more detail.
- I personally like the describe command, but for that I need to load a new package called Hmisc. The command library() loads packages.
  - *library(Hmisc)*
  - *describe(Ajay)*
- Suppose I want to add comments to my coding so it looks clean and legible. I just use the # sign, and anything after # looks commented out. This is mostly ignored but should be regularly done to make your code readable for future documentation and presentation.

4.4.4 Output

Finally, I want to save all my results. I can export them using menus in the R Commander GUI, using the menu within R, or *modifying the read.table statement to simply write.table with the new path*, which saves the dataset.

4.5 Comparison of R Graphical User Interfaces for Data Input

It is comparatively easy to use a GUI. The Rattle GUI offers the most extensive options in terms of types of inputs, except it needs data in a data frame and cannot accept data in other formats (such as R time series). Unlike other GUIs, Rattle is based mostly on tab and radio buttons and not drop-down menus.

Rattle

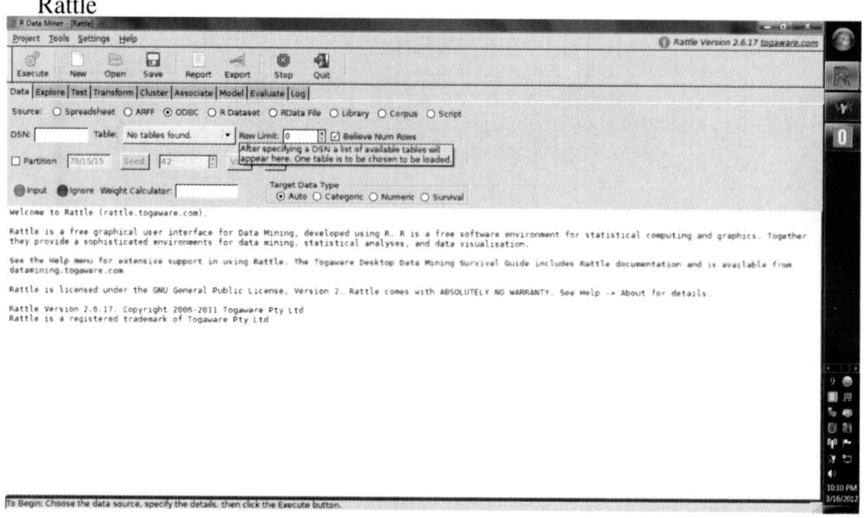

R Commander: It is quite easy to import data in R Commander with just a few clicks

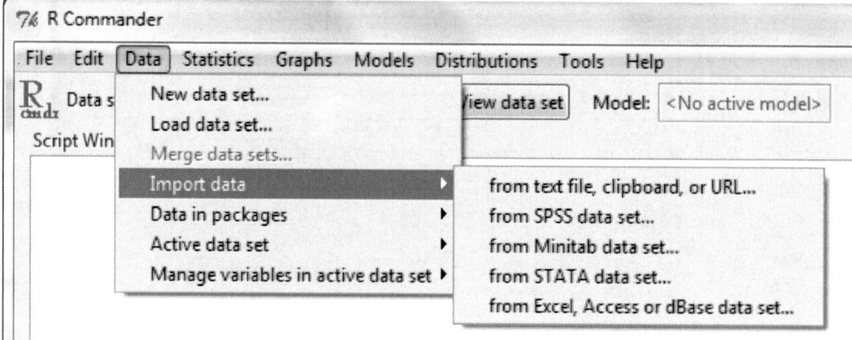

Deducer

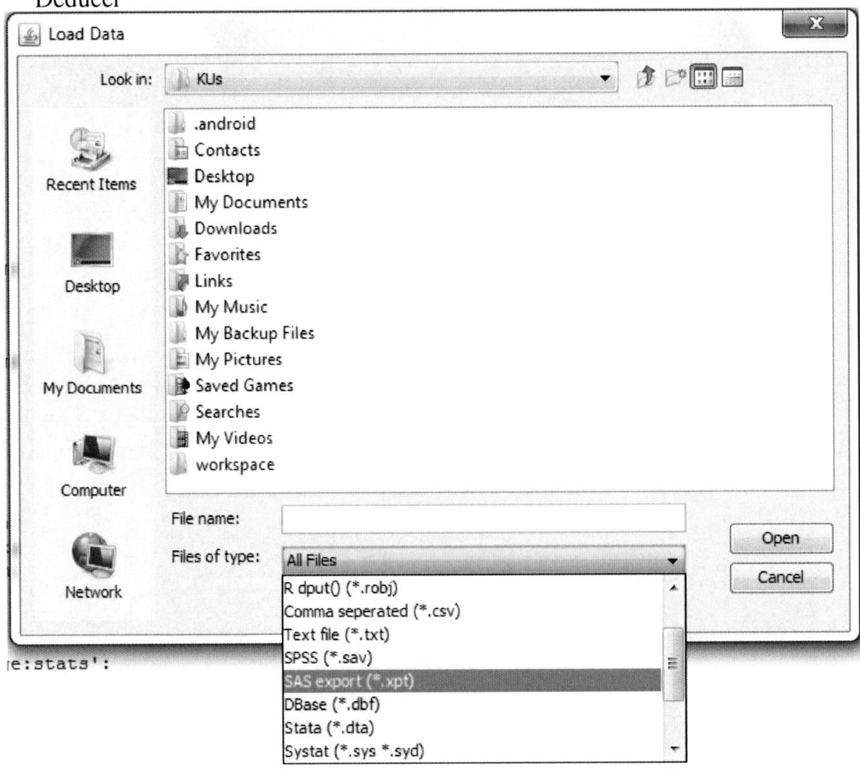

Name of R GUI	Read datasets from attached packages	Mixed R formats (*Non data frame*)
R Commander	Yes	Yes
Rattle	Yes	Not supported
JGR-Deducer	Not supported	Yes

4.6 Using R with Databases and Business Intelligence Systems

Business data are mostly stored in databases and accessed using querying tools. Databases are collectively given names of data marts, and many querying tools and associated analyses are called business intelligence (BI) tools.

Why do we use databases?

They are the cheapest, most efficient, and safest way to store data for concurrent access and storage.

From the official R documentation at http://cran.r-project.org/doc/manuals/R-data.html#Relational-databases:

Strengths of database management systems (DBMSs) and, in particular, relational DBMSs (RDBMSs):

- *Fast access to selected parts of large databases*
- *Powerful ways to summarize and cross-tabulate columns in databases*
- *Store data in more organized ways than in rectangular grid model of spreadsheets and R data frames; concurrent access from multiple clients running on multiple hosts while enforcing security constraints on access to data*
- *Ability to act as a server to a wide range of clients*

We give two case studies for databases, using MySQL and PostGreSQL. We also give information about software using Jaspersoft or Pentaho with R. We provide a brief listing of nonrelational DBMSs (noSQL) and Hadoop resources with R. Lastly, we give resources for using Google Prediction API and Google Fusion Tables within R.

4.6.1 RODBC

The RODBC package available at http://cran.r-project.org/web/packages/RODBC/index.html is the primary R package used for interfacing with databases. It was created by noted R experts Brian Ripley and Michael Lapsley.

From the documentation at http://cran.r-project.org/web/packages/RODBC/RODBC.pdf:

Two groups of functions are provided. The mainly internal odbc commands implement low-level access to the ODBC functions of a similar name. The sql* functions operate at a higher level to read, save, copy, and manipulate data between data frames and SQL tables. Many connections can be open at once to any combination of DSN/hosts.*

4.6.2 Using MySQL and R

There is a brief tutorial on working with R and MySQL. MySQL belongs to Oracle and is one of the most widely used databases.

- **Download mySQL** from http://www.mysql.com/downloads/mysql/ or (http://www.mysql.com/downloads/mirror.php?id=403831).

Click Install: Use the default options, and remember to write down the password that you use at this step=XX.

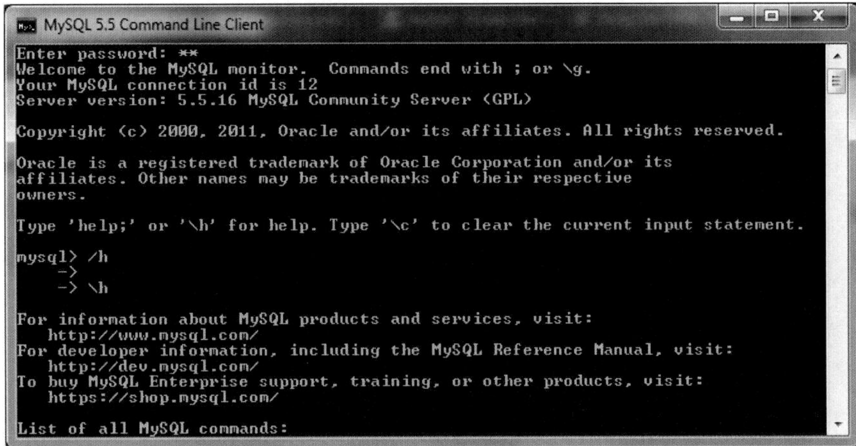

- **Download ODBC Connector** from http://www.mysql.com/downloads/connector/odbc/5.1.html.
- Locate the data sources (ODBC) in the Windows 7 control panel.

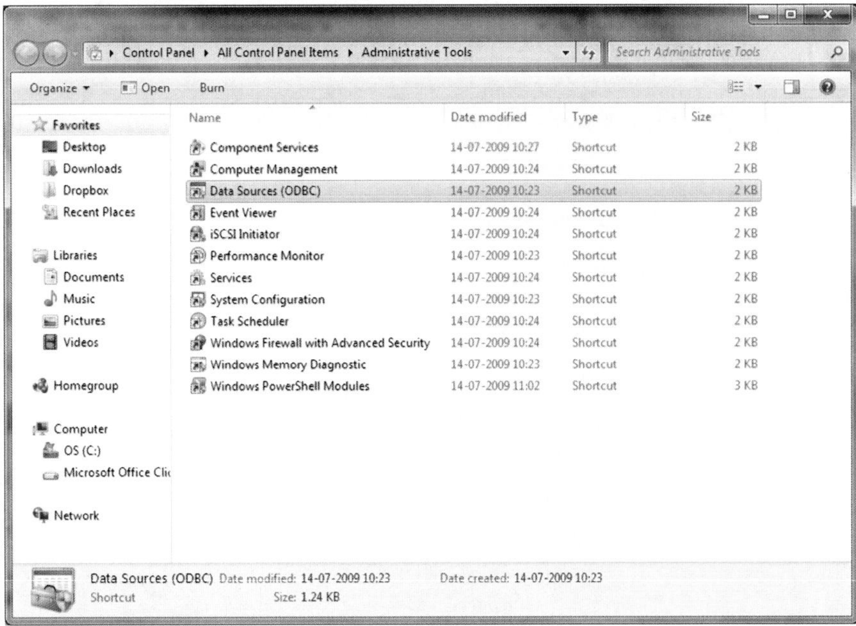

- Install **ODBC Connector** by double-clicking on the .msi file downloaded in Step 2.
- Check the following screenshot in ODBC Connectors to verify:

This is the Drivers tab in ODBC Data Source Administrator.

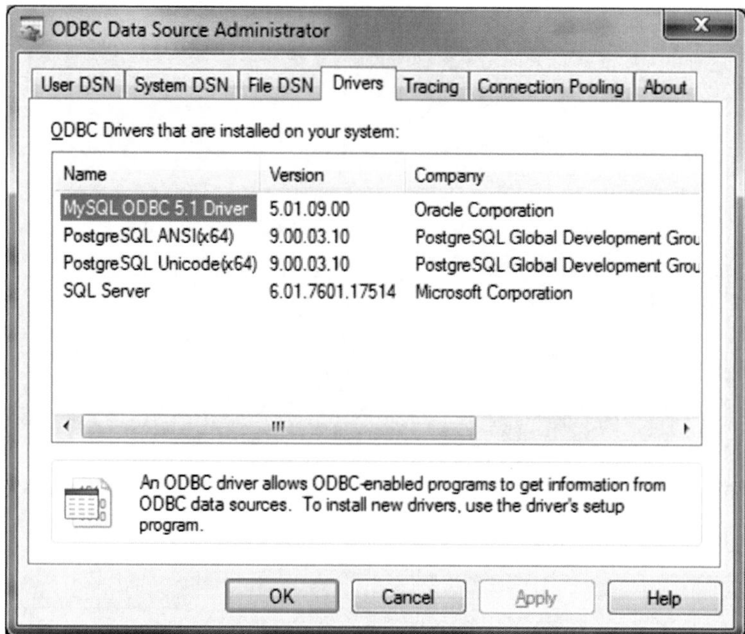

- Click **System DSN** and configure MySQL using the add button.

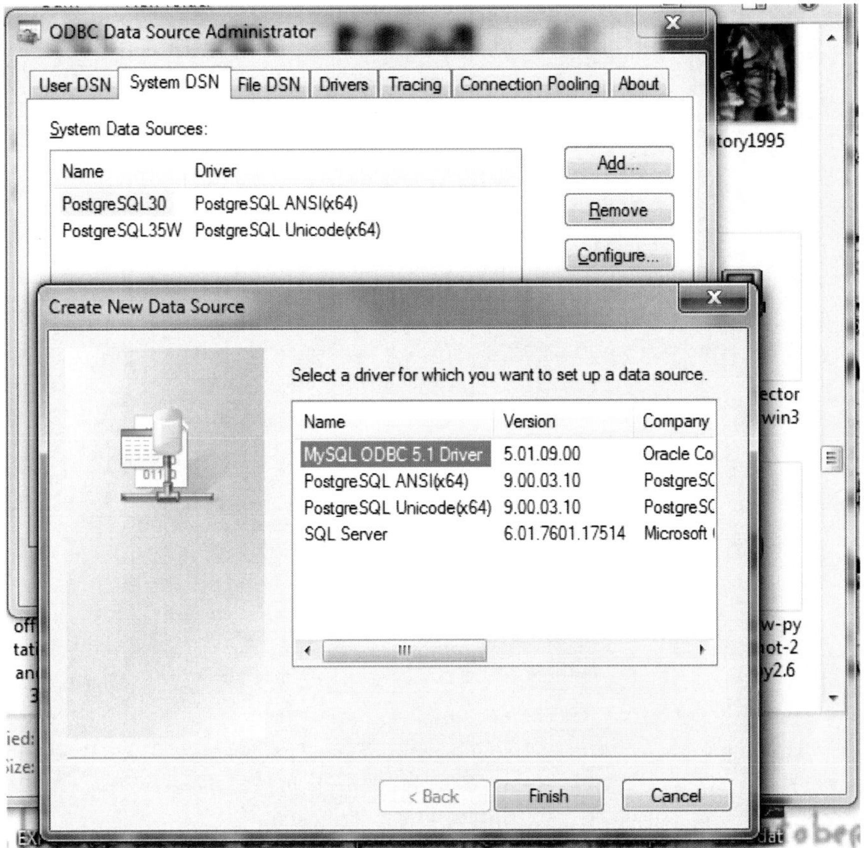

- Use the configuration options as shown here. The user is root, the TCP/IP Server is the local host (use the same password as in Step 1), and the database is MySQL. Test the connection. Click OK to finish this step.

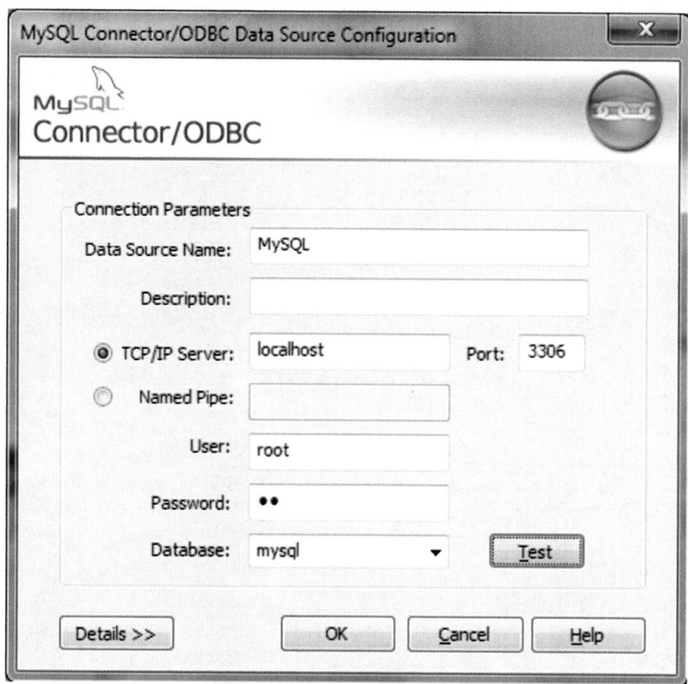

- Click the **User DSN** tab (and repeating the immediately preceding step. Add, and configure the connection using options The user is root, the TCP/IP Server is local host (use the same password as in Step 1), and the database is MySQL; test the connection and click OK to add the connection.

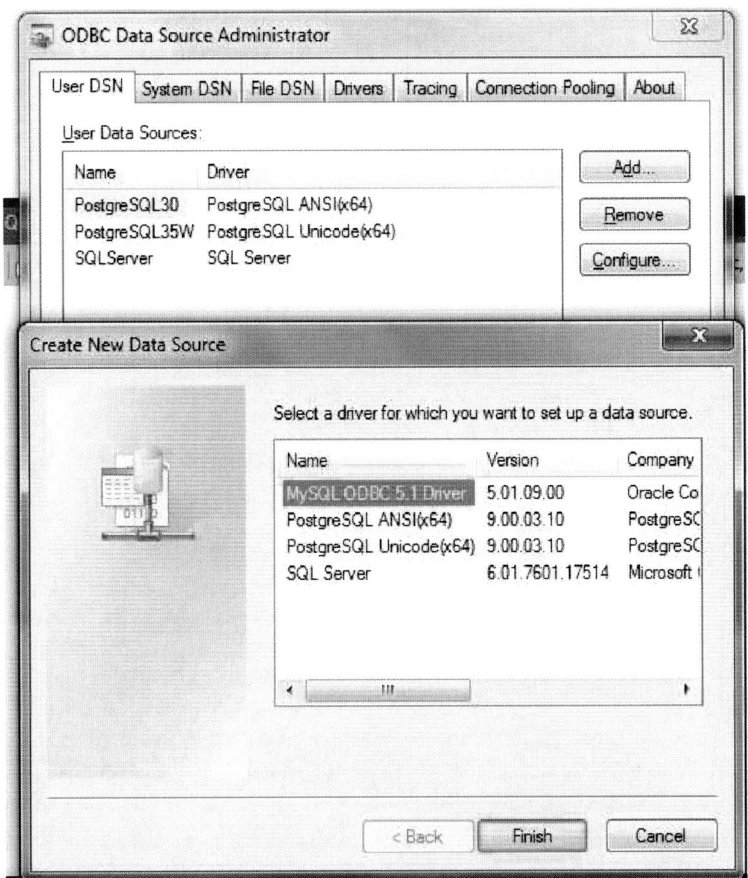

- Download MySQL Workbench from http://www.mysql.com/downloads/workbench/. This is very helpful in configuring the database: http://www.mysql.com/downloads/mirror.php?id=403983#mirrors.
- Assume a database is created as configuring databases is beyond the scope of this chapter.
- Start R.
- Type the commands in the screenshot below to create a connection to the database in MySQL:

```
> library(RODBC)
  > odbcDataSources()
  > ajay=odbcConnect("MySQL",uid="root",pwd="XX")
  > ajay
  > sqlTables(ajay)
  > tested=sqlFetch(ajay,"host")
```

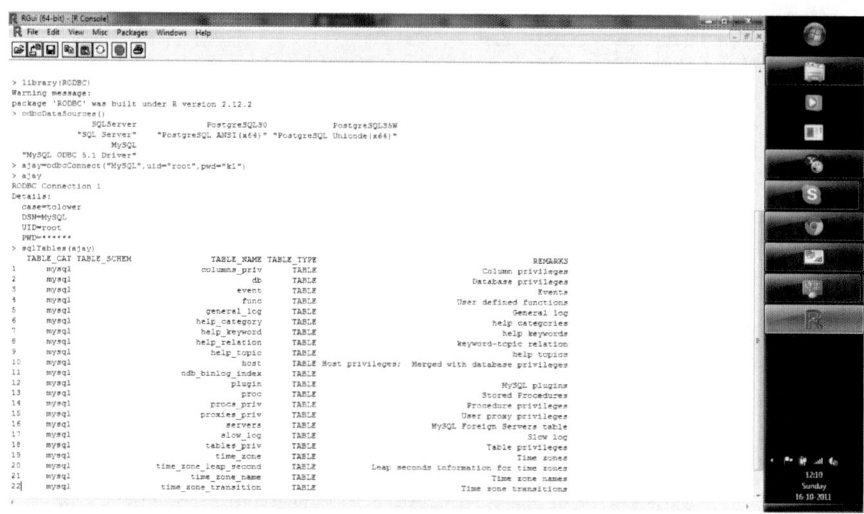

4.6.3 Using PostGresSQL and R

Here is a brief tutorial on using PostGreSQL and R

- Download PostgreSQL from http://www.postgresql.org/download/windows.
- Install PostgreSQL. Remember to store /memorize the password for the user postgres! Create a connection using the pgAdmin feature in the Start menu.

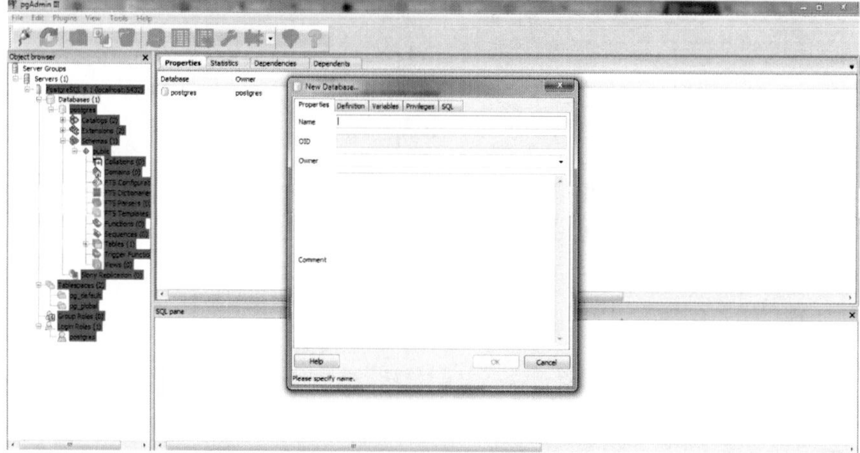

- Download the ODBC driver from http://www.postgresql.org/ftp/odbc/versions/msi/ and the Win 64 edition from http://ftp.postgresql.org/pub/odbc/versions/msi/psqlodbc_09_00_0310-x64.zip.
- Install the ODBC driver.

- Go to Start\Control Panel\All Control Panel Items\Administrative Tools\Data Sources (ODBC).
- Configure the following details in System DSN and User DSN using the ADD tabs.

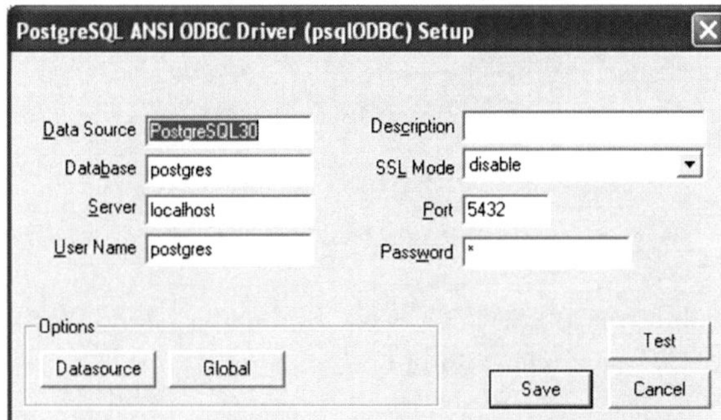

- Test the connection to check if it is working (this is the same as configuring DSN as was done for MySQL).
- Start R and install and load library RODBC.

> *library(RODBC)*

> *odbcDataSources(type = c("all", "user", "system"))*

SQLServer PostgreSQL30 PostgreSQL35W "SQL Server" "PostgreSQL ANSI(x64)" "PostgreSQL Unicode(x64)"

> *ajay=odbcConnect("PostgreSQL30", uid = "postgres", pwd = "XX")*

> *sqlTables(ajay)*

TABLE_QUALIFIER TABLE_OWNER TABLE_NAME TABLE_TYPE REMARKS 1 postgres public names TABLE

> *crimedat <- sqlFetch(ajay, "names")*

4.6.4 Using SQLite and R

SQLite is an alternative database system that can be used for moving data very quickly within R. But what is SQLite and when is it useful compared to other databases?

4.6.4.1 What Is SQLite

SQLite is an in-process library that implements a self-contained, serverless, zero-configuration, transactional SQL database engine. In addition, SQLite is an embedded SQL database engine. Unlike most other SQL databases, SQLite does not have a separate server process. SQLite reads and writes directly to ordinary disk files. A complete SQL database with multiple tables, indices, triggers, and views is contained in a single disk file.

Source: http://www.sqlite.org/about.html.

When to Use SQLite

Use SQLite in situations where simplicity of administration, implementation, and maintenance are more important than the countless complex features that enterprise database engines provide. For more information see http://www.sqlite.org/whentouse.html.

Using SQLite in R can facilitate reading larger files very fast.

Using SQLite with R

The RSQLite package within R is an interface to use SQLite within R. We also use the sqldf package to use SQL statements within R.

Example:

#Creating dummy random data in a dataset
bigdf <- data.frame(dim=sample(letters, replace=T, 4e7), fact1= rnorm(4e7), fact2=rnorm(4e7, 10, 20))
#Writing dataset to a file
write.table(bigdf, "C://Users//KUs//Desktop//bigdf.txt", sep=",", col.names= TRUE, row.names=TRUE, quote=TRUE, na="NA") #Removing all datasets in memory
rm(list=ls())
#Loading the sqldf
library(sqldf)
#Reading the data
a<- file("bigdf.csv")
*bigdf <- sqldf("select * from a",* ***dbname = tempfile(),*** *file.format = list(header = T, row.names = T))*

4.6.5 Using JasperDB and R

Jaspersoft is a leading open source BI platform that is integrated for use with R.

You can see this software at http://jasperforge.org/projects/rrevodeployrbyrevolutionanalytics.

RevoConnectR for JasperReports Server is a Java library interface between JasperReports Server and Revolution R Enterprise's RevoDeployR, a standardized collection of Web services that integrates security, APIs, scripts, and libraries for R into a single server.

JasperReports Server dashboards can retrieve R charts and result sets from RevoDeployR. RevoDeployR is a new Web service framework that can integrate dynamic R-based computations into Web applications. It was created by Revolution Analytics, the leading vendor of R-based commercial products. You can read more on RevoDeployR at http://www.revolutionanalytics.com/products/pdf/RevoDeployR.pdf.

4.6.5.1 July 2011 Interview with Mike Boyarski, Director of Product Marketing at Jaspersoft

Ajay: You announced a product partnership on the R Platform—RevoConnectR for JasperReports Server. How is that going and what are the initial results? Any other initiatives on business analytics?

Mike: We continue to see a steady stream of downloads and inquiries, roughly a hundred or more a month, of our RevoConnectR tool. The connector configures our JasperReports Server product to work seamlessly with the Revolution Analytics environment. As the predictive analytics and BI markets find tighter convergence for customers, we think our partnership with Revolution Analytics will provide the leading credible alternative for customers wanting this set of services without the costly price point of the leading proprietary vendors' products.

4.6.6 Using Pentaho and R

R is integrated into Pentaho's ETL tool.

4.6.6.1 Interview with James Dixon, Founder of Pentaho

Pentaho has been growing very rapidly. It makes open source BI solutions—basically, it currently has the largest share of enterprise software market.

Ajay: How would you describe Pentaho as a BI product for someone who is used to traditional BI vendors.

James: Pentaho has a full suite of BI software:

* ETL: Pentaho data integration
* Reporting: Pentaho reporting for desktop and Web-based reporting
* OLAP: Mondrian ROLAP engine, and Analyzer or Jpivot for Web-based OLAP client
* Dashboards: CDF and Dashboard Designer
* Predictive Analytics: Weka
* Server: Pentaho BI Server handles, e.g., Web-access, security, scheduling, sharing, report bursting. We have all of the standard BI functionality.

Ajay: Jaspersoft already has a partnership with Revolution Analytics for RevoDeployR (R on a Web server). Any R plans for Pentaho as well?

James: The feature set of R and Weka overlap to a small extent—both of them include basic statistical functions. Weka is focused on predictive models and machine learning, whereas R is focused on a full suite of statistical models. The creator and main Weka developer is a Pentaho employee. We have integrated R into our ETL tool.

4.7 Summary of Commands Used in This Chapter

4.7.1 Packages

- Doby
- Plyr
- Reshape2
- sqldf
- SQLite
- RODBC

4.7.2 Functions

- aggregate
- tapply
- merge
- diff
- cumsum
- rnorm
- scan
- sort
- melt
- acast
- reshape
- subset
- difftime
- sys.time
- strptime
- head
- tail
- ls
- rm
- names
- summary
- table
- str
- read.csv
- read.table
- write.table

4.8 Citations and References

- Example on using SQLite: http://code.google.com/p/sqldf/#Examples
- Frank E. Harrell Jr. with contributions from many other users. (2012). Hmisc: Harrell Miscellaneous. R package version 3.9-3. http://CRAN.R-project.org/package=Hmisc
- G. Grothendieck <ggrothendieck@gmail.com> (2011). sqldf: Perform SQL Selects on R Data Frames. R package version 0.4-2. http://CRAN.R-project.org/package=sqldf
- Tutorial on using SQLite with R: http://www.bioconductor.org/help/course-materials/2006/rforbioinformatics/labs/thurs/SQLite-R-howto.pdf
- Jonathan Lee (2011). RcmdrPlugin.doBy: Rcmdr doBy Plug-In. R package version 0.1-1. http://CRAN.R-project.org/package=RcmdrPlugin.doBy
- R FAQ How does R handle missing values? UCLA: Academic Technology Services, Statistical Consulting Group from http://www.ats.ucla.edu/stat/r/faq/missing.htm (accessed 16 March 2012)
- Additional tutorial on using R and databases: http://www.slideshare.net/jeffreybreen/accessing-databases-from-r
- Stack overflow R questions: http://stackoverflow.com/questions/4332976/how-to-import-csv-into-sqlite-using-rsqlite
- J.D. Long http://www.cerebralmastication.com/2009/11/loading-big-data-into-r/
- Using SQLite: http://datamining.togaware.com/survivor/Using_SQLite.html
- http://stackoverflow.com/questions/1299871/how-to-join-data-frames-in-r-inner-outer-left-right
- RSQLite package database interface R driver for SQLite: http://cran.r-project.org/web/packages/RSQLite/index.html
- SQLdf package. Manipulate R data frames using SQL: http://cran.r-project.org/web/packages/sqldf/index.html and http://code.google.com/p/sqldf/
- R Library: Advanced functions. UCLA: Academic Technology Services, Statistical Consulting Group: http://www.ats.ucla.edu/stat/r/library/advanced_function_r.htm (accessed 24 February 2012)
- Rtips. Paul E. Johnson: http://pj.freefaculty.org/R/Rtips.html#toc-Subsection-2.9

4.9 Additional Resources

A list of additional resources for using R with databases and BI software.

- Netezza: In-database analytics using R: http://www.r-project.org/conferences/useR-2010/slides/Hess+Chambers_1.pdf
- R 's role in BI architechture: **(Using R with Talend and Pentaho): http://www.r-project.org/conferences/useR-2010/slides/Colombo+Ronzoni+Fontana.pdf**
- OpenBI: Consulting on R in BI applications: http://www.openbi.com/demosarticles.html#Spotlight

- Mango Solutions: Consulting in analytics based on R: http://www.mango-solutions.com/r-validation.html
- Revolution Analytics: Consulting based on R: http://www.revolutionanalytics.com/products/consulting/

4.9.1 Methods for Larger Dataset Sizes

R is able to read in data in almost any format in any manner, list, array, vector, or data frame. This gives a considerable advantage to analysts in flexibility in handling data. However, business analytics sometimes needs to deal with datasets (or data frames) of large size in terms of observations (rows) or variables (columns). Multiple memory management techniques can help with size input issues including:

biglm, ff series of packages

MM RevoScaler package made by Revolution Analytics (which uses the XDF format)

Hardware optimization (RAM, 64-bit OS)

biglm, ff series of packages

R Enterprise and XDF format using RevoScalar

Using programming techniques

Note these packages should be used only by people with advanced analytical and computation needs.

Beginning with 2.14.0, R started offering direct support to parallelism. Many R packages can help with writing parallelized code that can use all the cores of a machine optimally. These packages are parallel, Rmpi, snowfall, foreach, and multicore. For datasets that are too big and out of memory, use the bigmemory, ff, HadoopStreaming, and Rhipe packages.

A complete list of all such resources is available in the CRAN High-Performance Computing and Parallel Computing task view: http://cran.r-project.org/web/views/HighPerformanceComputing.html.

Some tutorials on using these packages are at

- ByteMining tutorials Part 1: http://www.slideshare.net/bytemining/taking-r-to-the-limit-high-performance-computing-in-r-part-1-parallelization-la-r-users-group-727
- Part 2: http://www.slideshare.net/bytemining/r-hpc
- Introduction to HPC (2009): http://dirk.eddelbuettel.com/papers/ismNov2009intro-HPCwithR.pdf

Case Study: R and Oracle

The following brief interview on Oracle's initiatives in the R analytical field involves case studies and code examples authored with the help of Oracle employees Mark Hornick and Vaishnavi Sashikanth.

Ajay: What are the reasons that made you choose R for the Oracle Advanced Analytics option?

Mark/Vaishnavi: R is a powerful, extensible, and highly graphical open source language and environment for statistical computing. It has proven itself among statisticians and data analysts as a serious alternative to other proprietary statistical environments like SAS and SPSS. R also provides a rich ecosystem of open source packages, a vibrant community, and online resources that point to its popularity and usability. As such, R is highly desirable for statisticians and data analysts and an ideal interface for Oracle through which to provide transparent access to database-resident data. With the rapidly growing number of R users in enterprises, these users have been clamoring for an enterprise-ready R that seamlessly integrates with database data. We had been receiving requests for R integration with Oracle Database from our enterprise database customers, which prompted us to address this need.

That said, R has some well-known limitations in terms of scalability and performance that Oracle found could be overcome by coupling the R language and environment with Oracle Database as a high-performance and scalable computational engine. Enterprise users of R need such a solution since data marts and warehouses are the most common enterprise data storage layer. Since Oracle Database is distributed and massively parallel and has been evolving and tested for over 30 years, enterprise users know that Oracle understands how to manage and extract value from data—terabytes and petabytes of data—as well as how to leverage and optimize SQL. As a result, Oracle wanted to make the powerful Oracle Database available to R users, but to do so transparently, without requiring R users to learn a new language or environment. The core R functions and object types, such as data frames, matrices, and vectors, match database functionality and objects rather naturally. For **Oracle R Enterprise**

(ORE), Oracle has taken the best aspects of R—its language, environment, open source packages, graphics—and minimized or eliminated the limitations of base R—memory constraints and single-threaded execution.

Coupling with Oracle Database, R also provides an environment for leveraging parallelism in Oracle Database. Through the transparency layer, ORE allows R users to take advantage of in-database parallel execution of R-generated SQL queries. Through embedded R execution, ORE allows R users to not only operationalize R scripts in database applications but also to leverage database support for R-user-controlled data parallel execution of R scripts. Through the statistics engine, ORE allows R users to perform sophisticated statistical computations in Oracle Database.

Ajay: What are your views on other software in the analytics space, and how do you think in-database and big data analytics will evolve in the coming years?

Mark/Vaishnavi: At the enterprise level, data are normally stored in databases, and often Oracle databases. Oracle is in a unique position to bring advanced analytics to Oracle Database-resident data by incorporating algorithms in Oracle Database and providing transparent access, in this case from R. Other software vendors in the analytics space have also seen the wisdom in this approach. Some vendors try to execute their algorithms in Oracle Database; however, their code must still execute as nonnative code. This nonnative status inherently incurs performance penalties and is not transparently or seamlessly integrated with Oracle Database. Other vendors are also providing R interfaces to their products. From what we've seen so far, such vendors provide neither the level of transparency nor overall integration between R and their products. Transparency enables existing R scripts to remain largely intact while gaining the benefits of in-database execution. Embedded R execution enables, among other things, the ability to provide efficient data parallel execution at the database server.

In the coming years, we will no doubt continue to see an increase in the amount of analytical functionality available in DBMS software. However, this will also be supplemented by the recent upswing in interest in big data analytics. Oracle announced at Oracle OpenWorld 2011 the Big Data Appliance to address the needs of companies to analyze enterprise-level data stores, on the order of petabytes. These data often consist of low-density structured and unstructured data from heterogeneous sources like social media.

To address the needs of big data, Hadoop has been proven by a wide range of successful, data-intensive companies, e.g., Google, Facebook, eBay. The Oracle Big Data Appliance ships with a ready-configured Hadoop Cluster, based on Cloudera software. To facilitate the use of Hadoop by R users, Oracle provides the **Oracle R Connector for Hadoop** (ORCH) as one of its big data connector software products. ORCH allows R users not only to interact with Hadoop Distributed File System (HDFS) data seamlessly through an R interface but also to specify mapper and reducer functions in the R language and to invoke MapReduce jobs directly from R. The results, placed in HDFS, can be viewed in R or pushed to Oracle Database. (More on ORCH below in the examples.) Regarding big data in the coming years, Hadoop will spread across a much wider range of enterprises to meet the analytics challenges of ever-increasing dynamic and heterogeneous data.

Ajay: What are the differences between Oracle R Enterprise and other flavors of R including open source R, Revolution Analytics's version, and other software that integrates with R? How do you intend to support enterprise customers and academic customers?

Mark/Vashnavi: Oracle R Enterprise builds on Base R in several keys ways. As mentioned previously, ORE addresses the memory and parallel execution limitations of R by transparently leveraging Oracle Database as a computational engine. Before Oracle R Enterprise, R users would access database data using various connectors, such as ODBC or JDBC. These are relatively slow for pulling data from and pushing data to databases. With the Oracle R Enterprise transparency layer, tables and views are visible in the R environment through R objects serving as proxies for the corresponding database table or view. These are represented as a subclass of the R *data.frame* class, called *ore.frame*. Proxy objects enable manipulating database-resident data in Oracle Database. Oracle R Enterprise uses the ROracle package, which has been enhanced for use of Oracle's OCI interface for optimized performance. Oracle is now the maintainer of ROracle and has published this package back to the R open source community.

Oracle R Enterprise also enables embedded R execution, i.e., executing an R script at the database server machine, with data parallel execution managed by Oracle Database. Embedded R execution provides both an R and a SQL interface for executing R scripts through Oracle Database.

In addition, Oracle provides the Oracle R Distribution, an Oracle-supported distribution of open source R. Support for Oracle R Distribution is provided to customers of the Oracle Advanced Analytics (https://blogs.oracle.com/datawarehousing/entry/announcing_oracle_advanced_analytics option, Oracle Big Data Appliance (http://www.oracle.com/us/technologies/big-data/index.html), and Oracle Linux. Oracle R Enterprise and Oracle Data Mining are the two components of the Oracle Advanced Analytics option. Oracle Data Mining provides data mining functionality as native SQL functions within Oracle Database in combination with a workflow-based GUI enabled through SQL Developer.

The Oracle R Distribution facilitates enterprise acceptance of R since the lack of a major corporate sponsor has made some companies concerned about fully adopting R. With the Oracle R Distribution, Oracle plans to contribute bug fixes and enhancements to open source R. Oracle R Distribution also works with Intel's Math Kernel Library (MKL), which enables optimized, multithreaded math routines to provide relevant R functions with maximum performance on Intel hardware.

Support for Oracle R Distribution, Oracle R Enterprise, and Oracle R Connector for Hadoop is provided through standard Oracle support channels. Oracle also provides an R discussion forum (https://forums.oracle.com/forums/forum.jspa?forumID=1397) for customers of Oracle's R products.

With respect to other R offerings, we believe that no matter how cool a technology may be, if that technology requires the creation of a separate data tier, it's a nonstarter to keep pace with the upward spiraling growth of data volumes and velocity. Moreover, a separate data tier enables the reemergence of problems from the past that data warehouses tackled, namely, providing a single version of truth for analytic models. Putting solutions into production requires going back to the source data, so scoring, e.g., 300 million customers for a company with an established online presence in a separate data tier introduces unwelcome overhead and complexity.

Consider one such R offering, Revolution Analytics' Revolution R Enterprise (RRE). A key difference is in dealing with data representation. RRE requires working with a proprietary data format, an .xdf file. Extracting data from Oracle Database (or loading data from a file) and converting it to.xdf format can be expensive, incurring costs of data replication, extraction, and transport, in addition to the costs of overall system complexity, backup, recovery processes, and, perhaps most importantly, security. In contrast, the Oracle R Enterprise transparency layer enables data to remain in the database while requiring minimal changes to existing R code. Revolution's RevoScaleR package has its own functions, so R users need to rewrite their code. The Oracle R Enterprise embedded R execution capability greatly simplifies writing R code to enable data parallel execution. Revolution's HPC server functionality enables writing parallel R scripts but with much greater awareness on the part of the end user.

Ajay: Any plans to give something back to the R community? Will we see a more stepped presence by Oracle in the R and analytics space in the coming years?

Mark/Vashnavi: To support Oracle R Enterprise, Oracle has taken responsibility for and contributed modifications to ROracle (http://cran.r-project.org/web/packages/ROracle/index.html)—an Oracle database interface (DBI) driver for R, now based on OCI. As ROracle is LGPL and used for Oracle Database connectivity from R, Oracle is committed to ensuring this is the best package for Oracle connectivity.

With the Oracle R Distribution (ORD), Oracle plans to contribute bug fixes and enhancements to open source R. For example, the recently released version of ORD provides a step change to open source R in terms of its ability to handle chip-vendor-specific optimized blas and lapack libraries. Today, one needs to rebuild R for open source R to use a chip-vendor-specific optimized math library. The steps are nontrivial for R users, as evidenced by questions in many forums about how to build with Intel's MKL. ORD exploits MKL out of the box for x86 platforms. Moreover, ORD is integrated directly with the Fortran level interface for BLAS/LAPACK instead of at the C layer to squeeze out every ounce of performance. This also prepares ORD for other chip vendors' math libraries. Since MKL is proprietary and R is GPL, ORD is not allowed to prepackage

MKL. Instead, ORD detects the presence of MKL if installed and uses it. This enhancement is being made available to the R community.

- **Example: Connecting to Oracle Database**

To use Oracle R Enterprise from an R engine client, users first load the ORE package, which also loads several subpackages. At this point, the user connects to Oracle Database specifying a particular user, SID, and host. To synchronize the metadata in R with the tables and views available in the specified schema, we use the function ore.sync. This creates the proxy objects. To provide access to tables and views as though they were native R objects, we attach the specific schema to R's search path.

```
library(ORE)
ore.connect(user="rquser", sid="orcl", host="my-machine")
ore.sync(schema="rquser")
ore.attach (schema="rquser")
```

The function *ore.ls* returns the names of tables and views available in the attached schema. This set refers to R objects that act as proxies for the tables and views in the database. Suppose one of the tables is named ONTIME—a table that contains flight-related information for domestic US flights between 1988 and 2008. We can invoke standard R functions such as *names* and *dim* on the proxy object ONTIME to return the names of the columns of the table and the dimensions in terms of number of rows in number of columns, respectively.

```
ore.ls()
names(ONTIME)
dim(ONTIME)
```

- **Example: Transparency Layer**

The transparency layer allows users to invoke core R functions on instances of *ore.frame*, a subclass of *data.frame*, such that the corresponding SQL is generated and submitted to the database. From this SQL, a view is created and a corresponding proxy object returned. Whenever possible, results are not computed/materialized until the user explicitly requests the result, for example, using a function like head or print. Consider column selection in the following examples. Users can select columns by name using standard R syntax on the proxy object ONTIME. Columns can also be specified by index, vector of indexes, or a range of columns, which can be included or excluded. Although this is standard R, SQL does not allow the use of indices in the SELECT clause, nor does it allow the specification of columns to be excluded. The transparency layer enables this.

```
df <- ONTIME[,c("YEAR","DEST","ARRDELAY")]
head(df)
ONTIME[,c(1,4,23)]
ONTIME[,-(1:22)]
```

Similarly, in row selection, users can filter rows using a predicate, as well as select columns in combination. *df[df$DEST=="SFO",]*

```
df[df$DEST=="SFO" | df$DEST=="BOS",1:3]
```

R supports data transformations. For example, transformations such as recoding can be performed by specifying a function and applying that function to a column or using the transform function. This is standard R that works transparently on data stored in Oracle Database.

```
delayCategory_fmt <- function(x) {
ifelse(x>200,'LARGE',
ifelse(x>=30,'MEDIUM','SMALL')) }
attach(ONTIME)
ONTIME$ARRDELAY <- delayCategory_fmt(ARRDELAY)
ONTIME$DEPDELAY <- delayCategory_fmt(DEPDELAY)
detach(ONTIME)
ONTIME <- transform(ONTIME,
ARRDELAY = ifelse(ARRDELAY > 200, 'LARGE', ifelse(ARRDELAY >= 30,
'MEDIUM', 'SMALL')),
DEPDELAY = ifelse(DEPDELAY > 200, 'LARGE', ifelse(DEPDELAY >= 30,
'MEDIUM', 'SMALL')))
```

This next example involves aggregation. ORE overloads the function *aggregate* such that when invoked on an *ore.frame* proxy object it results in the corresponding SQL GROUP BY statement being generated and submitted to the database. In this particular example, we count the number of flights at each destination. The class of the result is also ore.frame. Only when we access the results through head, to get the first few results, do we compute and retrieve the data:

```
aggdata <- aggregate(ONTIME$DEST, by = list(ONTIME$DEST), FUN =
length)
class(aggdata)
head(aggdata)
```

This next example illustrates using the transparency layer to display a boxplot. We first split the data for arrival delay by day of the week. This result is passed to the *boxplot* function where the statistics required to draw the graph are computed in the database. The underlying data never leave the database, only the summary statistics.

```
bd <- split(ONTIME$ARRDELAY, ONTIME$DAYOFWEEK)
boxplot(bd, notch = TRUE, col = "red", cex = 0.5,
outline = FALSE, axes = FALSE, main = "Airline Flight Delay by Day of Week",
ylab = "Delay (minutes)", xlab = "Day of Week")
axis(1, at=1:7,
labels=c("Monday", "Tuesday", "Wednesday", "Thursday", "Friday", "Saturday",
"Sunday"))
axis(2)
```

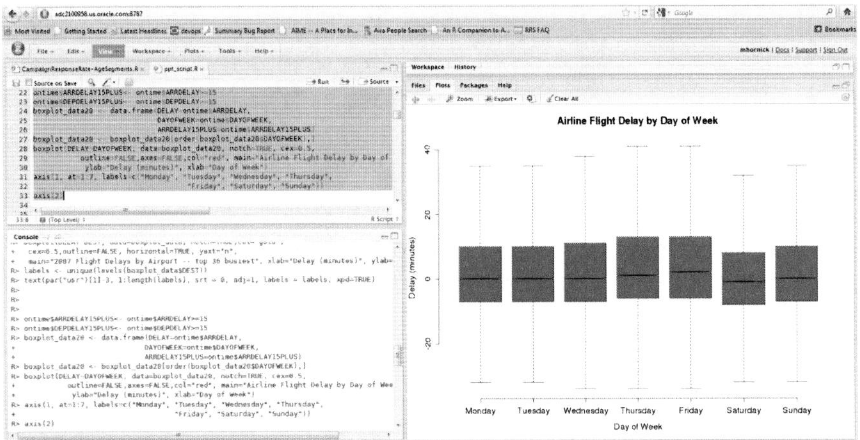

- **Examples: Embedded R Execution**

Embedded R execution enables users to submit an R function to be executed through Oracle Database by the spawning of one or more R engines at the database server machine. The following example uses the function *ore.groupApply*, one of several embedded R execution functions, to illustrate how R users can achieve data parallelism through the database. Here, we specify a column on which to partition the data. Each partition of the data is provided to the function through the first argument, in this case the function variable *dat*. This example also illustrates that embedded R execution enables the use of open source packages. Here we see the use of the R package biglm. Only when we want to see the results of these models do we need to retrieve them into R memory and perform, for example, the summary function.

```
modList <- ore.groupApply
( X=ONTIME,
INDEX=ONTIME$DEST,
function(dat) {
library(biglm)
biglm(ARRDELAY ~ DISTANCE + DEPDELAY, dat) });
modList_local <- ore.pull(modList)
modList_local$BOS
```

Whereas the previous example showed how to use embedded R execution from the R environment, we can also invoke R scripts from SQL. This next example illustrates returning a data frame from results computed in Oracle Database. We first create an R script in the database R script repository. The script is defined as a function that creates a vector of ten elements and returns a data frame with those elements in one column and those elements divided by 100 in a second column.

Once the script is created, we can invoke it through SQL. One of the SQL-embedded R execution table functions is *rqEval*. The first argument is NULL since we have no parameters to pass to the function. The second argument describes the structure of the result, which can be any valid SQL query that captures the name and

type of resulting columns. The third argument is the name of the script to execute.
begin

sys.rqScriptCreate('Example1',
'function() {
ID <- 1:10
data.frame(ID = ID, RES = ID / 100)
}');
end;
/
*select * from table(rqEval(NULL,*
'select 1 id, 1 res from dual',
'Example1'));

```
SQL> begin
  sys.rqScriptCreate('Example1',
'function() {
   ID <- 1:10
   res <- data.frame(ID = ID, RES = ID / 100)
   res}');
end;
/
select *
  from table(rqEval(NULL,
        'select 1 id, 1 res from dual',
        'Example1'));
  2    3    4    5    6    7    8
PL/SQL procedure successfully completed.

SQL>  2    3    4
        ID        RES
---------- ----------
         1        .01
         2        .02
         3        .03
         4        .04
         5        .05
         6        .06
         7        .07
         8        .08
         9        .09
        10         .1

10 rows selected.
```

R scripts may generate structured data, complex R objects, and graphs. Oracle R Enterprise embedded R execution enables returning results from an R script as an XML string. Consider the following example that creates a vector from the integers 1 to 10, plots 100 random normal points in a graph, and then returns the vector. After creating the script in the database R script repository, we invoke the script using *rqEval*, but instead of specifying the form of the result in a SQL query, we specify 'XML'.

```
begin
sys.rqScriptCreate('Example6',
'function(){
res <- 1:10
plot( 1:100, rnorm(100), pch = 21, bg = "red", cex = 2 )
res }');
end;
/
select value
from table(rqEval( NULL,'XML','Example6'));
```

While the actual graph looks as follows, the output from this query will be an XML string.

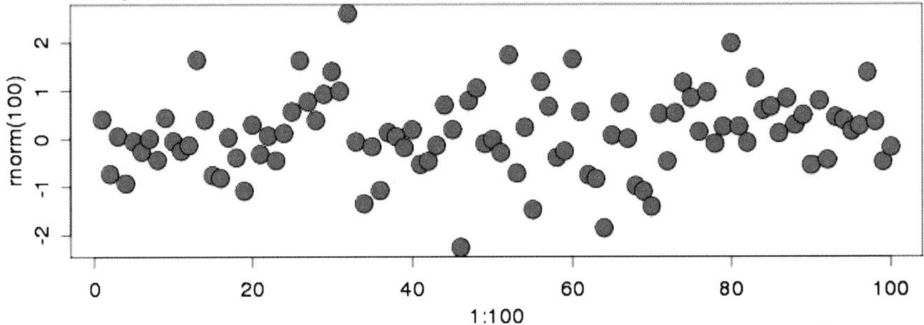

In the execution results shown below, the VALUE column returned is a string that contains first the structured data in XML format. Notice the numbers 1 through 10 set off by the <value> tags. This is followed by the image in PNG base 64 representation. This type of output can be consumed by Oracle Business Intelligence Publisher (BIP) to produce documents (templates) with R-generated graphs and structured content. This content can also be used to expose R-generated content in Oracle Business Intelligence Enterprise Edition (OBIEE) Web browser-based dashboards.

```
SQL> set long 20000
set pages 1000
begin
  sys.rqScriptCreate('Example6',
 'function(){
           res <- 1:10
           plot( 1:100, rnorm(100), pch = 21,
                 bg = "red", cex = 2 )
           res
           }');
SQL> end;
/
select    value
from      table(rqEval( NULL,'XML','Example6'));
SQL>  2   3   4   5   6   7   8   9   10
PL/SQL procedure successfully completed.

SQL>  2
VALUE
--------------------------------------------------------------------------------
```


```
<root><R-data><vector_obj> <ROW-vector_obj><value>1</value></ROW-vector_obj><ROW
-vector_obj><value>2</value></ROW-vector_obj><ROW-vector_obj><value>3</value></R
OW-vector_obj><ROW-vector_obj><value>4</value></ROW-vector_obj><ROW-vector_obj><
value>5</value></ROW-vector_obj><ROW-vector_obj><value>6</value></ROW-vector_obj
><ROW-vector_obj><value>7</value></ROW-vector_obj><ROW-vector_obj><value>8</valu
e></ROW-vector_obj><ROW-vector_obj><value>9</value></ROW-vector_obj><ROW-vector_
obj><value>10</value></ROW-vector_obj></vector_obj> </R-data><images><image><img
 src="data:image/pngbase64"><![CDATA[iVBORw0KGgoAAAANSUhEUgAAAeAAAAHgCAIAAADytin
CAAAgAElEQVR4nOzdZ1xT1x8G8CcMB6jgQq0IDnDVulsRBSKyZQjIUnCDKDhq3bvuValbcRYFFRFRUBF
ExYWrarGKA3GAgw0udvJ/wV8aTG5ESG4C/L4fXug9JzdPGL/c3hvJ/wVDJ1gcJdcG6b/wJc4VvU0Rw+nw9CCCHyROHWAQghhIhGBZo
QQuQUFWhCCJFTVKAJIUROUYEmhBA5RQWaEELkFBVoQgiRU1SgCSFET1GBJoQQQOUUFmhBC5BQVaEIIkVN
UoAkhRE5RgSaEED1FBZoQQuQUFWhCCJFTVKAJIUROUYEmhBA5RQWaEELkFBVoQgiRU1SgCSFET1GBJoQ
QOUUFmhBC5BQVaEIIkVNUoAkhRE5RgSaEED1FBZoQQuQUFWhCCJFTVKAJIUROUYEmhBA5RQWaEELkFBV
oQgiRU1SgCSFET1GBJoQQOUUFmhBC5BQVaEIIkVNUoAkhRE5RgSaEED1FBZoQQuQUFWhCCJFTVKAJIUR
```


- **Example: Oracle R Connector for Hadoop**

The Oracle R Connector for Hadoop (ORCH) provides transparent access to a Hadoop cluster—enabling manipulation of Hadoop Distributed File System (HDFS) resident data and the execution of MapReduce jobs. ORCH can be used on the Oracle Big Data Appliance or on non-Oracle Hadoop clusters and is part of the Oracle Big Data Connectors software suite. R users write mapper and reducer functions in R and execute MapReduce jobs from the R environment using a high-level interface. As such, R users are not required to learn a new language, e.g., Java, or environment, e.g., cluster software and hardware, to work with Hadoop. Moreover, functionality from R open source packages can be used in the writing of mapper and reducer functions. ORCH also gives R users the ability to test their MapReduce programs locally, using the same function call, before deploying on the Hadoop cluster.

ORCH includes functions for manipulating HDFS data. Users can move data in and out of HDFS with the file system, R data frames, and Oracle Database tables and views. This next example shows one such function, hdfs.push, which accepts an ore.frame object as its first argument, followed by the name of the key column, and then the name of the file to be used within HDFS. *ontime.dfs_D <- hdfs.push(ONTIME, key='DEST', dfs.name='ontime_DB')*

The following R script example illustrates how users can attach to an existing HDFS file object, getting a handle to the HDFS file. Then, using the *hadoop.run* function in ORCH, we specify the HDFS file handle, followed by the mapper and reducer functions. The mapper function takes the key and value as arguments, which correspond to one row of data at a time from the HDFS block assigned to the mapper. The function keyval in the mapper returns data to Hadoop for further processing by the reducer. The reducer function receives all the values associated with one key (resulting from the "shuffle and sort" of Hadoop processing). The result of the reducer is also returned to Hadoop using the keyval function. The results of the reducer tasks are consolidated in an HDFS file, which can be accessed using the *hdfs.get* function.

The following example computes the average arrival delay for flights where the destination is San Francisco Airport (SFO).

```
dfs <- hdfs.attach('ontime_DB')
res <- hadoop.run(
dfs,
mapper = function(key, value) {
if (key == 'SFO' & !is.na(x$ARRDELAY)) {
keyval(key, value)
}
},
reducer = function(key, values) {
sumAD <- 0
count <- 0
for (x in values) {
sumAD <- sumAD + x$ARRDELAY
count <- count + 1}
}
res <- sumAD / count
keyval(key, res)
}
)
hdfs.get(res)
```

Chapter 5
Exploring Data

While Chap. 4 dealt with getting your data in shape for processing (or, as it is commonly known, data preprocessing), in this chapter we actually start the process of looking at slices of data for generating various insights. We will emphasize the need for data visualization both as an acknowledgement of growing demands of data volume and easy understandability by business audiences. The fact that R currently has one of the most advanced graphical libraries also helps. We will be using basic graphical capabilities but will also briefly touch on advanced customization using the acclaimed ggplot2 package.

5.1 Business Metrics

Business metrics are important variables that are collected on a periodic basis to assess the health and sustainability of a business. They should have the following properties:

1. The absence of collection of regular updates of the business metric could cause business disruption by incorrect and incomplete decision making.
2. The costs of collecting, storing, and updating the business metric are less than the opportunity costs of wrong decision making caused by a lack of information of that business metric.
3. The business metrics are continuous in comparisons across time periods and business units—if necessary the assumptions for smoothing the comparisons should be listed in the business metric presentation itself.
4. Business metrics can be derived from other business metrics. If necessary, only the most important business metrics should be presented, or the metrics with the greatest deviation from past trends should be mentioned.
5. The scale of the business metric units should be comparable to other business metrics as well as significant in focus to emphasize any difference in numbers.

A. Ohri, *R for Business Analytics*, DOI 10.1007/978-1-4614-4343-8_5,

6. The dimension of the business metrics should be increased to enhance comparison and contrasts without increasing complexity.

5.2 Data Visualization

We list some of the principles of data visualization that we will use in our understanding of business analytics. Some of the guidelines for graphs are given at http://cran.r-project.org/doc/contrib/usingR.pdf, Sect. 3.8. One of the most useful guidelines is this: *draw graphs so that reduction and reproduction will not interfere with visual clarity.*

- Graphs that are sorted or ordered in terms of increasing data or decreasing data are easier to understand than graphs that do not have data sorted.
- Color shades are better than different colors for expressing or enhancing a change in magnitude of the same metric.
- Graphs that can be clicked or linked to more detail (or low-level versus high-level view) are better suited than lots of graphs not physically linked together by code or macros.
- Graphs should proportionally represent outliers as well as trends with appropriate scale.
- Typically, pie charts are used for market shares, line graphs for time-varying numbers, and histograms and density plots for frequency-varying numbers.
- The *X*- |*Y* scale of graphs needs to be appropriately matched to the variable for ease of analysis.
- A combination graphs or graphs in which two or more types of graphs are superimposed in the same area need to have a clear legend and scale and distinctly shaded without distracting colors or overlaps.

The R programming language offers multiple ways to achieve data visualization for business data. These range from the very simple (a GUI like GrapheR or R Commander) to the enhanced (a GUI like Red-R or Deducer). It offers a complete visual solution ranging from graphics in the base packages to newer enhanced packages like ggplot and jjplot. Corporations like Facebook, Google, and the New York Times have achieved stunning visualizations due to the R programming language, and part of this is due to its complete flexibility and range of visual options. In this chapter, we will explore a variety of these options, the context in which certain graphs are most appropriate, and the most efficient way to achieve those results.

A good reference manual for simpler data visualizations can be downloaded at http://cran.r-project.org/doc/contrib/Verzani-SimpleR.pdf.

For some examples of very good and very bad graphs, see http://www.datavis.ca/gallery/. Also see http://www.perceptualedge.com/examples.php for rectifying defects in data visualization.

5.3 Parameters for Graphs

We discuss a brief list of parameters that can be used for changing or customizing graphs.

- Title of graph: You can either change the main="Chart Title" parameter or use the title ("Chart Title") in a separate line. Note the main parameter will overwrite any previous value, while the title function will simply annonate or add in the title to an existing graph.
- The *x*- and *y*-axes are labeled using the *xlab*= "X AXIS LABEL" and *ylab*= "Y AXIS LABEL" parameter.
- The parameters *xlim=c(x,y)* and *ylim=c(x,y)* can help in defining the plotting area in terms of limits in the *x*- and *y*-axes.
- The thickness of plotted lines is changed using the parameter lwd, i.e., lwd=2.
- Colors are changed.

To change the colors, you need to add in the col= parameter within the specific graphical command. How many colors does R support? Just type *colors()* in the console. There are 657 colors from aliceblue to yellowgreen! You can choose any one of them as the value for your col= parameter.

> *hist(iris$Sepal.Length,col="blue")*

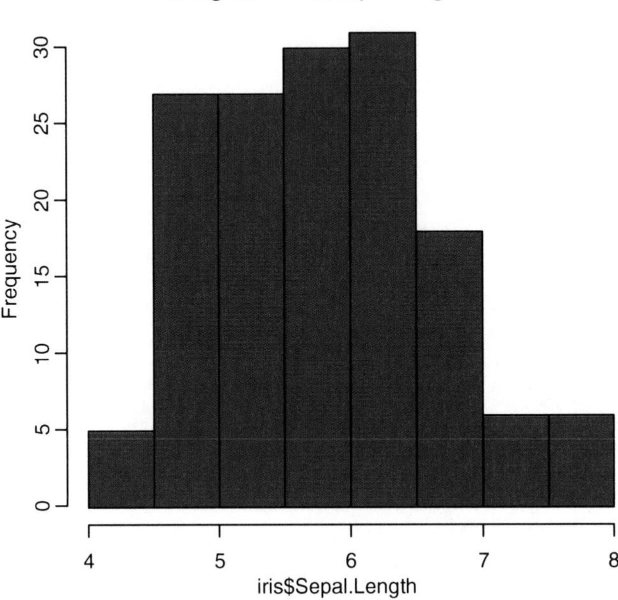

- For other parameters type *?par* in the console.
  - **To change the background color** to gray, change the bg parameter within the par command.

    > *par(bg="grey")*
  - **To fit in multiple graphs** within the same screen, use the mfrow and mfcol parameters within par. From the help /library/graphics/html/par.html we see mfcol, mfrow, a vector of the form c(nr, nc). Subsequent figures will be drawn in an nr-by-nc array on the device by columns (mfcol) or rows (mfrow).

5.4 Creating Graphs in R

R has one of the most celebrated graphical libraries and packages in analytics software. You can view an incredible range of graphs using R and the code associated with each at the beautiful R Graph Gallery Web site: http://addictedtor.free.fr/graphiques/.

5.4.1 Basic Graphs

We demonstrate some of the most commonly used graph types.

5.4.1.1 Histograms

Histograms are commonly used to show the distribution of data for variables and for probability estimation. Creating a histogram in R is as simple as typing *hist(*dataset$variable*)*:

> *data(iris)*
> *hist(iris)*
Error in hist.default(iris) : ‘x’ must be numeric
> names(iris) [1] "Sepal.Length" "Sepal.Width" "Petal.Length" "Petal.Width" "Species"
> str(iris) ’data.frame’: 150 obs. of 5 variables:
$ Sepal.Length: num 5.1 4.9 4.7 4.6 5 5.4 4.6 5 4.4 4.9 ...
$ Sepal.Width : num 3.5 3 3.2 3.1 3.6 3.9 3.4 3.4 2.9 3.1 ...
$ Petal.Length: num 1.4 1.4 1.3 1.5 1.4 1.7 1.4 1.5 1.4 1.5 ...
$ Petal.Width : num 0.2 0.2 0.2 0.2 0.2 0.4 0.3 0.2 0.2 0.1 ...
$ Species : Factor w/ 3 levels "setosa","versicolor",..: 1 1 1 1 1 1 1 1 1 1 ...
> *hist(iris$Sepal.Length)*

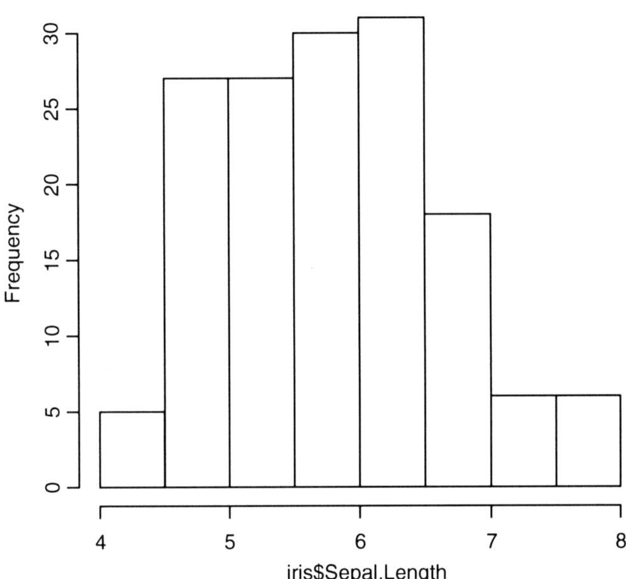

- There are multiple options for tweaking your histogram including *breaks* (for number of bins),*xlab* (for the *x*-axis label), freq (TRUE for counts and FALSE for plotting the probability density), and *col* (for color). To change colors, see also Sects. 5.3 and 5.9.

5.4.1.2 Stem and Leaf Plot

Stem produces a stem-and-leaf plot of the values in *x*. *stem(*dataset$variable*)*

```
> stem(iris$Sepal.Length)
The decimal point is 1 digit(s) to the left of the |
42 | 0
44 | 0000
46 | 000000
48 | 00000000000
50 | 0000000000000000000
52 | 00000
54 | 0000000000000
56 | 00000000000000
58 | 0000000000
60 | 000000000000
62 | 0000000000000
64 | 000000000000
66 | 0000000000
```

68 | 0000000
70 | 00
72 | 0000
74 | 0
76 | 00000
78 | 0

5.4.1.3 Pie Chart

To create a simple pie chart, use *pie(*dataset$variable*)*.

> ajay=c(12,5,35,67)
> names(ajay)=c("North","South","West","East")
> pie(ajay)

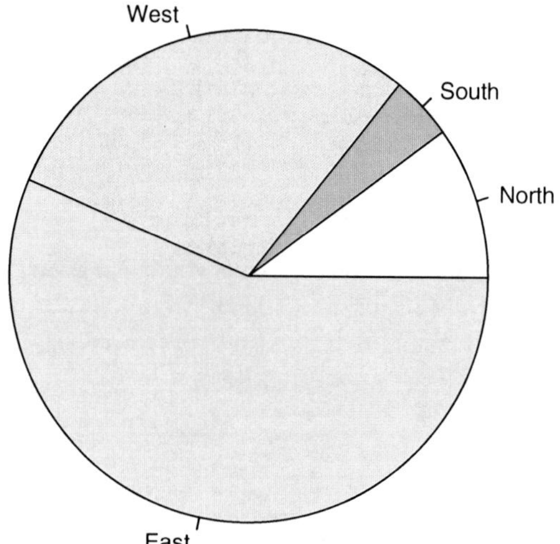

Many graphics experts discourage relying on pie charts based on perception issues. The following text is taken from http://stat.ethz.ch/R-manual/R-devel/library/graphics/html/pie.html to highlight this point: "*Pie charts are a very bad way of displaying information. The eye is good at judging linear measures and bad at judging relative areas. A bar chart or dot chart is a preferable way of displaying this type of data (Cleveland 1985); p. 264: "Data that can be shown by pie charts always can be shown by a dot chart. This means that judgements of position along a common scale can be made instead of the less accurate angle judgements." This statement is based on the empirical investigations of Cleveland and McGill as well as investigations by perceptual psychologists.*

5.4.1.4 Scatterplot

- To create a basic dot chart or a one-dimensional scatterplot of how variable values change, use stripchart(dataset).
  stripchart(iris)

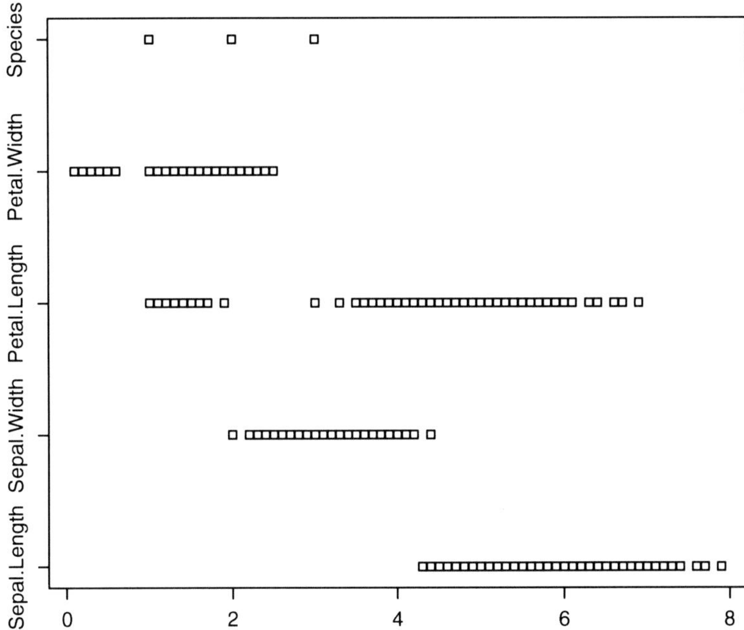

- To create a simple scatterplot where *x* and *y* are vectors of the *x* and *y* coordinates, use *plot(x,y)*.

  Note the different graphs created if we change the order of the *x–y* axis variables, when one variable is a factor with different levels.

  plot(iris$Species,iris$Sepal.Length)

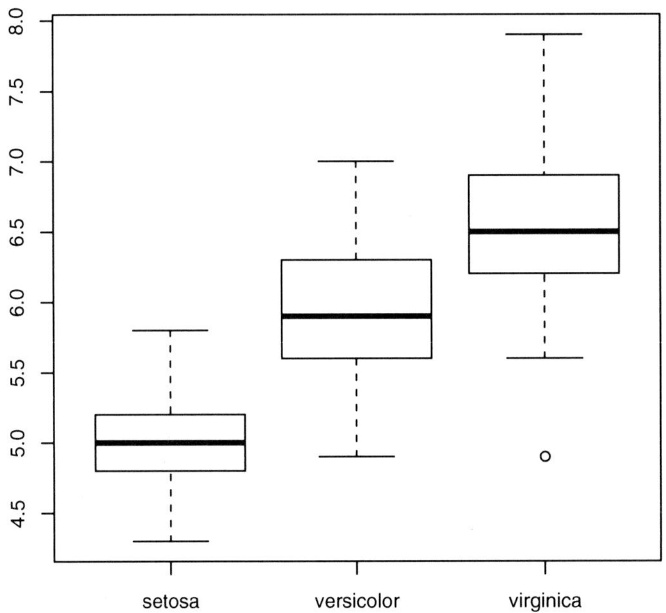

>plot(iris$Sepal.Length,iris$Species)

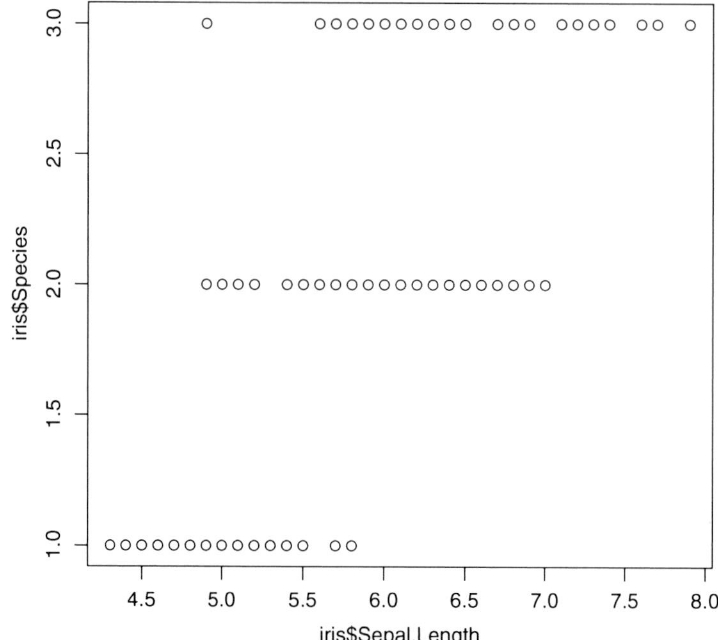

>plot(density(iris$Sepal.Length))

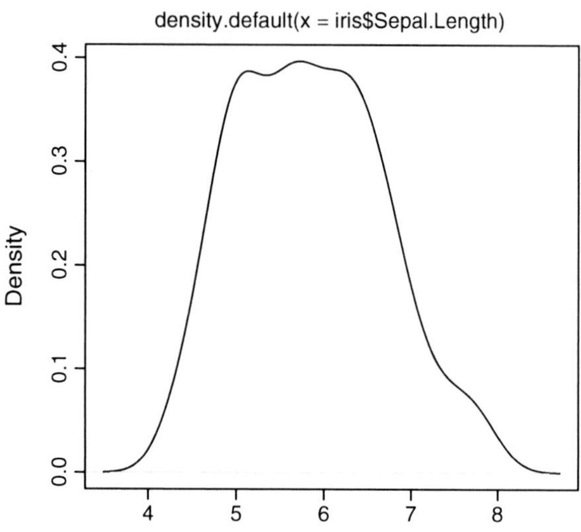

- To create a scatterplot matrix where you need to see how various numerical variables correlate or behave in relation to each other, use plot(dataset).

  plot(iris)

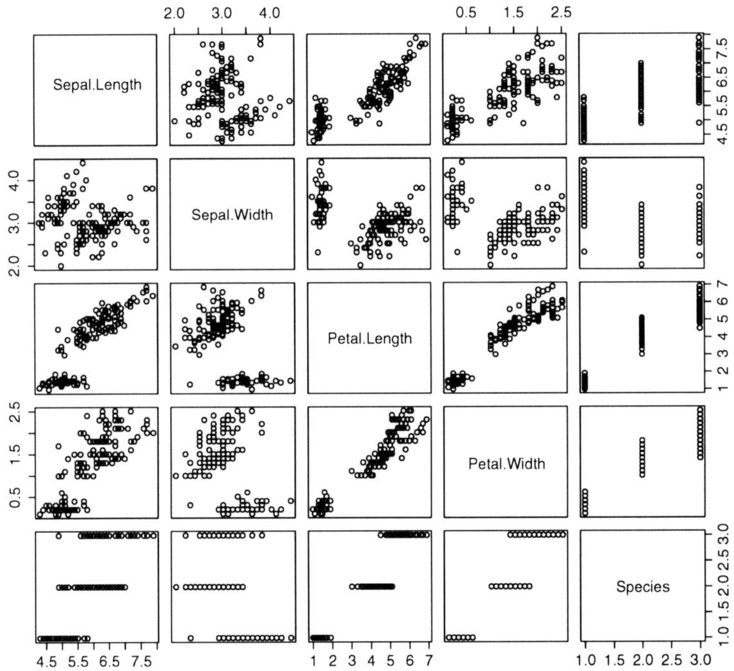

- The function rug(x) can show a rug plot for showing along the *x*-axis vertical bars that denote the distribution of *x*.
  - To make the rug plot on the *y*-axis, create the option side=2, and to make it on top, create side=3.

    > *plot(iris$Sepal.Length)*
    > *rug(iris$Sepal.Length,side=2)*

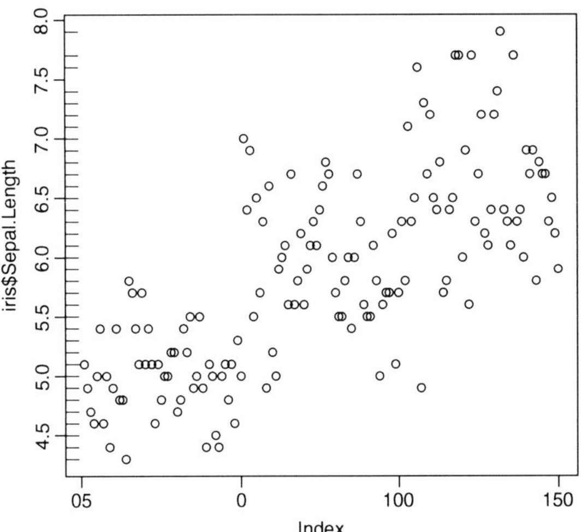

5.4.1.5 Line Graph

- The plot( ) command allows for a number of optional arguments, e.g., type="l", which sets the plot type to "lines".

  plot(iris$Sepal.Length,type="l")

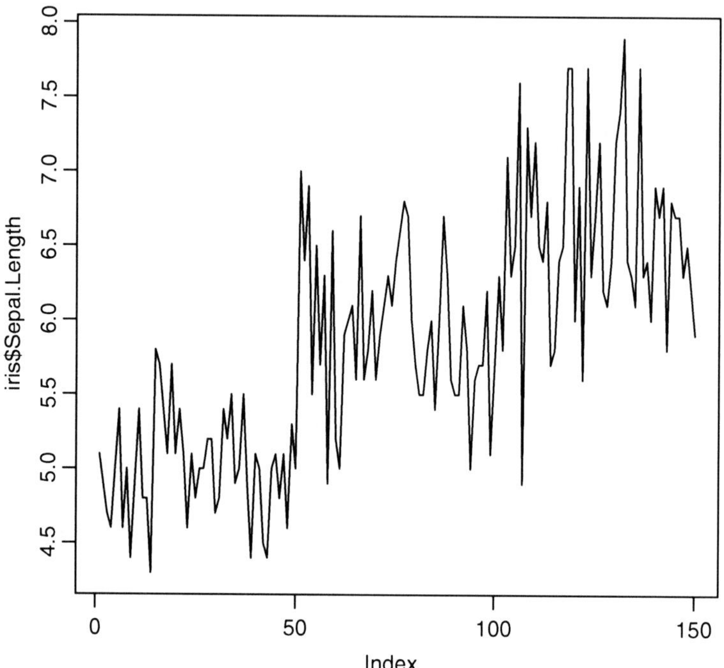

- For a simple line you can use the function *abline()*. It can also be used for plotting a trend line or a regression line using the reg= parameter, but care must be taken in plotting the parameters.

> data(cars)
> str(cars)
'data.frame': 50 obs. of 2 variables:
$ speed: num 4 4 7 7 8 9 10 10 10 11 ...
$ dist : num 2 10 4 22 16 10 18 26 34 17 ...
> head(cars)
speed dist
1 4
2 2
4 10
3 7
4 4
7 22
5 8

16 6
9 10
> tail(cars)
speed dist
45 23
54 46
24 70
47 24
92 48
24 93
49 24
120 50
25 85
> plot(cars)
> abline(v=0,h=0,reg=lm(cars$dist~cars$speed))
Note: The *v* and *h* values give values of the *x* and *y* intercept, while the reg parameter fits a regression or trend line.
This gives the following diagram:

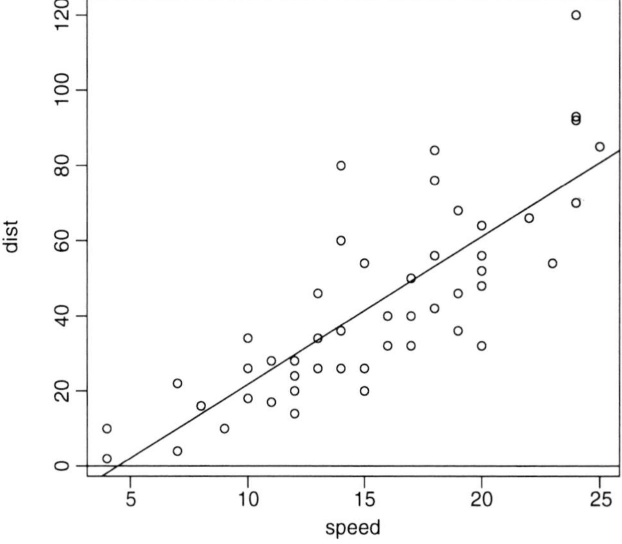

Now let us see what happens if we reverse the order of dependent and independent variables:
> plot(cars)
> abline(v=0,h=0,reg=lm(cars$speed~cars$dist))
> title("Wrong")
This gives a bad/incorrect trend line! We must be careful how we use **abline()**.
Note I used the title function (instead of main=) to give the title to the graph.

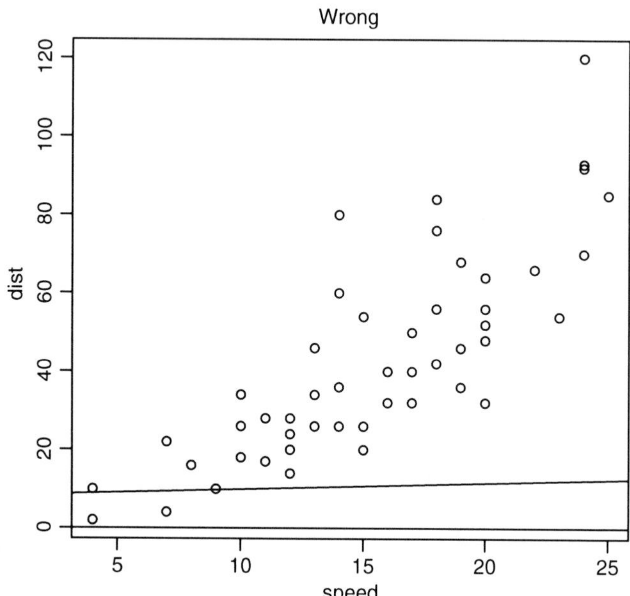

5.4.1.6 Bar Chart

For a simple bar chart to summarize variable *x* from a dataset:
> *barplot(table(iris$Sepal.Length))*.

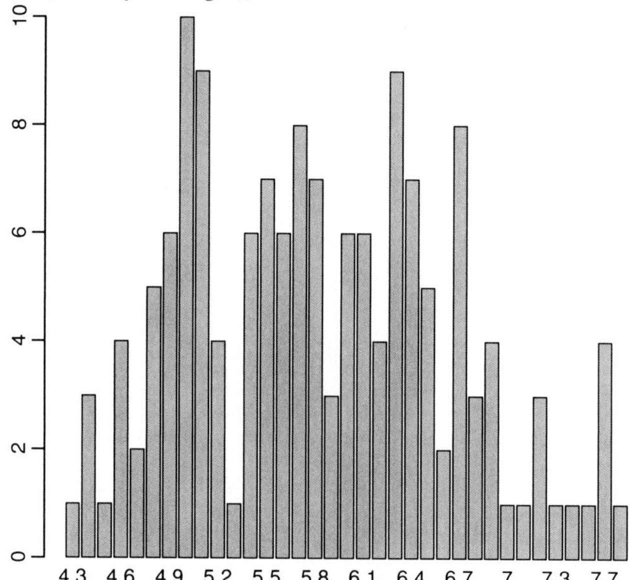

To make a horizontal bar plot, use the parameter horiz=TRUE:
>*barplot(table(iris$Sepal.Length),horiz=T)*

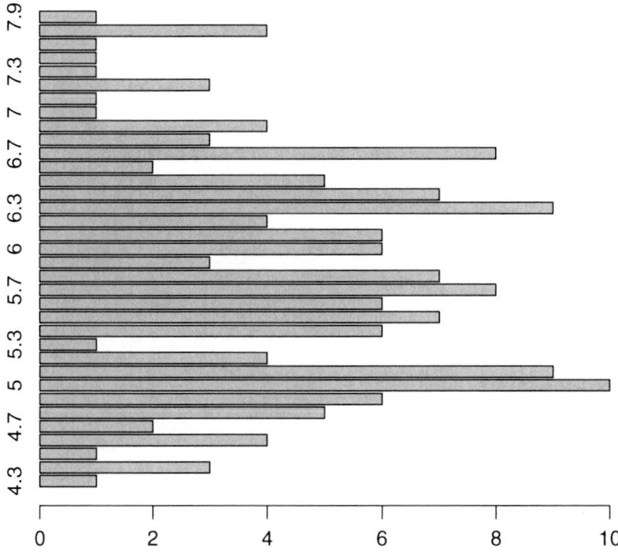

5.4.1.7 Sunflower Plot

An option for a scatterplot with significant point overlap is the sunflower plot. Multiple points are plotted as "sunflowers" with multiple leaves ("petals") such that overplotting is visualized instead of accidental and invisible. Another package for the same purpose (overlapping points) is hexbin.

Source: http://www.jstatsoft.org/v08/i03/paper

```
> par(mfrow=c(1,2) )
> plot(faithful)
> sunflowerplot(faithful)
```

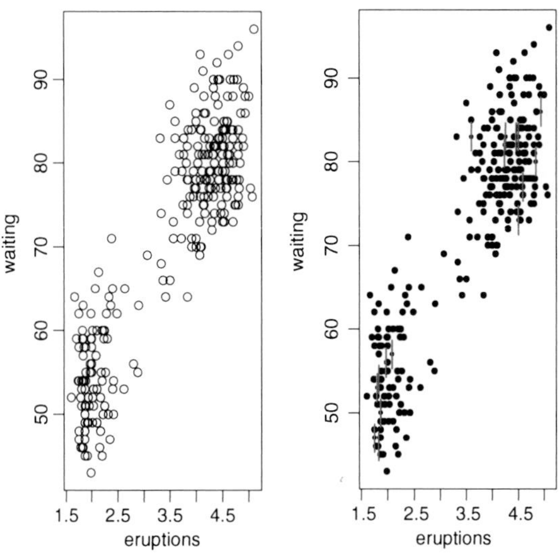

Another example
>library(HistData)
> data(Galton)
> par(mfrow=c(1,2))
> plot(Galton,main="Scatter Plot")
> sunflowerplot(Galton,main="Sunflower Plot")

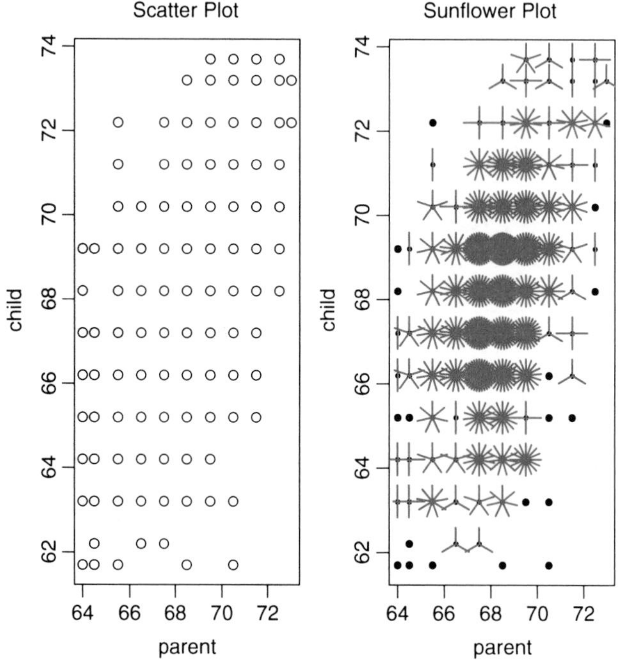

Two great resources for simple kinds of graphs are http://www.harding.edu/fmccown/r/#barcharts and http://www.statmethods.net/graphs/line.html.

5.4.1.8 Hexbin

- The hexbin(x, y) function in the hexbin package provides bivariate binning into hexagonal cells. It is useful for scatterplots when there is a large number of values that are overlapping. It is thus an alternative to the scatterplot or sunflower plot and is available at http://cran.r-project.org/web/packages/hexbin/hexbin.pdf.

library(hexbin)
ajay=hexbin(iris$Petal.Length,iris$Sepal.Length,xbins=30)
plot(ajay)

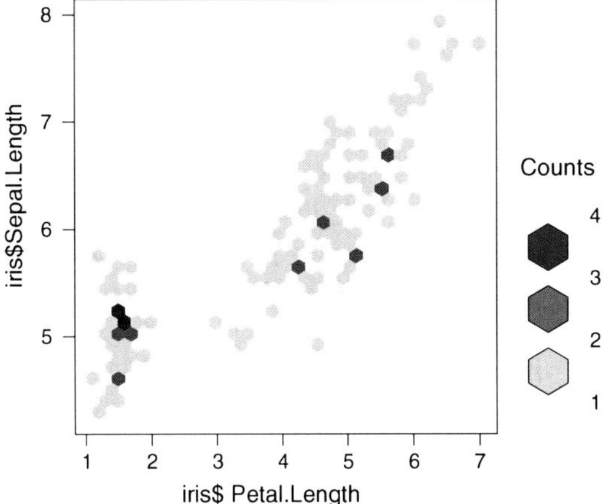

5.4.1.9 Bubble Chart

Using base graphics and symbols().

One of six symbols—circle, square, rectangle, star, thermometer, or boxplot—can be plotted at a specified set of *x*- and *y*-coordinates using the symbols() command. We use the inches command to limit the size of symbols and fg= and bg= parameters to control the color of the foreground and color of symbols.

> *radius <- sqrt(mtcars$qsec/ pi)*

> *symbols(mtcars[,6], mtcars[,1],* ***circles****=mtcars$qsec,inches=0.15,*
+fg="grey", bg="black", xlab="Wt", ylab="Miles per Gallon")

Note we calculated the parameter plots radius of circles, as circles is a vector giving the radii of the circles, if you need to change the area, then you need to use the radius as the value calculated. Using the text function to plot text at certain *x*- and *y*-values and using cex to limit the size of the font

>text(mtcars[,6],mtcars[,1],mtcars[,7],cex=0.5,col="white").

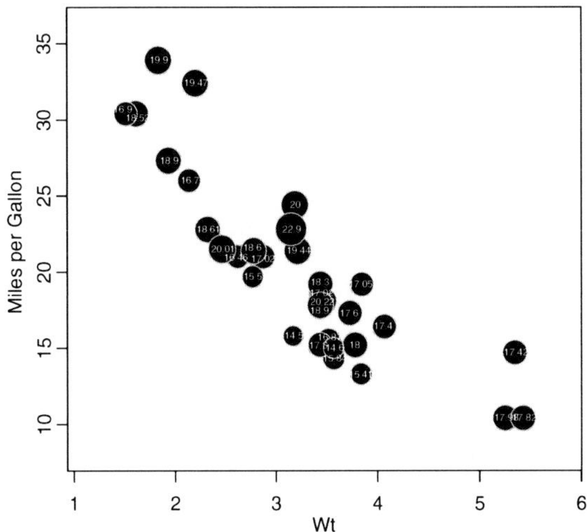

There is an elaborate example of ggplot2 at http://had.co.nz/ggplot2/geom_point.html.

Clearly ggplot2 can help us create even better graphs. We examine ggplot later in this chapter.

library(ggplot2)

ggplot(mtcars, aes(wt, mpg)) + geom_point(aes(size = qsec))

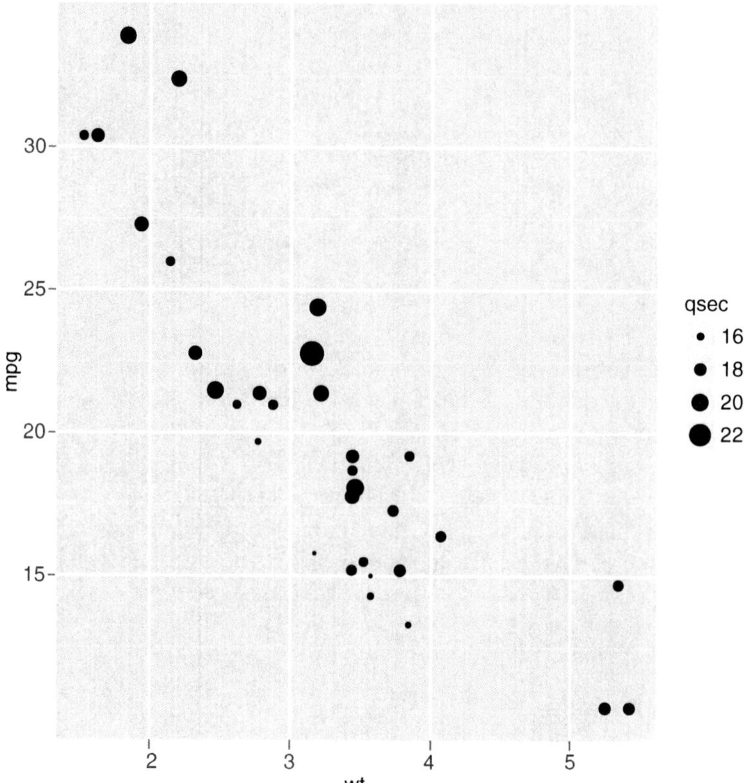

5.4.2 Summary of Basic Graphs in R

Let us summarize the graphs we have learned.

Graphing in R requires just one or two words for each type of basic graph.

> par(mfrow=c(3,3))

> plot(iris$Sepal.Length, main="Scatter Plot with Rug ")

> rug(iris$Sepal.Length,side=2)

> barplot(table(iris$Sepal.Length), main="Bar Plot")

```
> plot(iris$Sepal.Length,type="l", main="Line Plot")
> plot(iris$Sepal.Length, main="Scatter Plot")
> boxplot(iris$Sepal.Length, main="Box Plot")
> stripchart(iris$Sepal.Length, main="Strip Chart")
> sunflowerplot(iris$Sepal.Length, main="Sunflower Plot")
> hist(iris$Sepal.Length, main="Histogram")
> plot(density(iris$Sepal.Length), main="Density Plot")
```

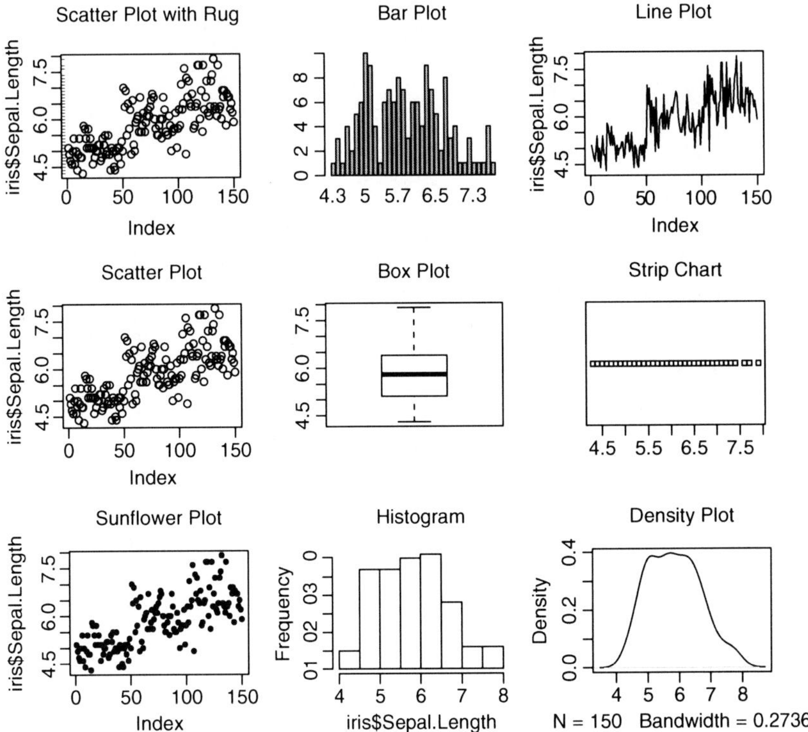

5.4.3 Advanced Graphs

5.4.3.1 Table Plot

A table plot is a visualization of a (large) multivariate dataset. Each column represents a variable, and each row bin is an aggregate of a certain number of records. Numeric variables are depicted as bar charts (bar lengths are determined by mean values per row bin) and categorical variables as stacked bar charts. Missing values are taken into account. Table plots can be designed with a GUI. They also support large datasets from the ff package.

```
> data(iris)
> library(tabplot)
> tableplot(iris)
```

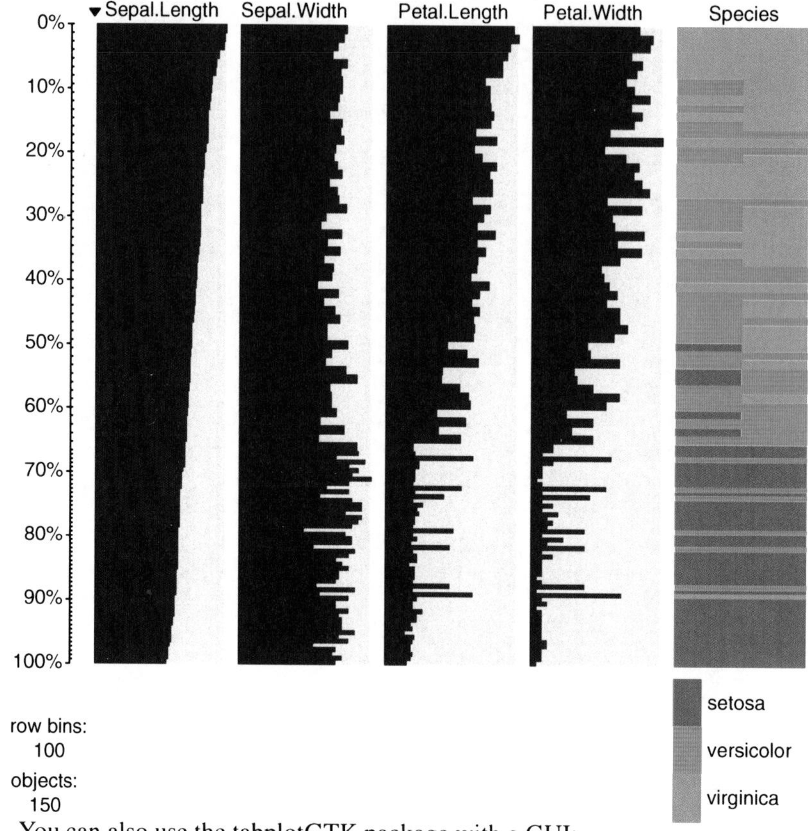

You can also use the tabplotGTK package with a GUI:
http://cran.r-project.org/web/packages/tabplotGTK/tabplotGTK.pdf

```
library(tabplotGTK)
data(iris)
tableGUI(iris) # This invokes the GUI for Tableplot.
```

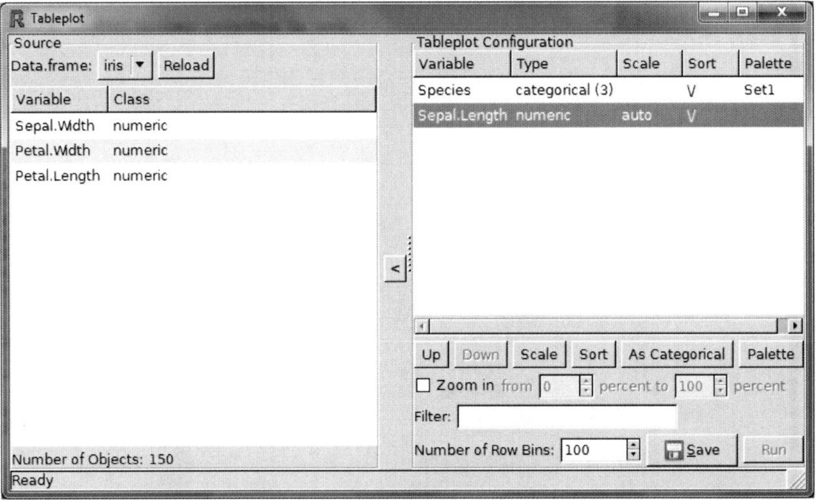

#Notice how we can keep the variables we want, sort the variables, and even choose a custom color palette.

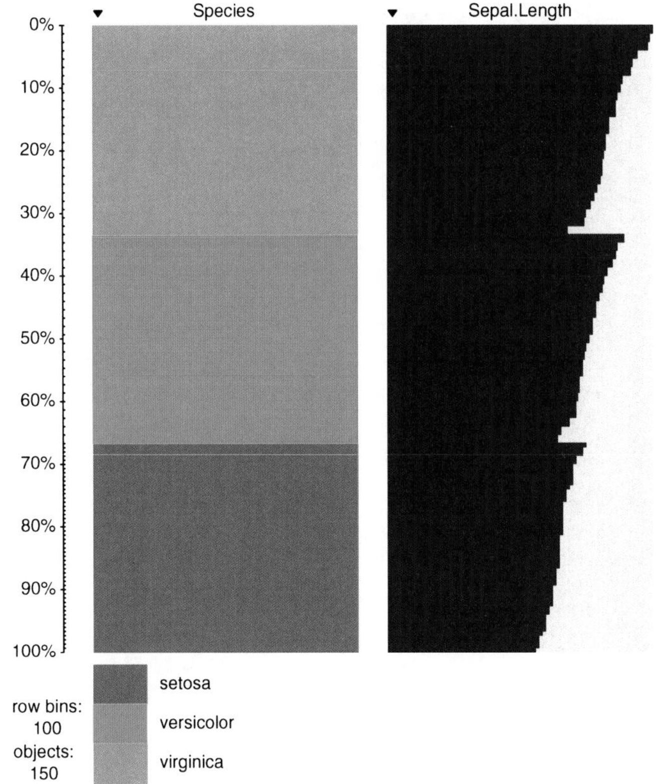

5.4.3.2 Mosaicplot and Treemaps

This is basically a contingency table; it is also known as a treemap. The area represents the number of observations. A treemap is used for visualizing categorical data.

mosaicplot(HairEyeColor,col=terrain.colors(2)) #Note we use color palettes here.

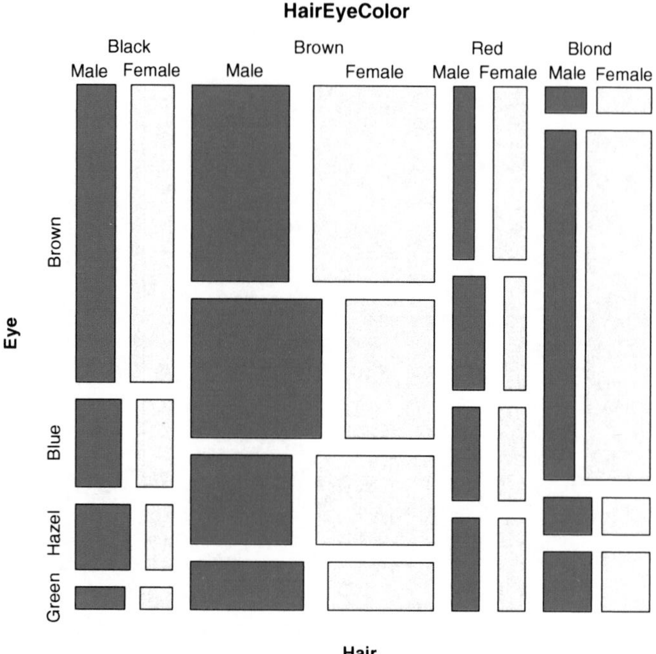

mosaicplot(HairEyeColor,shade=T) #Note we use shade=TRUE here.

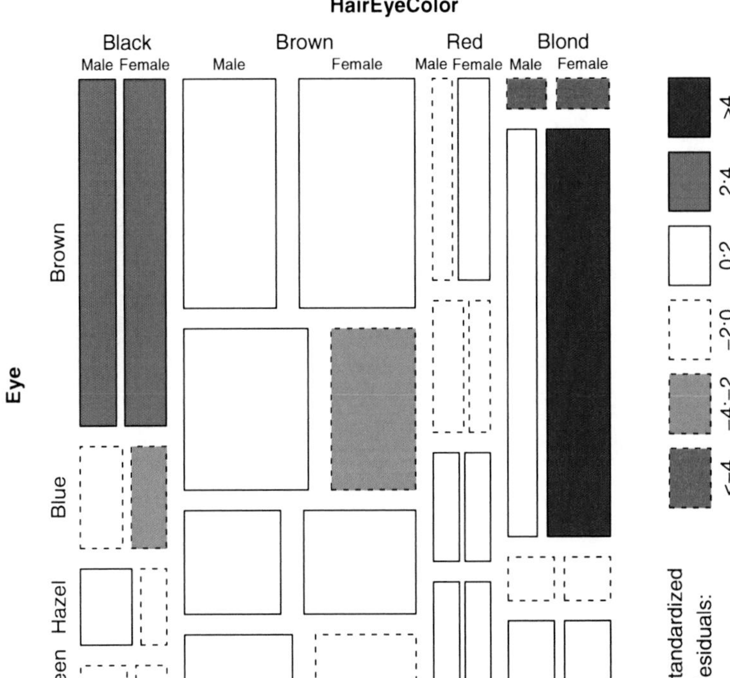

- To visualize a lot of categorical data, see the vcd package in R. The strucplot framework in the R package vcd, used for visualizing multiway contingency tables, integrates techniques such as mosaic displays, association plots, double-decker plots, and sieve plots: http://cran.r-project.org/web/packages/vcd/vcd.pdf.
- Sunburst graphs: The treemap uses a rectangular, space-filling, slice-and-dice technique to visualize objects in the different levels of a hierarchy. The area and color of each item correspond to an attribute of the item as well.

The sunburst technique is an alternative, space-filling visualization that uses a radial rather than a rectangular layout. For more details, visit http://www.cc.gatech.edu/gvu/ii/sunburst/.

The difference between sunburst and treemap is shown below (these are screenshots using the disk utlization in Ubuntu Linux):

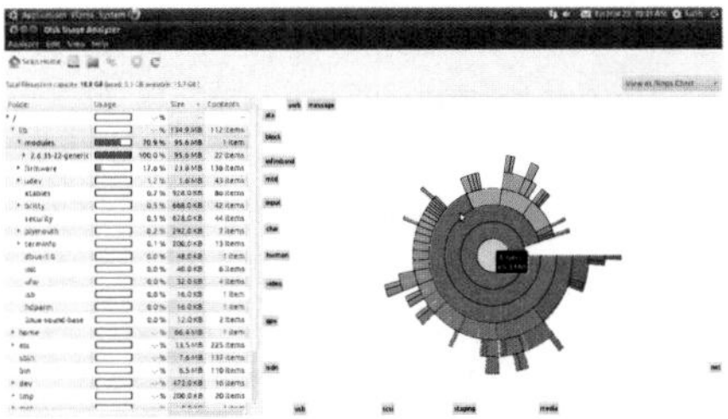

and

Note: Sunburst graphs can be visually similar to coxcomb graphs (Florence Nightingale). Coxcomb graphs are based on code from the histdata package and http://rgm2.lab.nig.ac.jp/RGM2/func.php?rd_id=HistData:Nightingale.

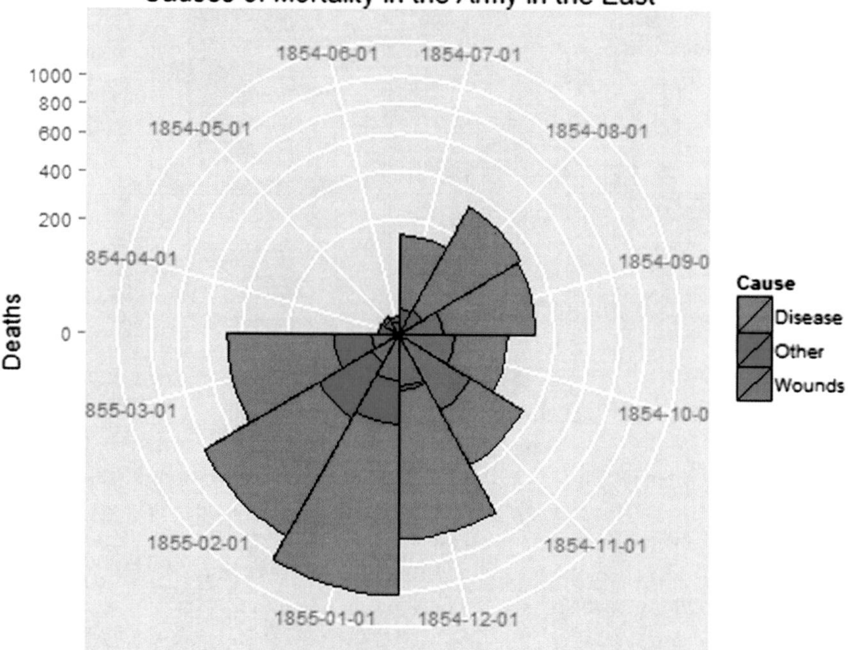

5.4.3.3 Heat Map

A heat map is a false color image with a dendrogram added to the left side and to the top. The function heatmap() can be used to make a heat map of a matrix. Heat maps are also used in sort analytics or to compare several categorical values and how they perform relative to many numerical values.

```
> ajay = matrix(rnorm(200), 20, 10)
> heatmap(ajay)
```

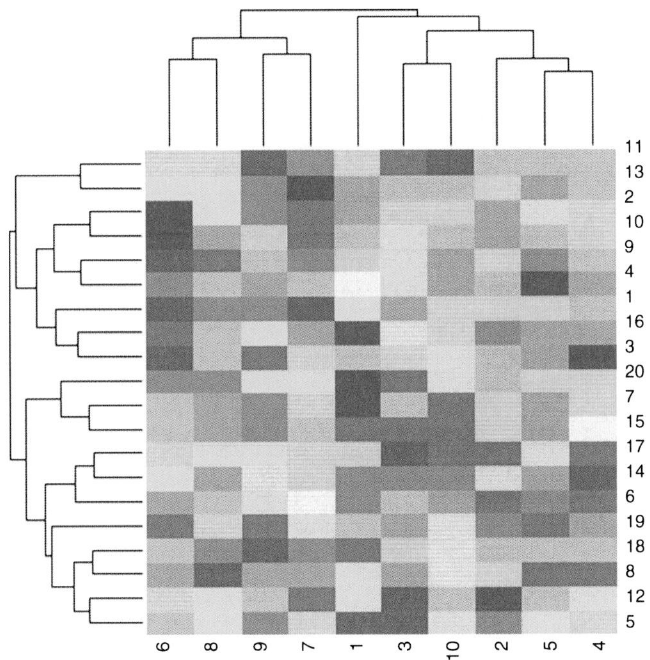

Another example

```
> data(mtcars)
> a=as.matrix(mtcars)
> heatmap(a,col=terrain.colors(256))
```

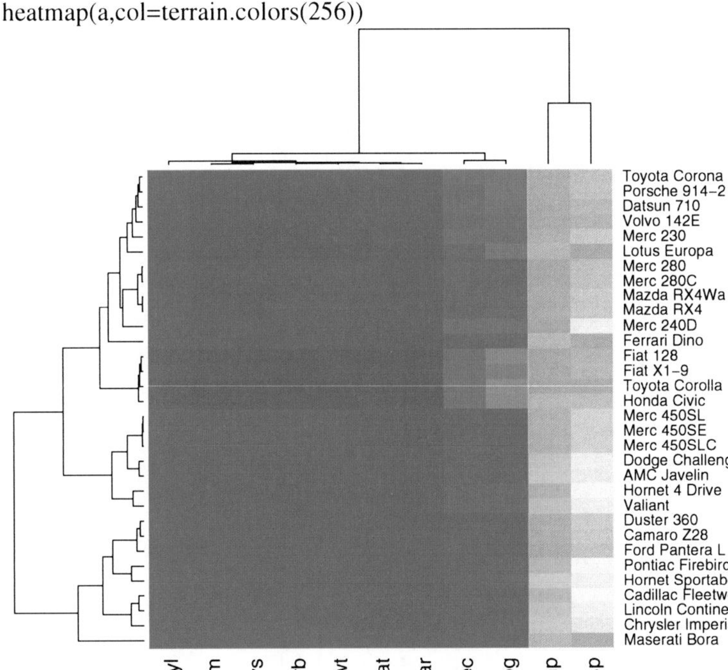

5.4.3.4 Plots to Show Distributions

Alternatives to boxplot graphs are violin plots and bean plots. Violin plots show the density of the distribution plotted next to the strip chart of the values. Bean plots show individual values as small lines next to the density and the average values.

- **Violin plot**

```
>library(vioplot)
>vioplot(iris$Sepal.Length,iris$Sepal.Width,iris$Petal.Length,iris$Petal.Width,
col="grey")
```

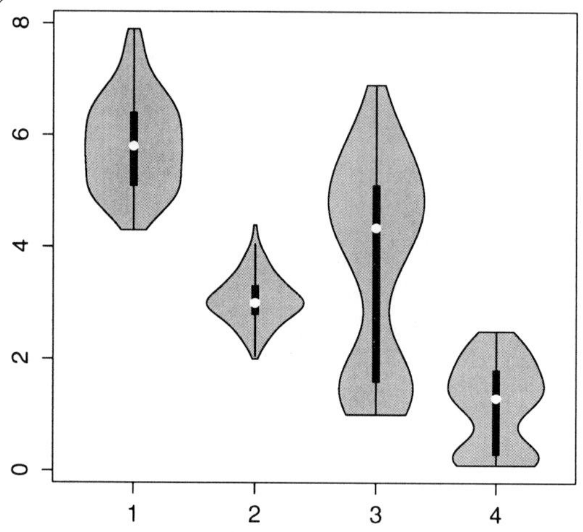

- **Beanplot**

```
> library(beanplot)
> beanplot(iris$Sepal.Length,iris$Sepal.Width,iris$Petal.Length,iris$Petal.Width)
```

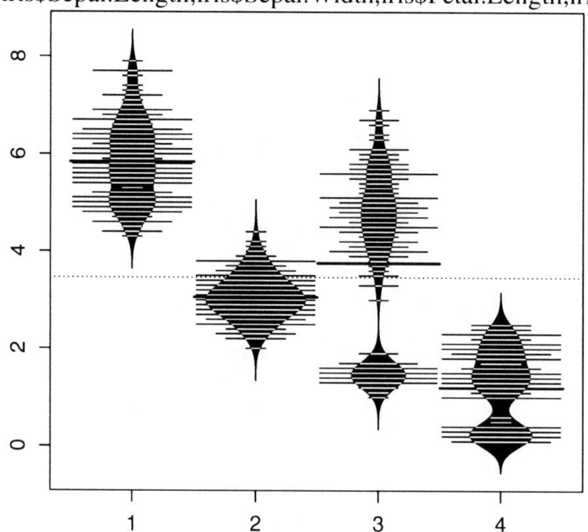

- Source: http://www.jstatsoft.org/v28/c01/paper

5.4.3.5 3D Graphs

- 3D Scatterplot: 3D scatterplot graphs can be built in just three clicks in Rcmdr or three lines of code in R! It uses the function scatter3d(x, y, z) in the Rcmdr package.

  . *# 3d Scatterplot*
  library(Rcmdr)
  data(iris)
  scatter3d(iris[,1],iris[,2]iris[,3])
  #Note the notation Dataset[,c] refers to the cth columnor variable in a data frame.

Using the GUI we can easily also plot by group. There are multiple options for 3D graphs; such graphs are easily made by the R Commander GUI.

Step 1: Navigate in R Commander to 3D Graphs

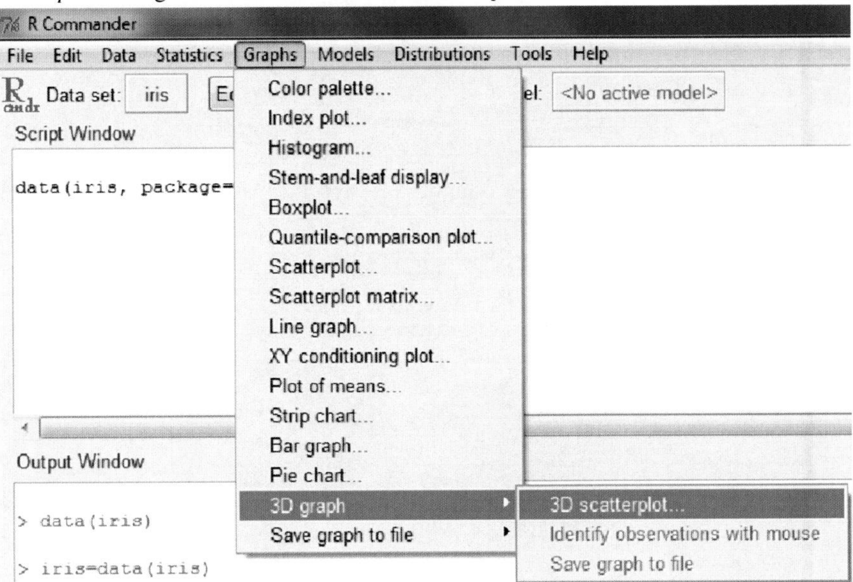

Step 2: Choose options for 3D graphs. We have chosen a linear least squares to fit as a surface and at the bottom we have chosen Plot by Species (which is a categorical variable).

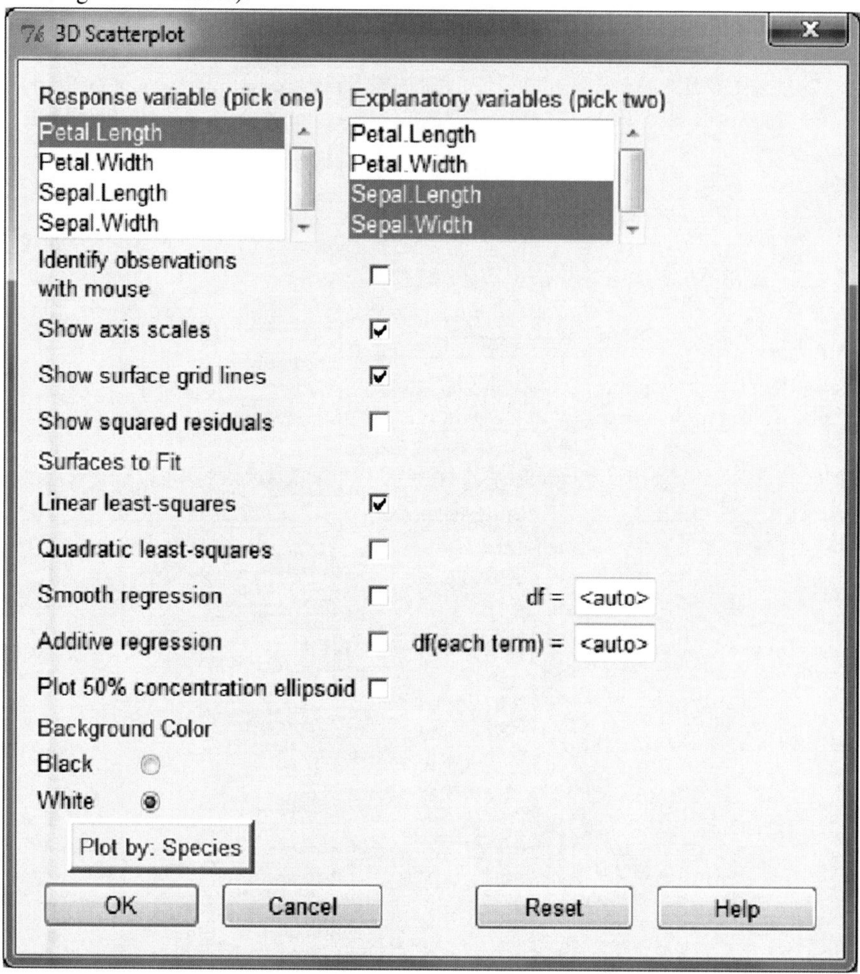

Step 3: Review the graph and use the mouse to rotate the graph interactively.

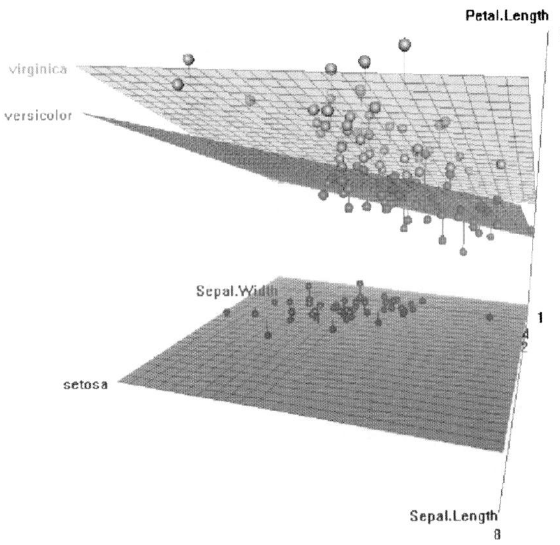

- 3D WireFrame

You can build stunning 3D graphs using the cloud and wireframe functions in the "lattice" package.

library(lattice)
wireframe(volcano, shade = TRUE, aspect = c(61/87, 0.4), light.source = c(10,0,10))

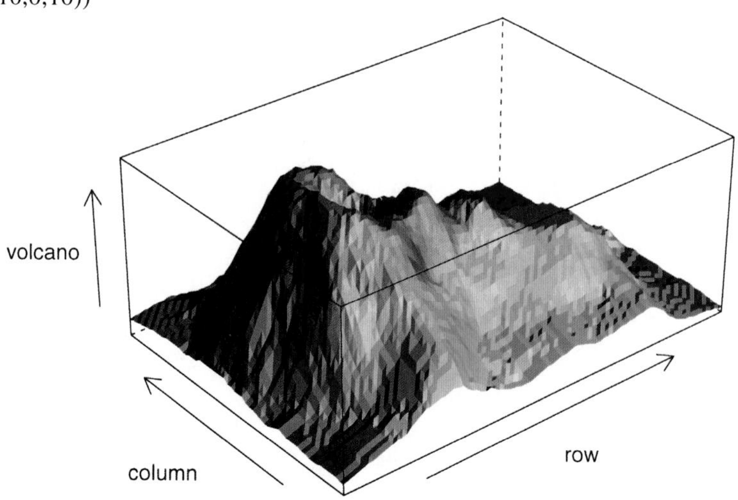

cloud
cloud(Sepal.Length ~ Petal.Length * Petal.Width | Species, data = iris, screen = list(*x* = −90, *y* = 70), distance = .4, zoom = .6)

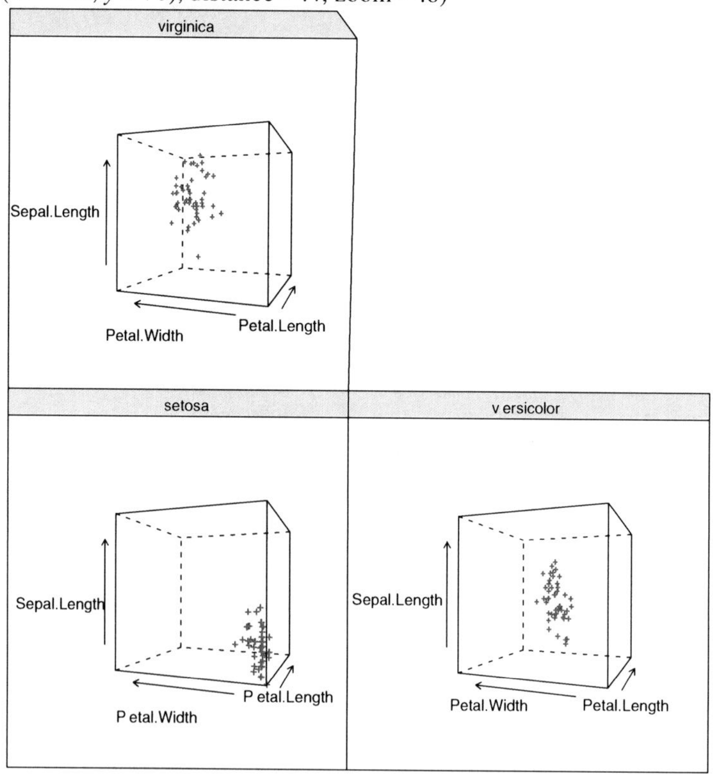

Suppose we don't want to have seperate 3D wireframes for each species but have just one 3D wireframe.
*cloud(Sepal.Length ~ Petal.Length * Petal.Width, data = iris, screen = list(x = −90, y = 70), distance = .4, zoom = .6)*

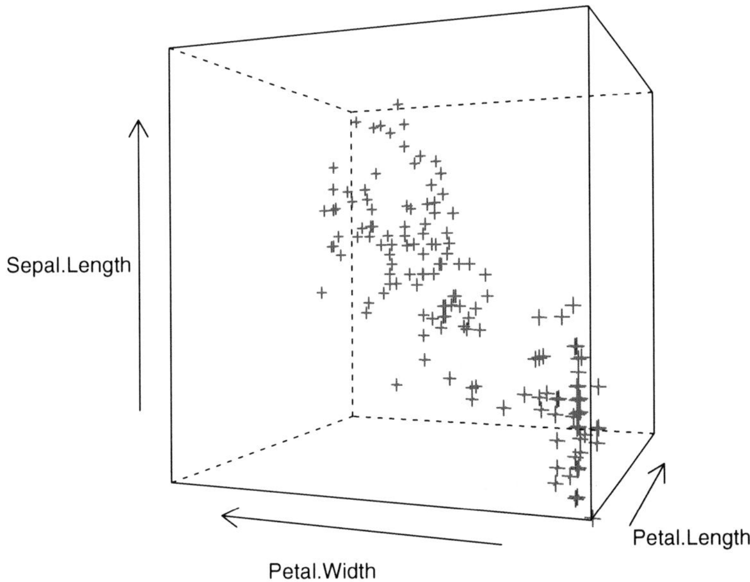

5.4.3.6 Sparkline

Sparklines are compact graphs based on Edward Tufte's work. They are graphs used for conveying information with high data density to see comparative trends and variations. Sparklines can be created in R by use of either YaleToolkit or SparkTable packages. Sparklines are roughly the same height as the accompanying text.

```
> library(YaleToolkit)
>data(EuStockMarkets)
> EuStockMarkets2=as.data.frame(EuStockMarkets)
#Note how we use the as.data.frame function above to transform a time series to a data frame.
> sparklines(EuStockMarkets2,
+ sub=c(names(EuStockMarkets2)),
+ yaxis = TRUE, xaxis=TRUE,
+ main = 'EuStockMarkets')
```

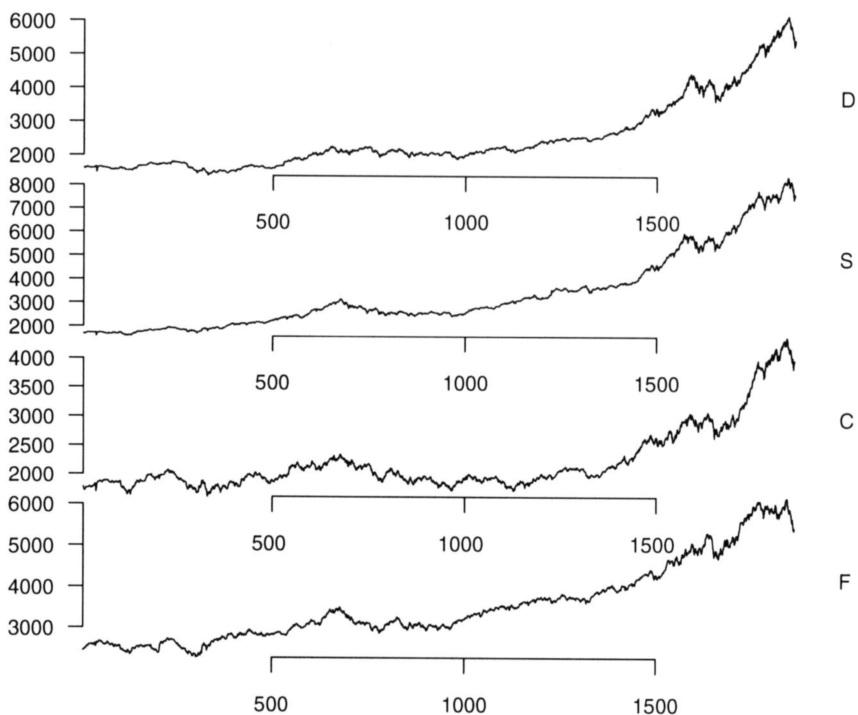

But if we want sparklines to look better, we use the following code. Go to http://rss.acs.unt.edu/Rdoc/library/YaleToolkit/html/sparklines.html to see how the parameters can be elaborated. or *?sparklines*

```
> sparklines(EuStockMarkets2,
+ sub=c(names(EuStockMarkets2)),
+ outer.margin = unit(c(2,4,4,5), ‘lines’),
+ outer.margin.pars = gpar(fill = ‘lightblue’),
+ buffer = unit(1, "lines"),
+ frame.pars = gpar(fill = ‘lightyellow’),
+ buffer.pars = gpar(fill = ‘lightgreen’),
+ yaxis = TRUE, xaxis=FALSE,
+ IQR = gpar(fill = ‘grey’, col = ‘grey’),
+ main = ‘EuStockMarkets’)
```

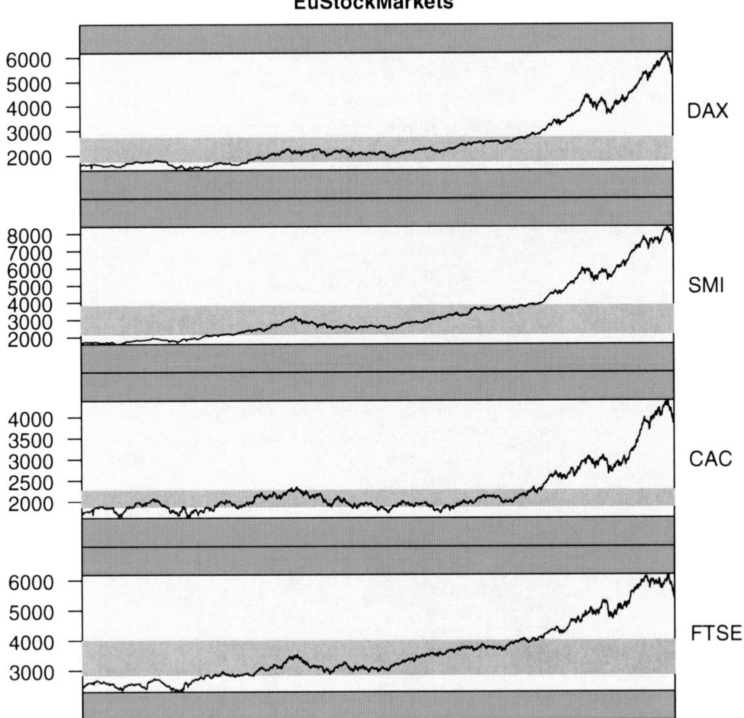

5.4.3.7 Radial Plots

Radial plots call for the use of the plotrix package. The following plots are covered here:

- Radial Plot
  - Polar chart
  - Circular plot
  - Clock plot
  - Windrose

The radial.plot family of plots is useful for illustrating cyclic data such as wind direction or speed (but see oz.windrose for both), activity at different times of day, and so on. A wind rose is displayed at oz.windrose in the style used by the Australian Bureau of Meteorology.

See http://rss.acs.unt.edu/Rdoc/library/plotrix/html/00Index.html.

```
> library(plotrix)
> testlen<-c(rnorm(36)*2+5)
> testpos<-seq(0,350,by=10)
> polar.plot(testlen,testpos,main="Test Polar Plot",lwd=3,line.col=4)
> radial.plot(testlen,testpos,rp.type="p",main="Test Radial Plot",line.col="blue")
> clock24.plot(testlen,testpos,main="Test Clock24 (lines)",show.grid=FALSE, line.col="green",lwd=3)
```

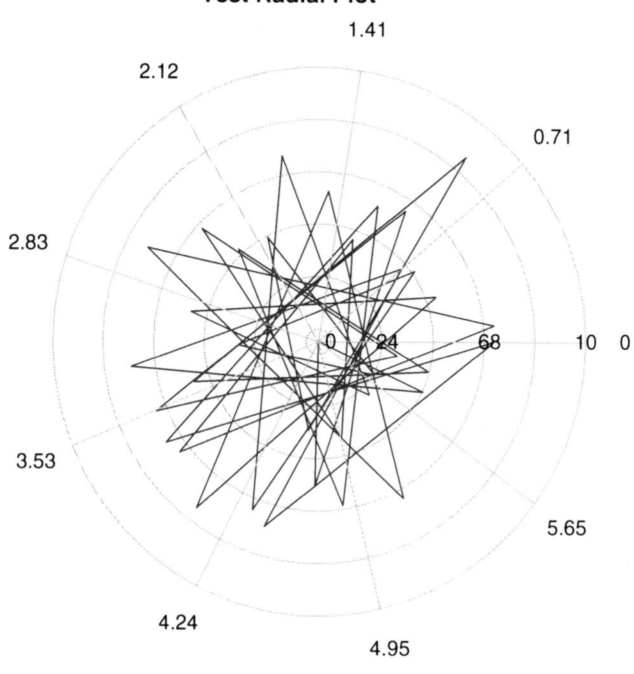

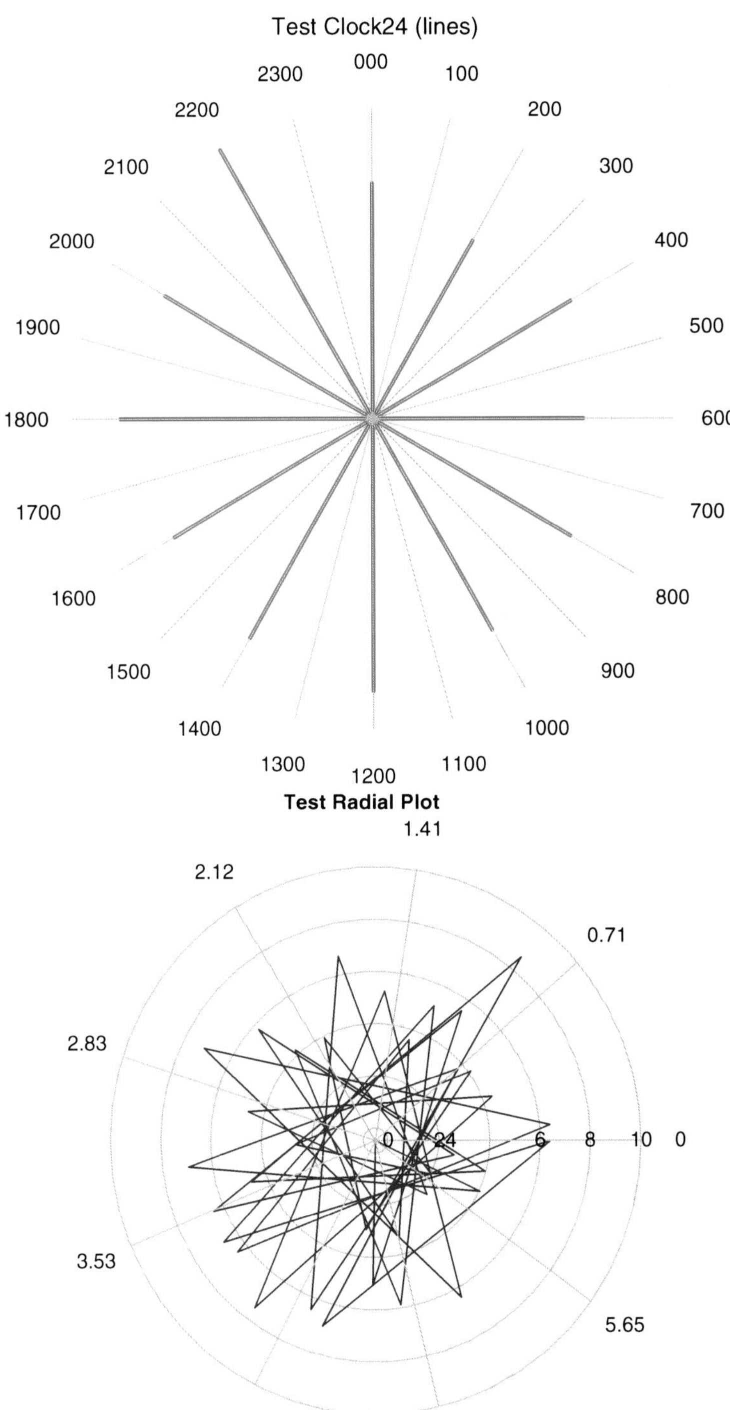
Test Clock24 (lines)
000
100
200
300
400
500
600
700
800
900
1000
1100
1200
1300
1400
1500
1600
1700
1800
1900
2000
2100
2200
2300
Test Radial Plot
1.41
0.71
0
2.12
2.83
3.53
4.24
4.95
5.65
0
2
4
6
8
10

```
>#Note: we are also demonstrating the use of a matrix data structure here.
>windagg<-matrix(c(8,0,0,0,0,0,0,0,4,6,2,1,6,3,0,4,2,8,5,3,5,2,1,1,5,5,2,4,1,
4,1,2,1,2,4,0,3,1,3,1), nrow=5,byrow=TRUE)
> windagg
[,1] [,2] [,3] [,4] [,5] [,6] [,7] [,8]
[1,] 8 0 0 0 0 0 0 0
[2,] 4 6 2 1 6 3 0 4
[3,] 2 8 5 3 5 2 1 1
[4,] 5 5 2 4 1 4 1 2
[5,] 1 2 4 0 3 1 3 1
>oz.windrose(windagg)
```

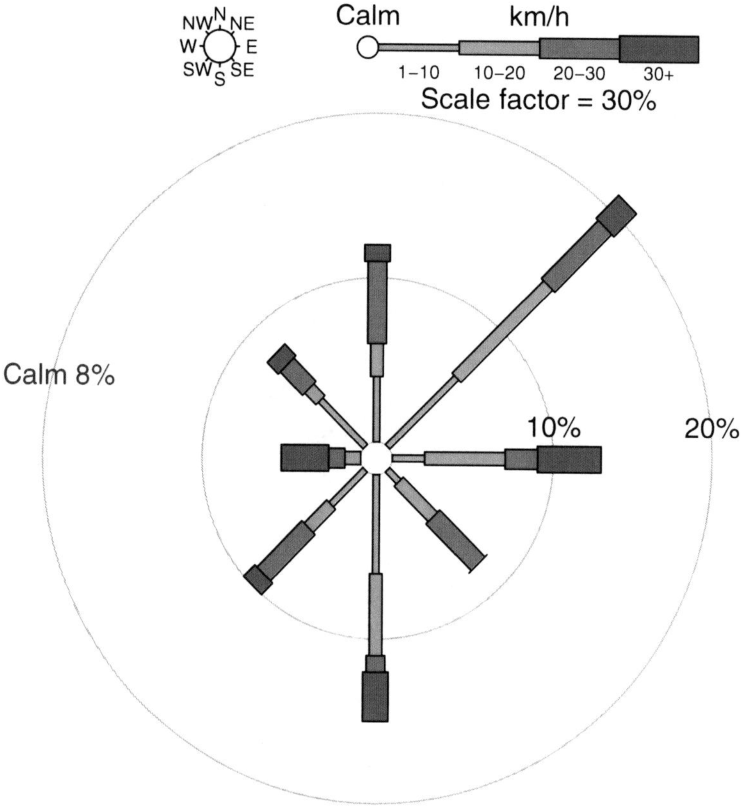

5.4.3.8 Bullet Charts

Bullet charts were proposed by Stephen Few. They can be created in R using the ggplot2 package (Sect. 5.5). Basically, ggplot uses a series of layers to create objects for data visualization in R.

Source: http://www.perceptualedge.com/blog/?p=217

> fake.data2 <- data.frame(measure=letters[15:24], value=round(rnorm(10,5,3)),
+ mean=round(rnorm(10,4,4)),target=round(rnorm(10,12,32)))
> p <- ggplot(fake.data, aes(measure, value)) #*Note:* The measure is on the *x*-axis and the value on the *y*-axis
> p <- p + geom_bar(fill="grey", width=0.5) # barchart for measure and plot is wide gray bar
> p <- p + geom_bar(aes(measure, mean), width=0.2)
barchart for measure (x) and mean (y) is narrow black bar
> p <- p + geom_point(aes(measure, target), colour="red")
This marks the red point for the target (y) whereas the measure is x
> p <- p + geom_errorbar(aes(y = target,x = measure, ymin = target,ymax + = target), width = .45)
#width of error bars for target
> p <- p + coord_flip() #The axis is flipped now to make them horizontal barplots
> p

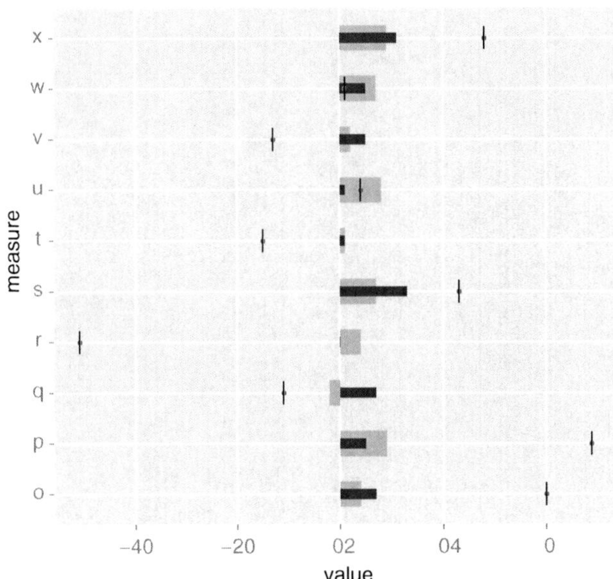

5.4.3.9 Word Cloud

A word cloud represents the frequency of words occurring in a document, with the size of the font for each word being proportional to its frequency. Thus a more frequently occurring word will look larger, and this can help visualize long texts.

We copy and paste President Barack Obama's "Yes We Can" speech in a text document and read it in. To create a word cloud, we need a data frame with two columns, one with words, the other with frequency. We read in the transcript from http://politics.nuvvo.com/lesson/4678-transcript-of-obamas-speech-yes-we-can and paste in the file located in the local directory: C:/Users/KUs/Desktop/new.

(Note tm is a powerful package and will read ALL the text documents within the particular folder.)

```
> library(tm)
>library(wordcloud)
> txt2="C:/Users/KUs/Desktop/new"
> b=Corpus(DirSource(txt2), readerControl = list(language = "eng"))
> b<- tm_map(b, tolower) #Changes case to lowercase
> b<- tm_map(b, stripWhitespace) #Strips white space
> b <- tm_map(b, removePunctuation) #Removes punctuation
> tdm <- TermDocumentMatrix(b)
> m1 <- as.matrix(tdm)
> v1<- sort(rowSums(m1),decreasing=TRUE)
> d1<- data.frame(word = names(v1),freq=v1)
> wordcloud(d1$word,d1$freq)
```

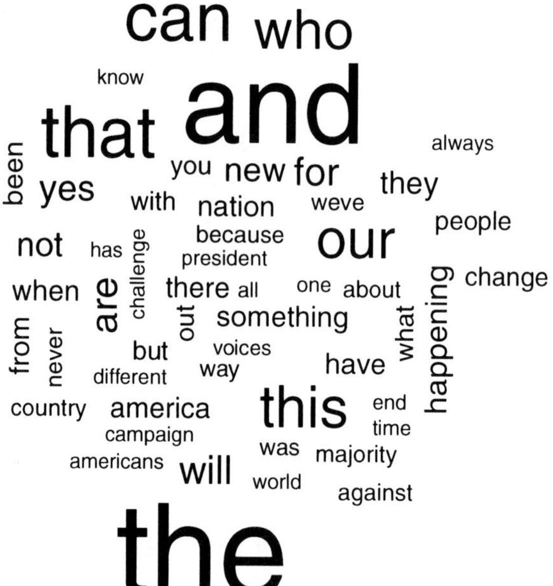

We need to remove some of the commonly occurring words like “the” and “and.” We do not use the standard stop words in English (the tm package provides that; see Chap. 13) as “we” and “can” are also included.

```
> b <- tm_map(b, removeWords, c("and","the")) # Remove “and” and “the”.
> tdm <- TermDocumentMatrix(b)
> m1 <- as.matrix(tdm)
> v1<- sort(rowSums(m1),decreasing=TRUE)
> d1<- data.frame(word = names(v1),freq=v1)
> wordcloud(d1$word,d1$freq)
```

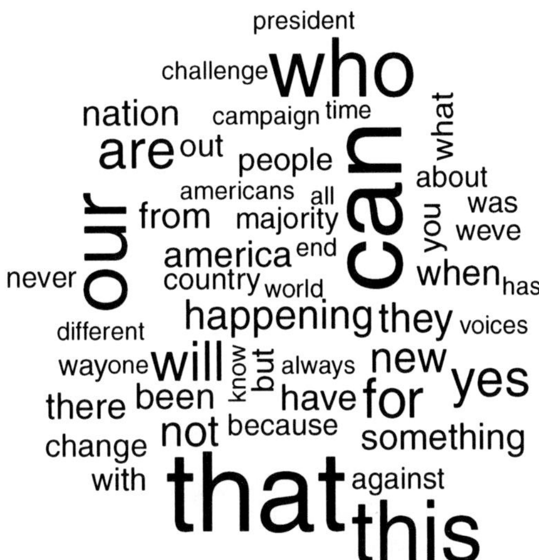

But let us see how the word cloud changes if we remove all English stop words

> b <- tm_map(b, removeWords, stopwords("english")) # remove stop words

and then rerun the code:

> tdm <- TermDocumentMatrix(b) > m1 <- as.matrix(tdm) > v1<- sort(rowSums(m1),decreasing=TRUE) > d1<- data.frame(word = names(v1),freq=v1) > wordcloud(d1$word,d1$freq)

change
nation
happening
people
americans
voices
weve
challenge
world
time
campaign
president
country
america
majoityr

You can draw your own conclusions from the content of this famous speech based on your political preferences!

5.4.4 Additional Graphs

Some specialized graphs to specific domains and graphs not covered previously are added in this subsection.

5.4.4.1 Financial Charts

For specialized financial charting see the quantmod package and especially http://www.quantmod.com/examples/charting/.

```
library(quantmod)
getSymbols("AAPL")
chartSeries(AAPL)
```

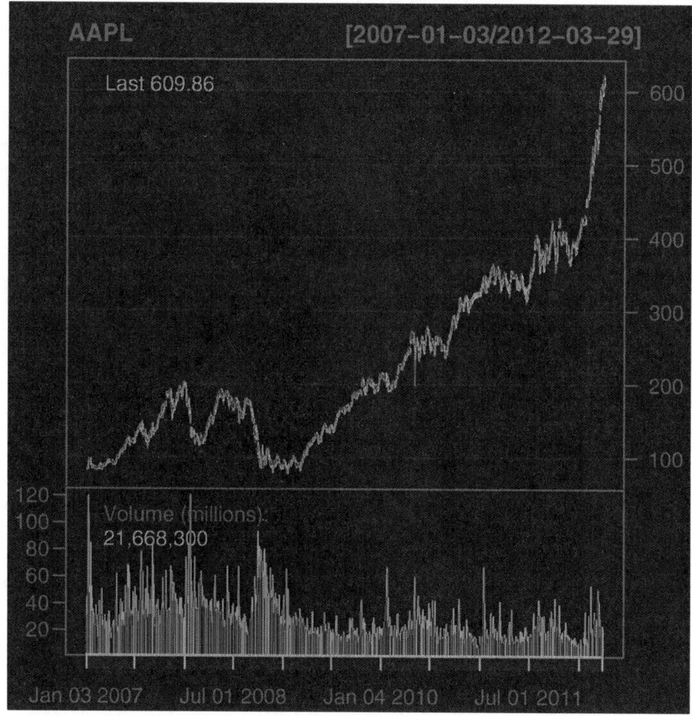

barChart(AAPL,theme='white.mono')

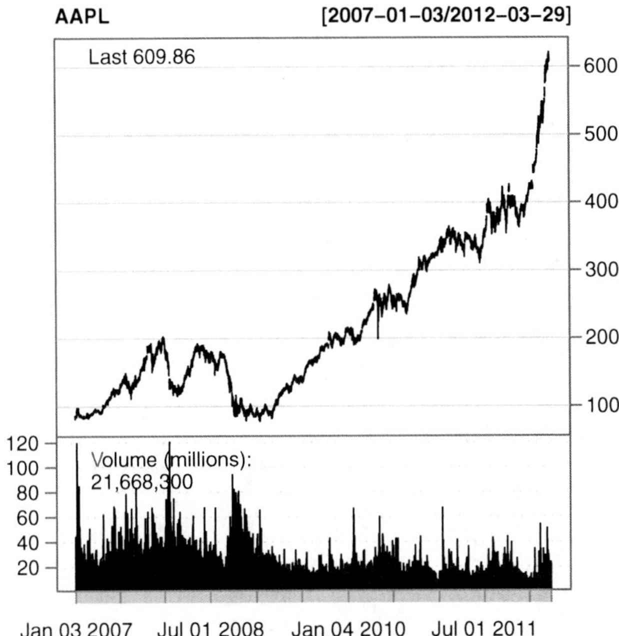

candleChart(AAPL)

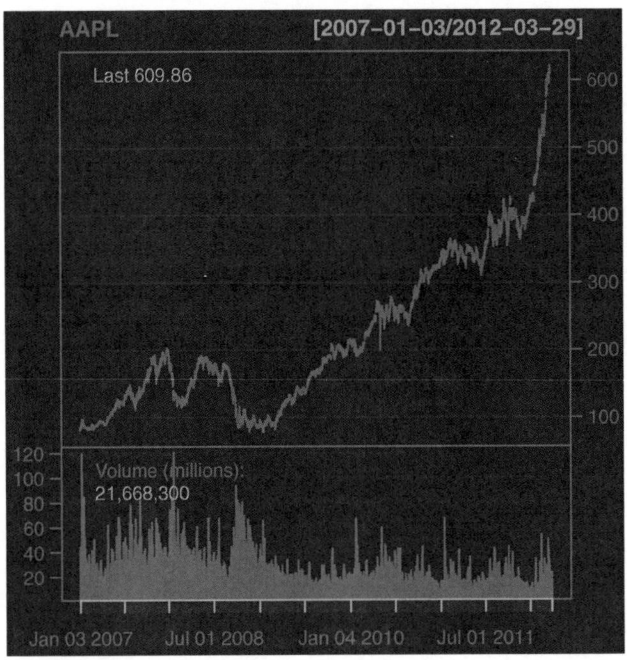

lineChart(AAPL)

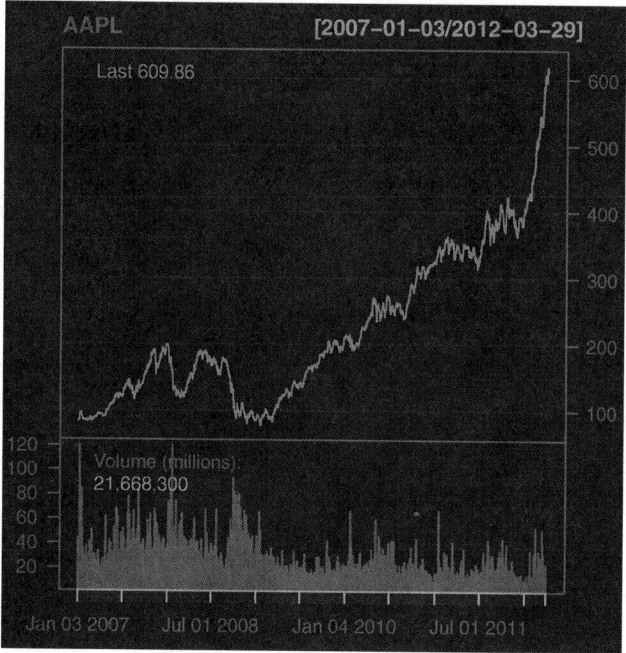

5.4.4.2 Consultant Chart

A consultant chart is basically a barplot along a circular radial axis. It is considered more compact than a standard barchart. Note that the consultant chart is almost visually similar to the coxcomb charts of Florence Nightingale (nineteenth century).

```
library(ggplot2)
#In ggplot2, barcharts are made by geom_bar, polar coordinates are made by coord_polar
#aes is used for the x- and y-axis parameters
ajay <- data.frame(variable = 1:20, value = sample(10,replace = T)) #random values
p=ggplot(ajay,aes(factor(variable),value))+coord_polar()+geom_bar() p
```

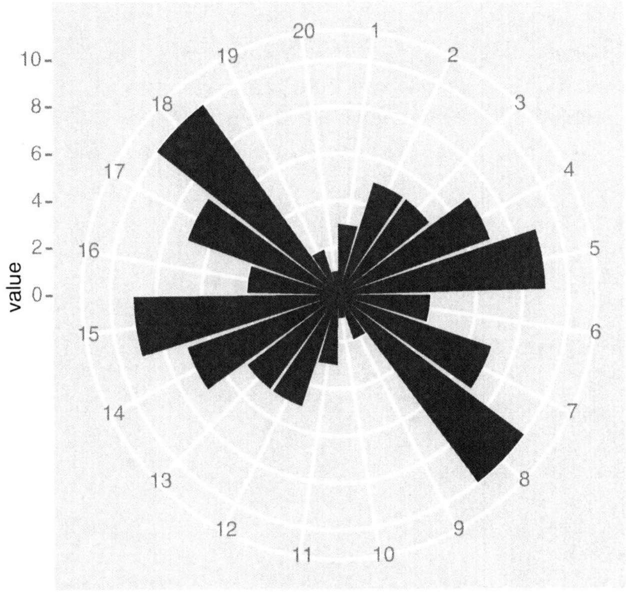

#*Note:* We are trying to show the effect of using color in ggplot by changing the fill parameter to a factor.

p=ggplot(ajay,aes(factor(variable), value, ***fill = factor (variable)))****+coord_polar()+geom_bar()*

p

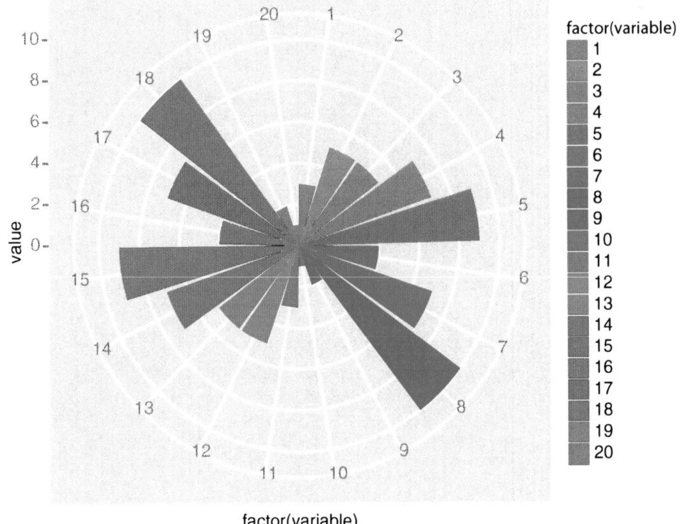

5.4.4.3 Cascade or Waterfall Chart

To show the effect of various metric values introduced in a sequence, we can use cascade or waterfall charts, also known as McKinsey charts after the noted consulting company. The chart shows the cumulative effect of these sequentially introduced values, particularly the effect of positive and negative values. They use the R package waterfall.

```
library(waterfall)
data(jaquith)
waterfallchart(jaquith$factor~jaquith$score,jaquith)
```

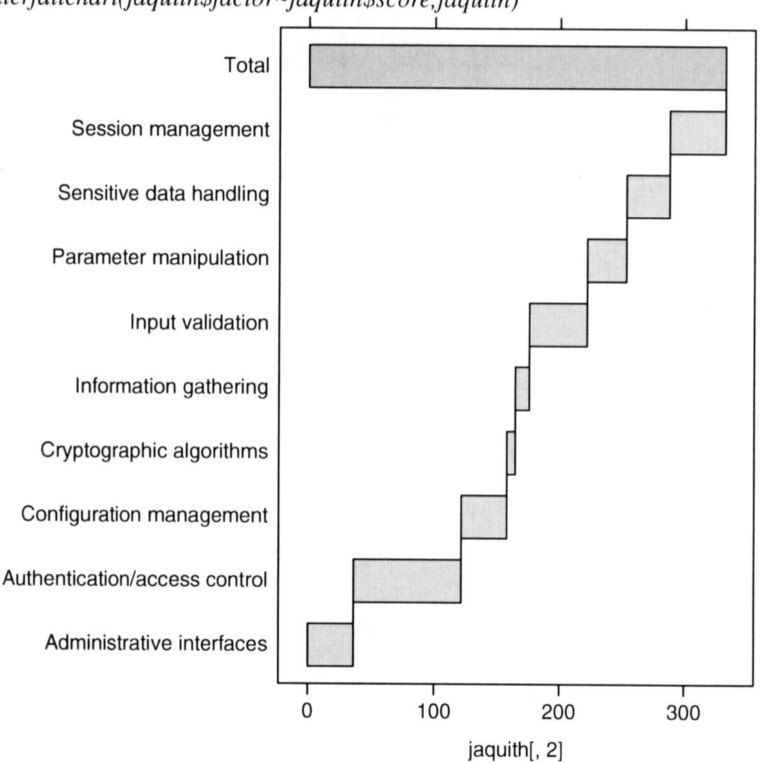

and

```
library(waterfall)
data(rasiel)
b=rasiel
waterfallchart(b[,1]~b[,2],b,col=terrain.colors(3))
#Note we use square brackets to refer to the first and second columns
```

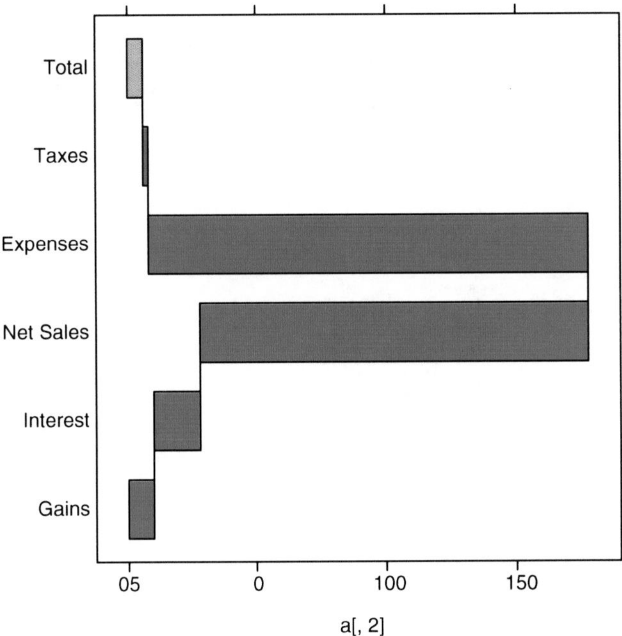

5.4.4.4 Geographic Maps

The spatial view shows many of the packages available in R for creating maps and visualizing spatial data: http://www.cran.r-project.org/web/views/Spatial.html.

Choropleth: A map in which areas are shaded or colored in proportion to a variable shown on the map. A special case is a prism map, which is a 3D map where the height of an area on the map is proportional to the variable's value for that regional area. Download shape files in the Rdata format from http://gadm.org/country.

The packages maps, mapdata, and sp can help with the creation of geographic maps in R.

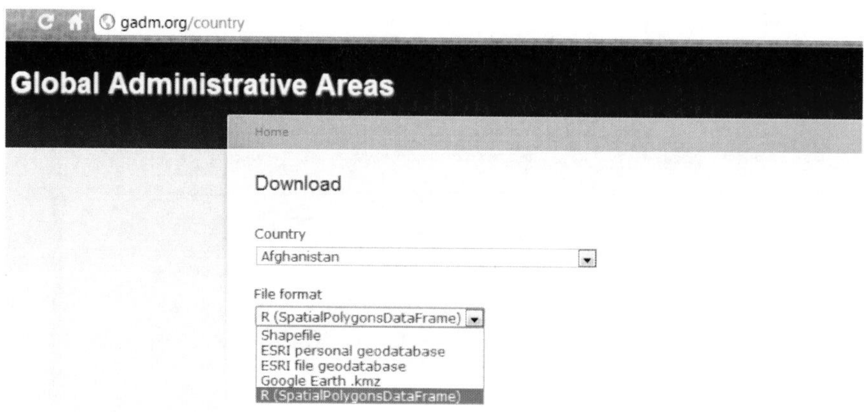

GADM is a spatial database of the location of the world's administrative boundaries. R objects can be plotted directly with the spplot function (from the sp package) or can be downloaded to a local file and then loaded within R.

```
> library(sp)
> con <- url("http://gadm.org/data/rda/IND_adm1.RData")
> print(load(con))
[1] "gadm"
> close(con)
>spplot(gadm[1])
```

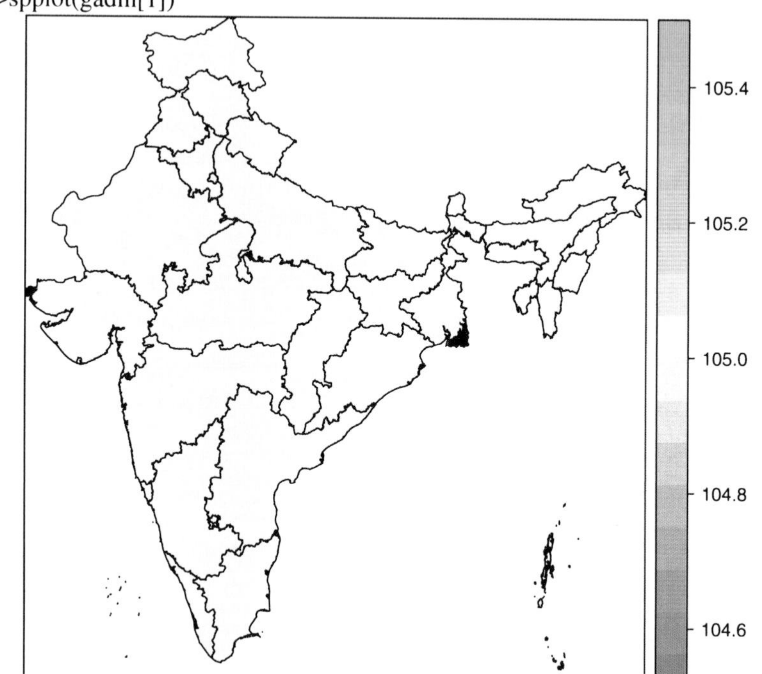

mapdata is an R package for extra map databases, a supplement to the maps package, providing a larger or higher-resolution database.

- **Cartogram** is used to produce spatial plots where the boundaries of regions can be transformed or distorted to be proportional to density/counts/populations. An R package is available at http://www.omegahat.org/Rcartogram/, but it is not updated. Cartograms have yet to have a dedicated package in the R language, and it is best to use the software tools at Scape Toad http://scapetoad.choros.ch/.
- **Contour maps** can also be created in R. They are 2D representations of 3D objects (like the heights of mountains for a map). See an example at http://addictedtor.free.fr/graphiques/RGraphGallery.php?graph=11.
- A gallery of spatial graphs is available at http://r-spatial.sourceforge.net/gallery/.
- You can use the **Deducer spatial plugin** for a GUI-based solution to maps. For more details on this very easy-to-use solution, visit http://www.deducer.org/pmwiki/index.php?n=Main.SpatialPlotBuilder (see figure below).

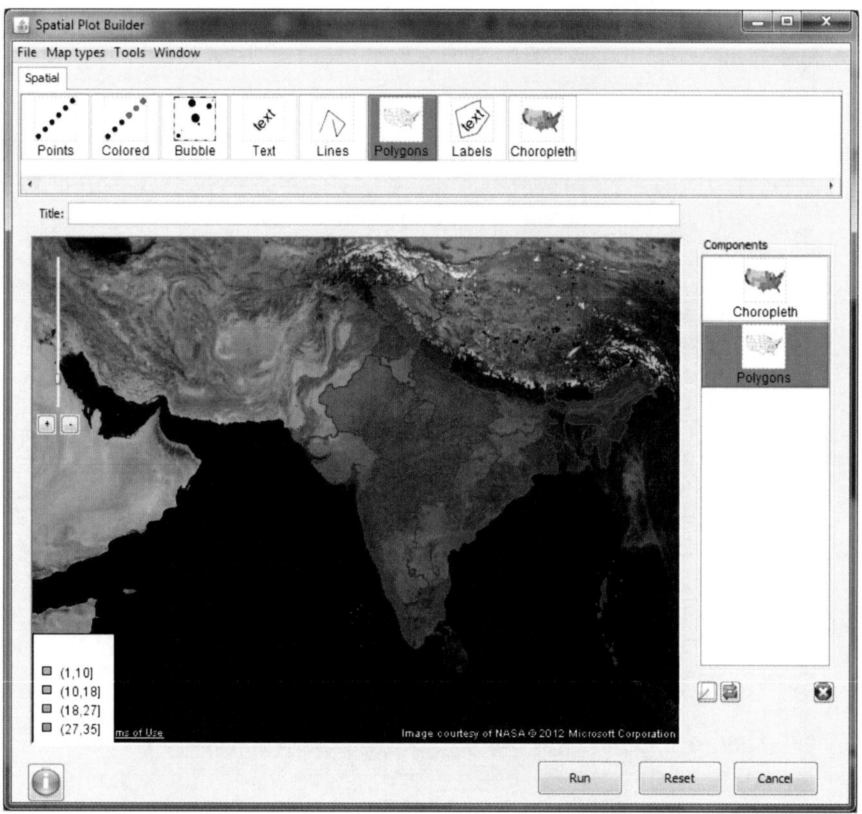

5.4.4.5 Calendar Heat Map

Calendar heat maps try to represent time series data using colors in a calendar. Thus, they can be considered time series equivalents of chloropleths. Based on functions created by Paul Bleicher, Chief Medical Officer of Humedica, and a blog post by David Smith at the Revolution Analytics blog, we plotted Apple's stock price for the last couple of years.

```
source("http://blog.revolution-computing.com/downloads/calendarHeat.R")
stock <- "AAPL"
start.date <- "2010-01-01"
end.date <- Sys.Date()
quote <- paste("http://ichart.finance.yahoo.com/table.csv?s=", stock,
"&a=", substr(start.date,6,7),
"&b=", substr(start.date, 9, 10),
"&c=", substr(start.date, 1,4),
"&d=", substr(end.date,6,7),
"&e=", substr(end.date, 9, 10),
"&f=", substr(end.date, 1,4),
"&g=d&ignore=.csv", sep="")
stock.data <- read.csv(quote, as.is=TRUE)
calendarHeat(stock.data$Date, stock.data$Adj.Close, varname="Apple Price")
```

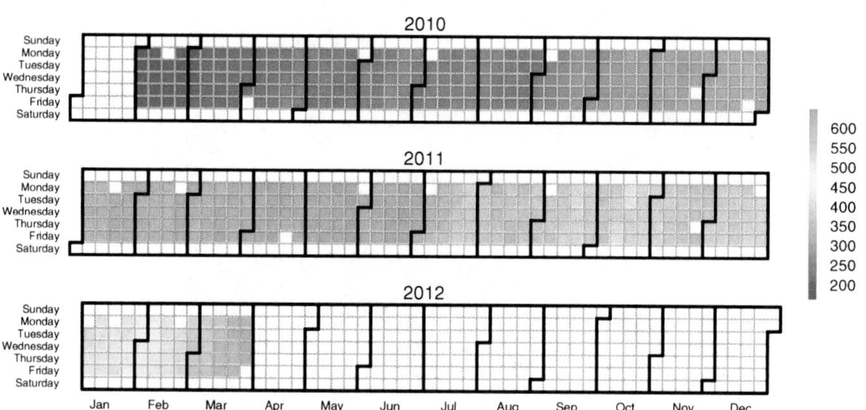

5.4.4.6 Quality Management Charts in R

You can use the qcc package for specific charts on quality management including Pareto charts, cause-and-effect diagrams, and control charts. You can see examples of charts on quality management created with the qcc package, as well as read more about the qcc package at http://stat.unipg.it/~luca/Rnews_2004-1-pag11-17.pdf

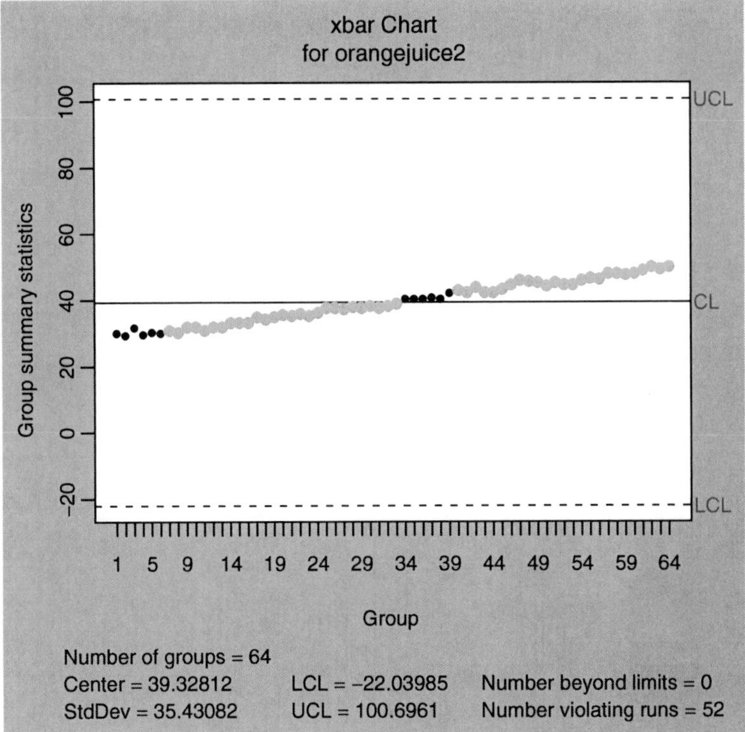

An R Commander plugin is available for qcc, so if quality management is what you need to do, you have a GUI available in R.

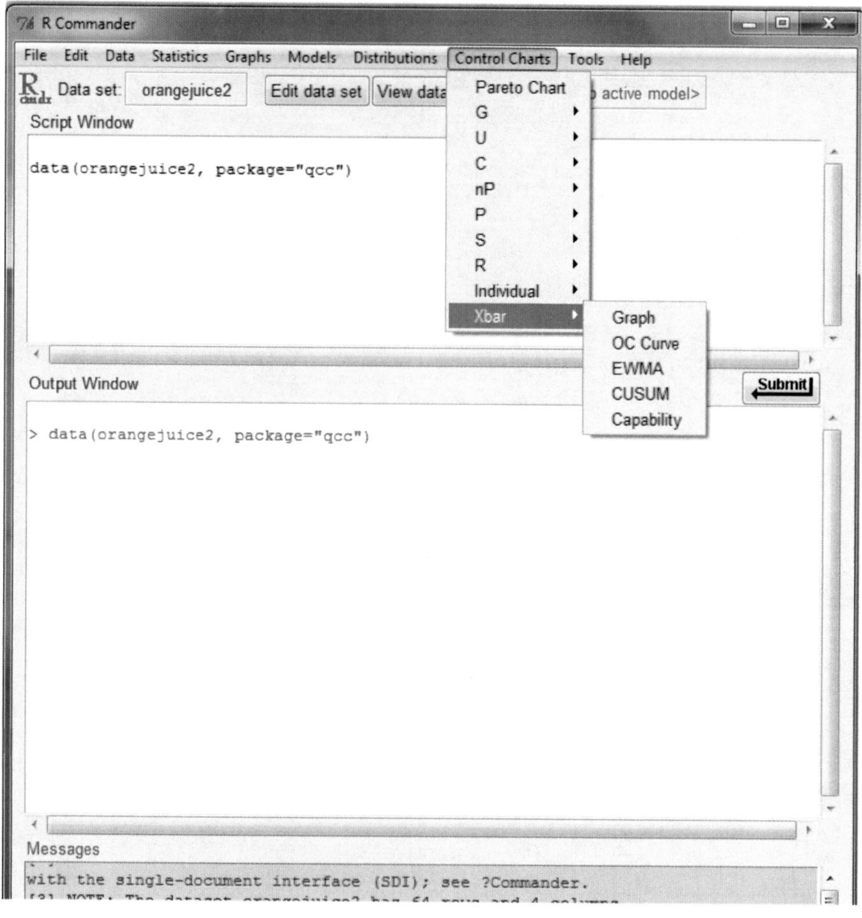

5.4.4.7 Venn Diagrams

Venn diagrams are used to determine the overlap between two or more sets of objects. It is quite easy to draw Venn diagrams in R using the VennDiagram package available at http://cran.r-project.org/web/packages/VennDiagram/.

```
library(VennDiagram)
venn.plot<- venn.diagram
( x = list( MBAs= c(1:240), Engineers= c(1:180), Doctors=c(238:240), Humanities=c(180:238)),
fill=c("grey","red","blue","green"),
filename = NULL );
grid.draw(venn.plot);
```

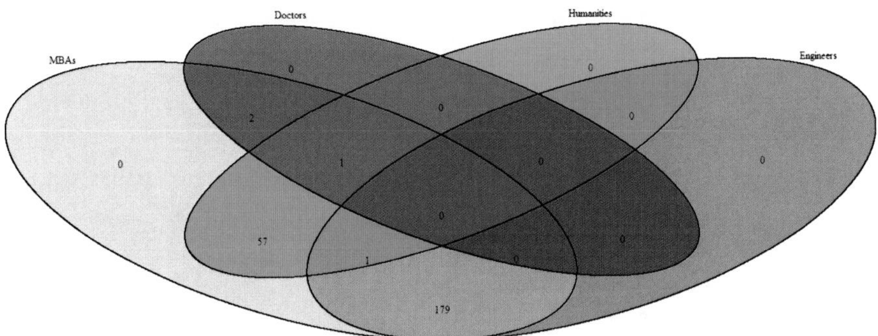

For better examples see /library/VennDiagram/html/venn.diagram.html using *??venn* at the command terminal.

5.5 Using ggplot2 for Advanced Graphics in R

The package ggplot2 has two basic graphing functions, qplot and ggplot. You are advised to use the Deducer GUI for starting points of ggplot for business analytics. Fortunately, even though ggplot can be a bit complex for the beginner, it has an excellent documentation Web site at http://had.co.nz/ggplot2/, which elaborates the terminology given below with graphical examples.

A very basic ggplot object can be represented by
ggplot(dataset_name, mapping) + layer(stat = "", geom = "", position = "").
Here mapping = *aes(x = dataset_name$VarX, y =dataset_name$VarY)*, where varX and varY are the variables for plotting on the x- and y-axes, respectively.
The graphing functions in ggplot have the following parameters for layers:

- geom_: Geoms, short for geometric objects, describe the type of plot.

The most commonly used are geom_point, geom_histogram, geom_bar, and geom_line.

- stat_: These are statistical functions that transform the data in the plot before plotting.

The most commonly used are stat_abline, stat_smooth, and stat_box_plot.

- scale_: These control how data are mapped to the plot on the basis of esthetics.

The most commonly used are scale_size and scale_gradient.

- coord_: These change the coordinate systems of plots to the 2D plane of the computer screen.

While coord_cartesian is the default option, other coord_ options used are coord_flip for flipping the axes and coord_polar for polar coordinates.

- facet_: These faceting functions allow subsets of data to be displayed in different panels for emphasis or to focus on particular parts of data and graph.

The most commonly used facet is facet_grid, and its typical use is facet_grid(Var1 ~ Var2) for plotting multiple graphs across two different variables (and their changing values).

- position_: These adjust the position of points in a plot and allow for fine-tuning. The effects used are like dodging, jittering, and stacking.

5.6 Interactive Plots

An interactive plot is a graph in which you can interact and edit with the layers and elements within it. R has many packages that offer interactivity with graphs. For some business purposes static graphs and pictures may not be good enough; the solution is interactive plots. The following R packages offer interactivity:

- Iplots: Iplots gives interactive plots for R: http://cran.r-project.org/web/packages/iplots/index.html. It has a new development version called Acinonyx whose main focus is speed and scalability to support large data (it uses OpenGL, optimized native code, and object sharing to allow visualization of millions of datapoints): http://www.rforge.net/Acinonyx/.
- playwith: playwith is an R package that provides a GTK+ GUI for editing and interacting with R plots: http://code.google.com/p/playwith/.

library(playwith)
library(zoo)
playwith(xyplot(sunspots ~ yearmon(time(sunspots)), xlim = c(1900, 1930), type = "l"), time.mode = TRUE)

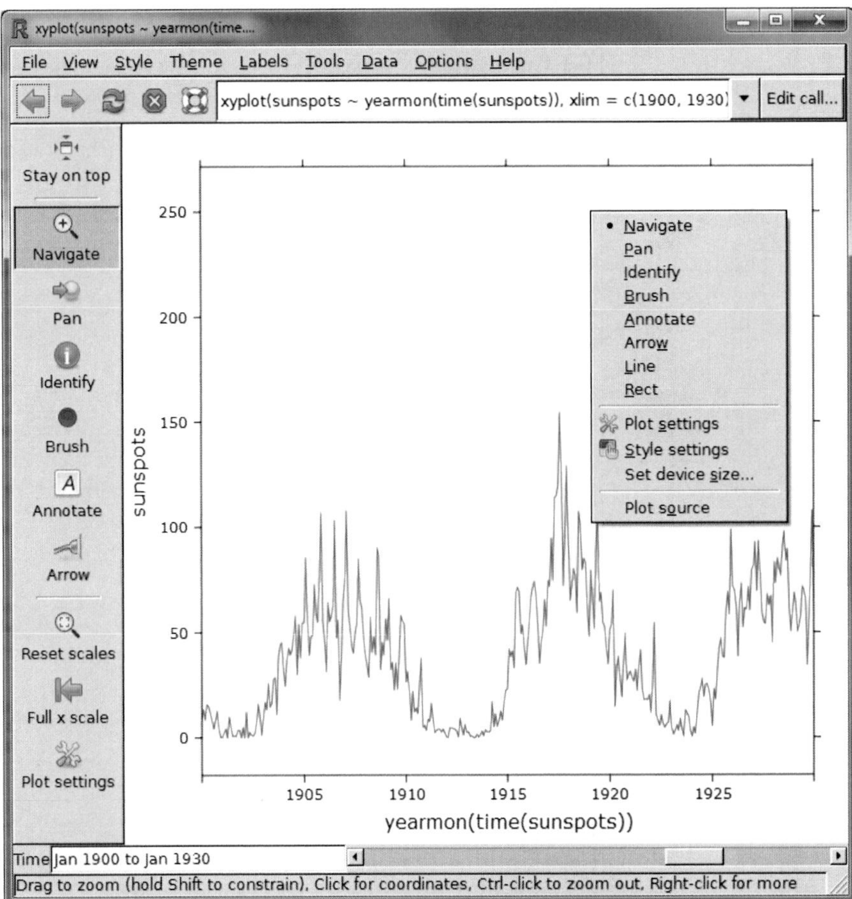

- rggobi: The rggobi package provides a command-line interface to GGobi, an interactive and dynamic graphics package. rggobi complements GGobi's GUI, providing a way to fluidly transition between analysis and exploration and automate common tasks: http://www.ggobi.org/rggobi/introduction.pdf.
- latticist: latticist is an R package that provides a GUI for exploratory visualization. It is primarily an interface to the lattice graphics system but also produces displays from the vcd package for categorical data: http://code.google.com/p/latticist/.

Both rggobi and latticist are present in the Rattle GUI.

Note: To create animations with R, you can use the RPackage animation. This package contains various functions for animations in statistics. It includes several approaches to saving animations in various formats: http://cran.r-project.org/web/packages/animation/index.html

An example is here: http://robjhyndman.com/researchtips/animations/.

5.7 GrapheR: R GUI for Simple Graphs

This is a basic GUI for plotting six types of graphs. It is quite easy to use and understand, as can be seen from the screenshot below.

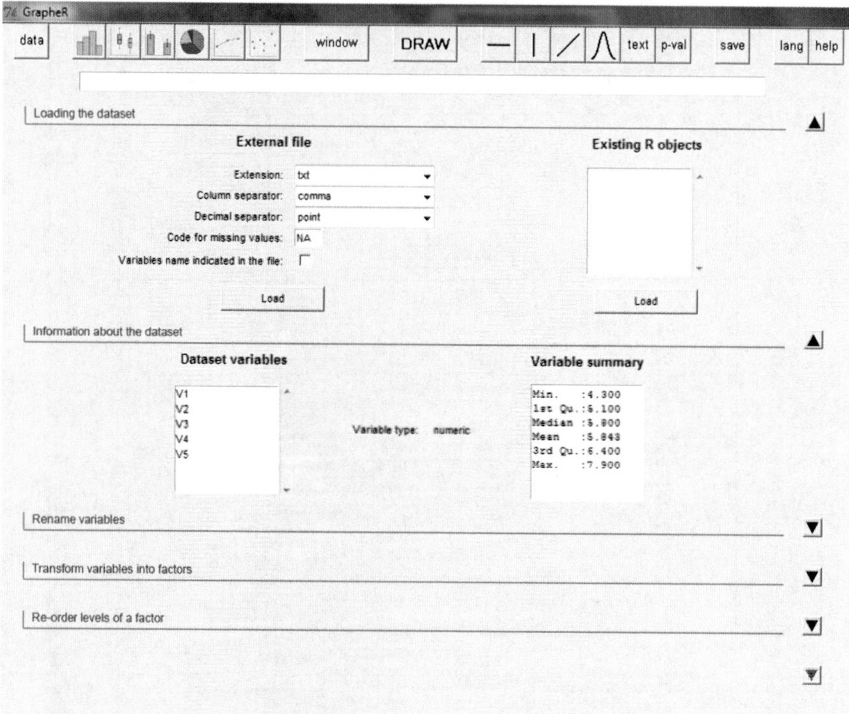

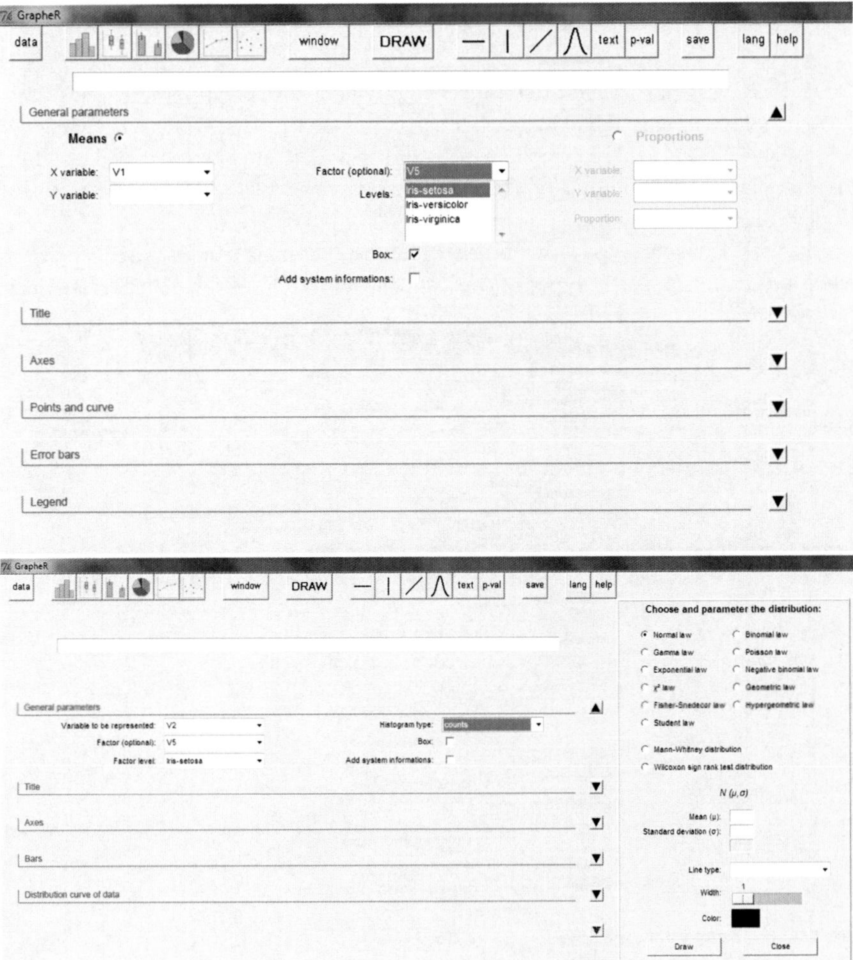

5.7.1 Advantages of GrapheR

It has an easy-to-use interface that is quite intuitive in creating basic graphs.

5.7.2 Disadvantages of GrapheR

It is not suitable for anything except data visualization. Also, it lacks a capability for advanced data visualization.

5.8 Deducer: GUI for Advanced Data Visualization

Deducer is the appropriate GUI for advanced data visualization and other uses. JGR is a Java-based GUI. Deducer is recommended for use with JGR. Deducer has basically been made to implement ggplot in a GUI; it is an advanced graphics package based on Grammar of Graphics and was part of the Google Summer of Code project.

The JGR Console is as shown below (note the automatic syntax suggestion even when a function is partly typed). You can download JGR from https://rforge.net/JGR/files/.

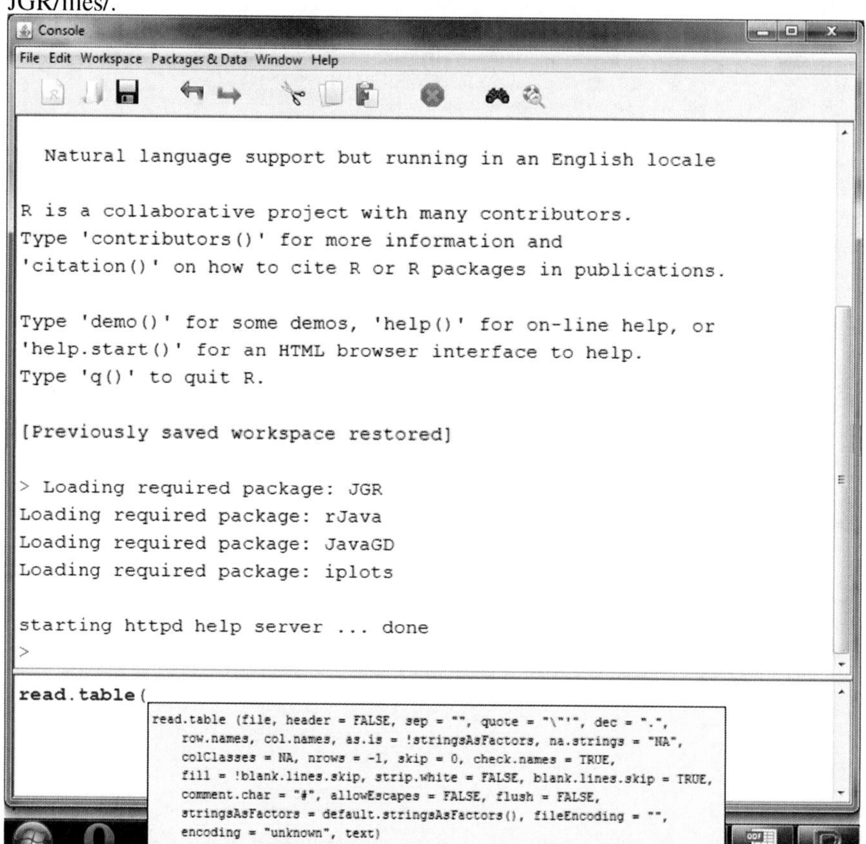

Deducer first asks you to either open an existing dataset or load a new one with just two icons. It has two initial views in Data Viewer: a Data view and a Variable view, which is quite similar to Base SPSS. The other Deducer options are loaded within the JGR console. Note the difference in menu tabs after loading Deducer.

Deducer can be downloaded from http://www.deducer.org/manual.html.

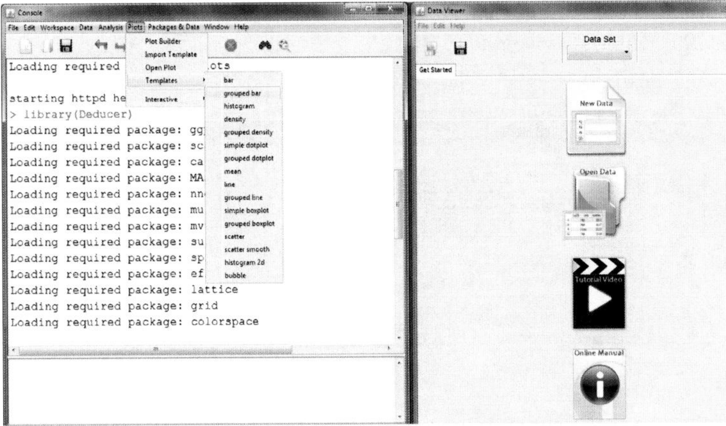

5.8.1 Advantages of JGR and Deducer

- Deducer has an option for factor as well as reliability analysis that is missing in other GUIs like R Commander and Rattle.
- The plot builder option gives very good graphics, perhaps the best in all the GUIs for R. This includes a color-by option that allows you to shade colors based on a variable value.

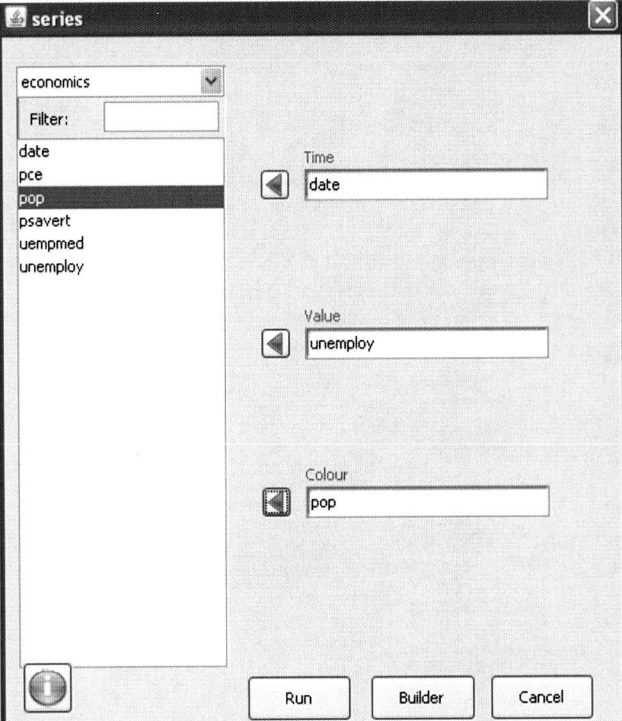

- An additional innovation is the form of templates, which enables even users unfamiliar with data visualization to choose among various graphs and click and drag them to the plot builder area.

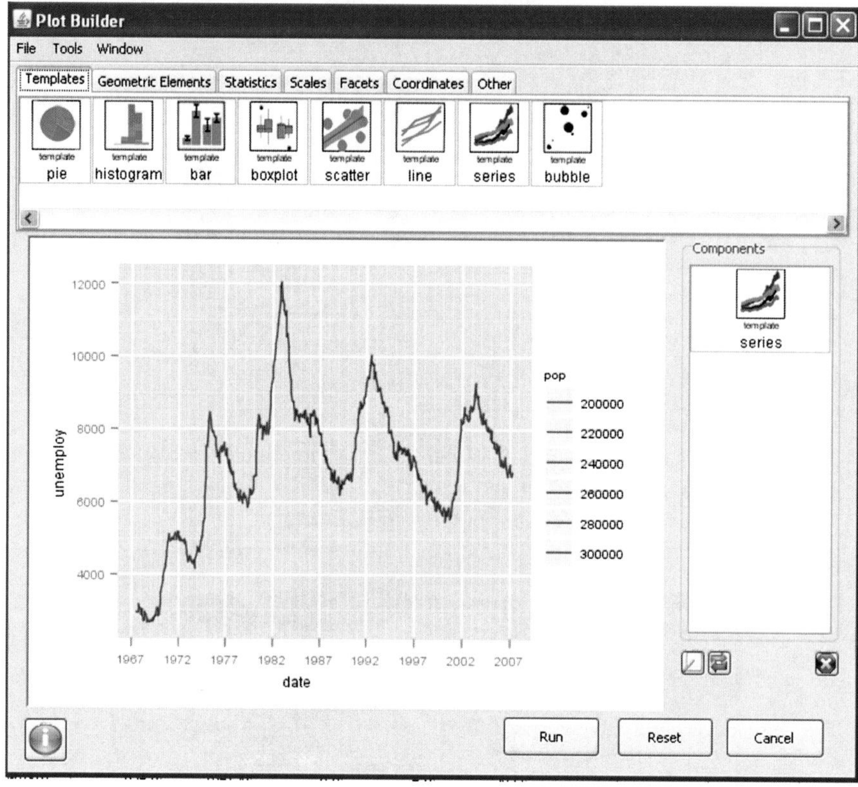

- In addition, you can plot multiple aspects of a graph using geometric aspects, statistics, scales, facets, coordinates, and other parameters.

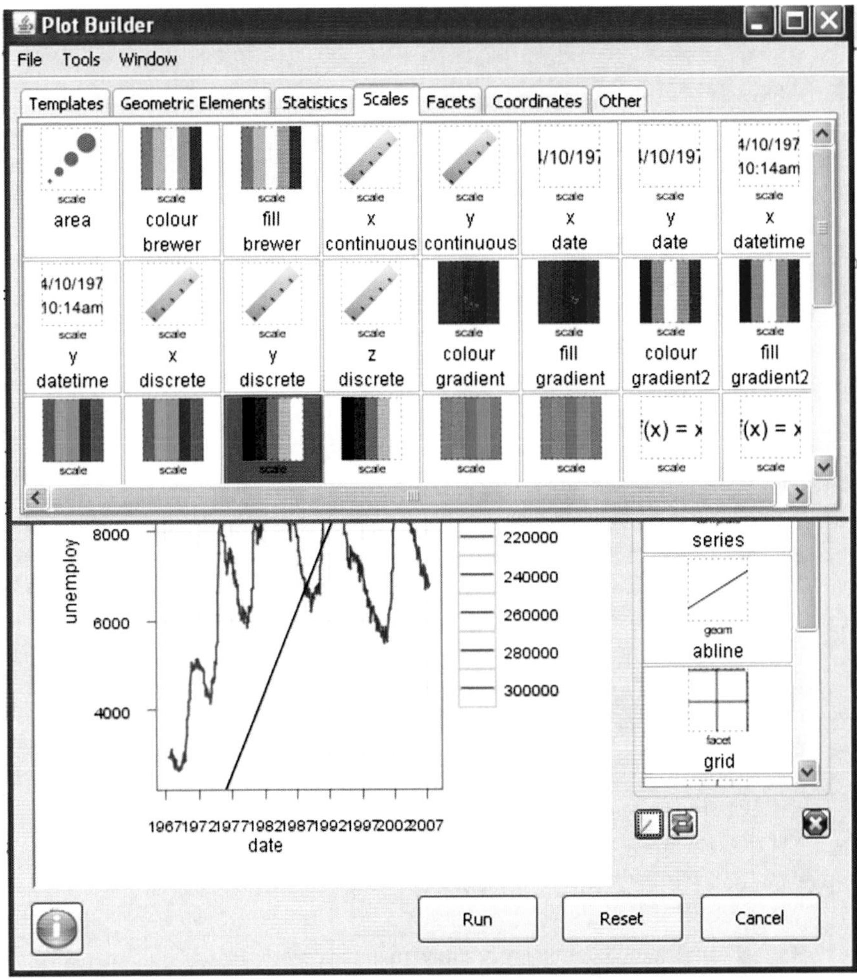

- You can set the Java GUI for R (JGR) menu to automatically load some packages by default using an easy checkbox list.
- Deducer offers a way to build other R GUIs using Java Widgets.
- The overall feel is of SPSS (Base GUI) SPSS users should be more comfortable using this.

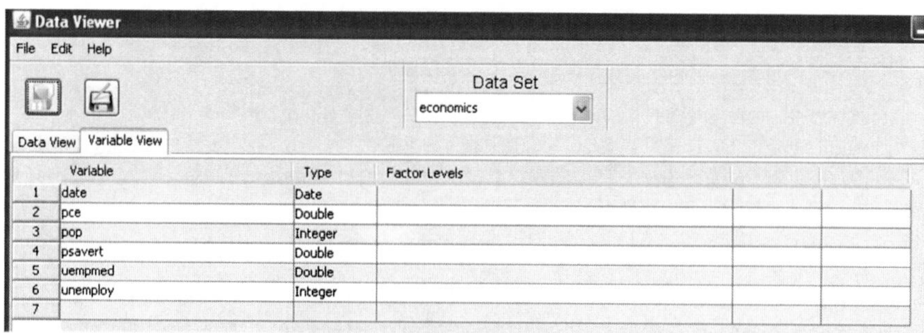

- It is very easy to move between two or more datasets using a dropdown list.
- It is the most convenient GUI for merging two datasets using a common variable.
- Deducer has a number of plugins to act as additional packages for additional functionality.
  - DeducerExtras: An add-on package containing a variety of additional analysis dialogs. These include distribution quantiles, single/multiple sample proportion tests, paired t-test, Wilcoxon signed-rank test, Levene's test, Bartlett's test, k-means clustering, hierarchical clustering, factor analysis, and multidimensional scaling.
  - DeducerPlugInScaling: This allows for reliability and factor analysis.
  - DeducerMMR: This allows for moderated multiple regression and simple slope analysis.
  - DeducerSpatial: A GUI for spatial data analysis and visualization.
  - gMCP (experimental): A graphical approach to sequentially rejective multiple test procedures.
  - RGG (experimental): A GUI generator
  - DeducerText (experimental): Text mining.
- JGR offers automatic syntax suggestions that can be helpful for beginners in R.
- It has a great dedicated Web site for help including video tutorials as well as practical examples.
- In summary, the Deducer GUI is a great way to introduce R and ggplot in the business analytics workplace due to ease of use.
- Deducer can also be used to build a customized GUI (see http://www.deducer.org/pmwiki/pmwiki.php?n=Main.Development).

5.8.2 *Disadvantages of Deducer*

- There are no menu options, unlike R Commander and Rattle, for reading in datasets from attached packages.
- It is not suitable for data mining.

5.8.3 *Description of Deducer*

- The data menu gives options for data manipulation including recoding variables, transforming variables (binning, mathematical operations), sorting dataset, transposing datasets, and merging two datasets.
- The analysis menu gives options for frequency tables, descriptive statistics, cross tabs, one-sample tests (with plots), two-sample tests (with plots), k sample tests, correlation, linear and logistic models, and generalized linear models.
- The plot builder menu allows plots of various kinds to be created in an interactive manner.

To read more on Deducer visit http://blog.fellstat.com/.

5.9 Color Palettes

If you are unable to decide whether blue or brown is a better color for your graph, the color palettes in R are a big help in providing esthetically acceptable alternatives. Using the same graphs, we choose the five main kinds of color palettes; using them is as easy as specifying the col= parameter in graphical display in Base Graphs. I modified the n parameter for number of colors to be used; you can specify more or fewer depending on the desired gradient or difference in colors.

An example follows, and each graph is labeled appropriately. Note in line two below that we use the mfrow in par to make a two row × three column grid of graphs in the same screen to conserve space.

```
data(VADeaths)
par(mfrow=c(2,3))
hist(VADeaths,col=rainbow(3),main="rainbow 3 colors")
hist(VADeaths,col=rainbow(7),main="rainbow")
hist(VADeaths,col=cm.colors(7),main="cm")
hist(VADeaths,col=topo.colors(7),main="topo")
hist(VADeaths,col=heat.colors(7),main="heat")
hist(VADeaths,col=terrain.colors(7),main="terrain")
```

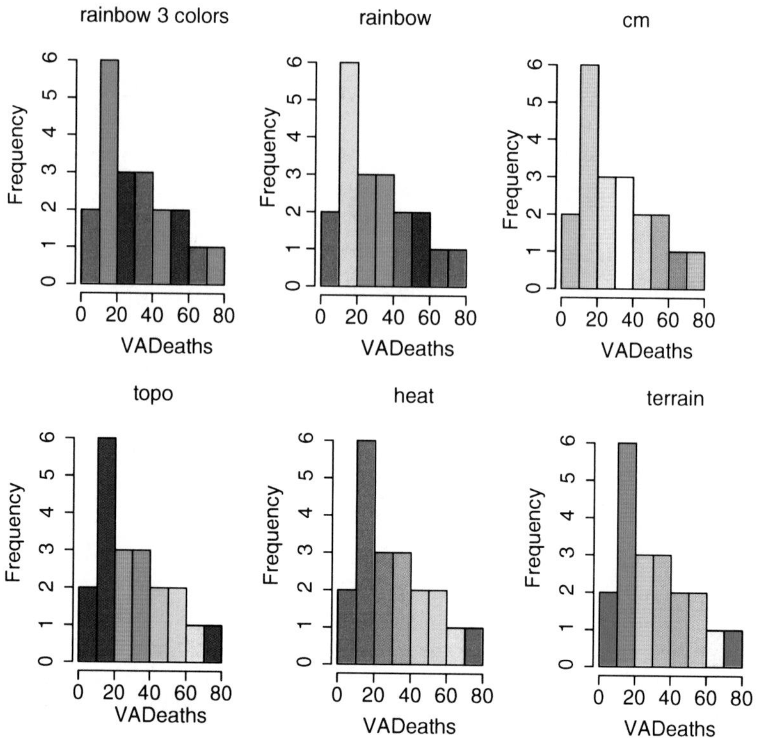

But we can also use more color palettes than the default ones in R using the RColorBrewer package.

library(RColorBrewer)

display.brewer.all()#This displays all the new palettes available in R using Color Brewer.

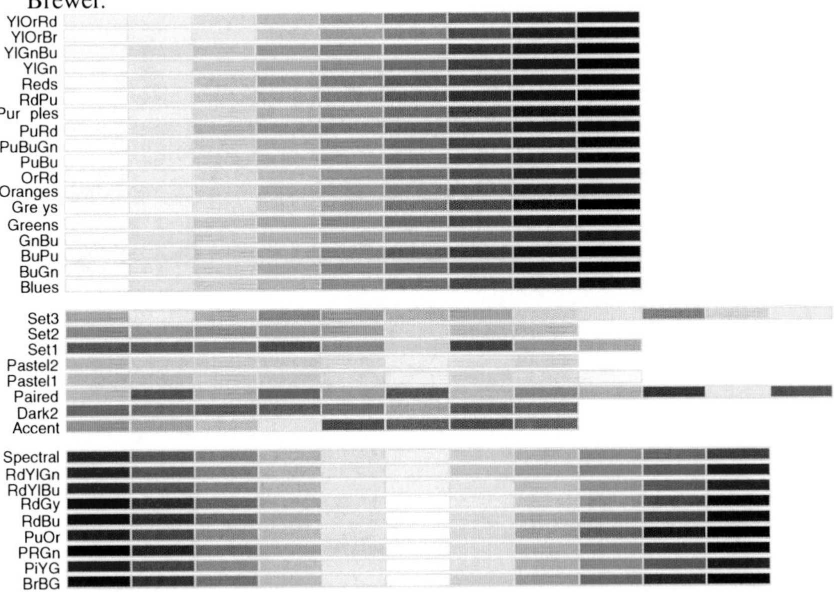

We use the brewer.pal(N,"Name") as the color parameter for the new color palette where we can see the name from the list above.

data(VADeaths)
library(RColorBrewer)
par(mfrow=c(2,3))
hist(VADeaths,col=brewer.pal(3,"Set3"),main="Set3 3 colors")
hist(VADeaths,col=brewer.pal(3,"Set2"),main="Set2 3 colors")
hist(VADeaths,col=brewer.pal(3,"Set1"),main="Set1 3 colors")
hist(VADeaths,col=brewer.pal(8,"Set3"),main="Set3 8 colors")
hist(VADeaths,col=brewer.pal(8,"Greys"),main="Greys 8 colors")
hist(VADeaths,col=brewer.pal(8,"Greens"),main="Greens 8 colors")

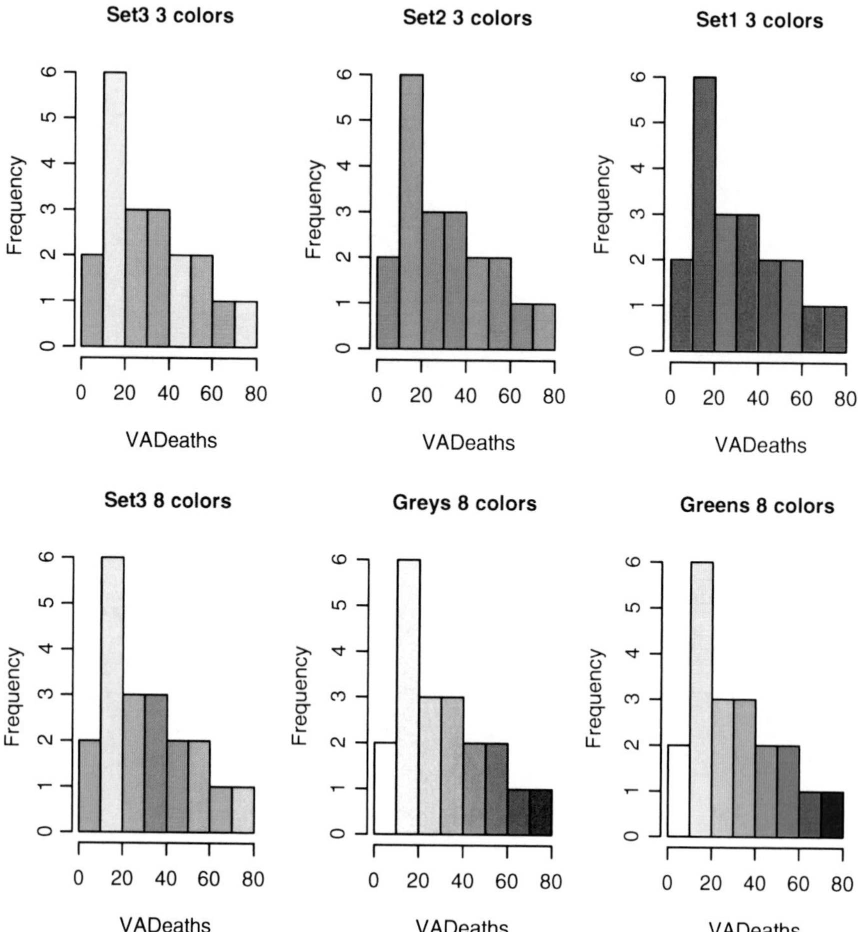

You can also visit http://www.colorbrewer.org for more details on color palettes.

5.10 Interview: Hadley Wickham, Author of *ggplot2: Elegant Graphics for Data Analysis*

Ajay: You have created almost ten R packages as per your Web site http://had.co.nz/. Do you think there is a potential for a commercial version for a data visualization R software? What are your views on the current commercial R packages?

Hadley: I think there's a lot of opportunity for the development of user-friendly data visualization tools based on R. These would be great for novices and casual users, wrapping up the complexities of the command line into an approachable GUI—see Jeroen Oom's site at http://yeroon.net/ggplot2 for an example. Developing these tools is not something that is part of my research endeavors. I'm a strong believer in the power of computational thinking and the advantages that programming (instead of pointing and clicking) brings. Creating visualizations with code makes reproducibility, automation, and communication much easier—all of which are important for good science. Commercial packages fill a hole in the R ecosystem. They make R more palatable to enterprise customers with guaranteed support, and they can offer a way to funnel some of that money back into the R ecosystem. I am optimistic about the future of these endeavors.

Ajay: Clearly, with your interest in graphics, you seem to favor visual solutions. Do you also feel that R Project could benefit from better R GUIs or GUIs for specific packages?

Hadley: See above—while GUIs are useful for novices and casual users, they are not a good fit for the demands of science. In my opinion, what R needs more are better tutorials and documentation so that people don't need to use GUIs. I'm very excited about the new dynamic html help system—I think it has huge potential for making R easier to use. Compared to other programming languages, R currently lacks good online (free) introductions for new users. I think this is because many R developers are academics and the incentives aren't there to make freely available documentation. Personally, I would love to make (for example) the ggplot2 book available openly available under a creative common license, but I would receive no academic credit for doing so.

Ajay: Describe the top three to five principles explained in your book *ggplot2: Elegant Graphics for Data Analysis*. What are other important topics that you cover in the book?

Hadley: The ggplot2 book gives you the theory to understand the construction of almost any statistical graphic. With this theory in hand, you are much better equipped to create visualizations that are tailored to the exact problem you face, rather than having to rely on a canned set of premade graphics. The book is divided into sections based on the components of this theory, called the layered grammar of graphics, which is based on Lee Wilkinson's excellent *Grammar of Graphics*. It's quite possible to use ggplot2 without understanding these components, but the better you understand, the better your ability to critique and improve your graphics.

5.11 Summary of Commands Used in This Chapter

5.11.1 Packages

- wordcloud
- waterfall
- histdata
- maps
- tm

- scatter3D
- rattle
- mapdata
- sp
- SparkTable
- Rcolorbrewer
- quantmod
- qcc
- R Commander and plugins
- JGR
- Deducer and plugins
- GrapheR
- ggplot2
- RGGobi
- lattice
- playwith
- Iplots
- VennDiagram
- YaleToolkit
- tabplot
- tabplotGTK
- Plotrix
- Hexbin
- VioPlot
- vcd
- BeanPlot

5.11.2 Functions

- colors()
- hist()
- stem()
- plot()
- boxplot()
- barplot()
- pie()
- symbols()
- sunflowerplot()
- par(bg= "")
- par(mfrow=" ")
- heatmap()
- mosaicplot()
- cloud(lattice)

- wireframe(lattice)
- hexbin(hexbin)
- tableGUI(tabplotGTK)
- ggplot(ggplot2)
- coord_polar (ggplot2)
- geom_bar(ggplot2)
- geom_point(ggplot2)
- qplot(ggplot2)
- coord_flip(ggplot2)
- brewer.pal(RColorBrewer)
- display.brewer.all(RColorBrewer)
- calenderHeat()
- corpus(tm)
- tdm(tm)
- wordcloud(wordcloud)
- venn.diagram(VennDiagram)
- grid.draw(VennDiagram)
- spplot(sp)
- waterfallchart(waterfall)
- getSymbols(quantmod)
- barChart(quantmod)
- candleChart(quantmod)
- lineChart(quantmod)
- chartSeries(quantmod)

Citations and References

- Color Palettes in R. http://decisionstats.com/using-color-palettes-in-r/
- Jeffrey A. Ryan (2011). quantmod: Quantitative Financial Modelling Framework. R package version 0.3-17. http://CRAN.R-project.org/package=quantmod
- Colors from http://www.ColorBrewer.org by Cynthia A. Brewer, Geography, Pennsylvania State University
- Erich Neuwirth (2011). RColorBrewer: ColorBrewer palettes. R package version 1.0-5. http://CRAN.R-project.org/package=RColorBrewer
  - *Note:* ColorBrewer is Copyright (c) 2002 Cynthia Brewer, Mark Harrower, and The Pennsylvania State University. All rights reserved. The ColorBrewer palettes have been included in the R package with permission of the copyright holder.
- Felix Andrews (2010). playwith: A GUI for interactive plots using GTK+. R package version 0.9-53. http://CRAN.R-project.org/package=playwith
- Scrucca, L. (2004). qcc: an R package for quality control charting and statistical process control. R News 4/1, 11–17.

- On creating Slopegraphs in R: https://github.com/bobthecat/codebox/blob/master/table.graph.
- On creating Voronoi diagrams: http://rgm2.lab.nig.ac.jp/RGM2/func.php?rd_id=alphahull:plot.delvor
- Hanbo Chen (2012) VennDiagram: Generate high-resolution Venn and Euler plots. R package version 1.1.3. http://CRAN.R-project.org/package=VennDiagram
- Fox, J. (2005). The R Commander: A Basic Statistics Graphical User Interface to R. Journal of Statistical Software, 14(9): 1–42.
- John W. Emerson and Walton A. Green (2012). YaleToolkit: Data exploration tools from Yale University. R package version 4.1. http://CRAN.R-project.org/package=YaleToolkit
- Alexander Kowarik, Bernhard Meindl and Matthias Templ (2012). sparkTable: Sparklines and graphical tables for tex and html. R package version 0.9.3. http://CRAN.R-project.org/package=sparkTable
- Sarkar, Deepayan (2008) Lattice: Multivariate Data Visualization with R. Springer, New York. ISBN 978-0-387-75968-5
- Martijn Tennekes and Edwin de Jonge (2011). tabplot: Tableplot, a visualization of large datasets. R package version 0.11-1. http://CRAN.R-project.org/package=tabplot
- James P. Howard, II, Waterfall Charts in R
- William D. Dupont* and W. Dale Plummer Jr. Vanderbilt University School of Medicine. Density Distribution Sunflower Plots. Journal of Statistical Software http://www.jstatsoft.org/v08/i03/paper
- Peter Kampstra (2008). Beanplot: A Boxplot Alternative for Visual Comparison of Distributions. Journal of Statistical Software, Code Snippets 28(1): 1–9. URL http://www.jstatsoft.org/v28/c01/
- Daniel Adler (2005) vioplot: Violin plot. R package version 0.2. http://wsopuppenkiste.wiso.uni-goettingen.de/~dadler
- Friendly, Michael (2001) Gallery of Data Visualization. Electronic document: http://www.datavis.ca/gallery/. Accessed: 03/13/2012 17:40:28
- Lemon, J. (2006) Plotrix: a package in the red light district of R. R-News, 6(4): 8–12.
- Ian Fellows (2012) wordcloud: Word Clouds. R package version 2.0. http://CRAN.R-project.org/package=wordcloud
- Dan Carr, ported by Nicholas Lewin-Koh and Martin Maechler (2011). hexbin: Hexagonal Binning Routines. R package version 1.26.0. http://CRAN.R-project.org/package=hexbin
- H. Wickham. A layered grammar of graphics. Journal of Computational and Graphical Statistics, 19(1): 3–28, 2010. [http://books.google.co.in/books?id=F_hwtlzPXBcC&lpg=PA3&ots=FsATdmyYoH&dq=A%20Layered%20Grammar%20of%20Graphics&pg=PA10#v=onepage&q=A%20Layered%20Grammar%20of%20Graphics&f=false]
- Some URLs used for researching this chapter:

- http://www.ats.ucla.edu/stat/r/gbe/histogram.htm
- http://www.ats.ucla.edu/stat/R/dae/logit.htm
- http://maths.anu.edu.au/~johnm/r/rgraphics.pdf
- http://www.gardenersown.co.uk/Education/Lectures/R/graphs2.htm#pie
- http://chem-eng.utoronto.ca/~datamining/dmc/data_mining_map.htm
- http://www.infovis.net/printMag.php?num=179&lang=2
- http://www.statmethods.net/graphs/creating.html
- http://daphne.palomar.edu/design/gestalt.html
- http://hci.stanford.edu/jheer/files/2010-MTurk-CHI.pdf

Chapter 6
Building Regression Models

One of the most common uses of statistical software is for building models, specifically logistic regression models for propensity in the marketing of goods and services. Within the R Project, regression packages are shown in the documentation in both the Econometrics view—http://cran.cnr.berkeley.edu/web/views/Econometrics.html—and the Finance view. A basic summary of all the R functions used for building regression models can be seen at http://cran.r-project.org/doc/contrib/Ricci-refcard-regression.pdf. A very good textbook on the basics of regression is *Practical Regression and Anova Using R* by Julian J. Faraway (available for free at http://cran.r-project.org/doc/contrib/Faraway-PRA.pdf).

6.1 Linear Regression

The basic theory of linear regression is assumed to be or known by the reader of this chapter. In any case, this book strives to be a practical handbook for the use of R in business analytics.

For a more Web-oriented discussion on linear regression, the reader is referred to the excellent article at http://en.wikipedia.org/wiki/Linear_regression.

This is what the official view on econometrics says about building linear regression models:

From http://cran.r-project.org/web/views/Econometrics.html:

1. Linear models can be fitted (via OLS) with lm() (from stats) and standard tests for model comparisons are available in various methods such as summary() and anova().
2. Analogous functions that also support asymptotic tests (z instead of t tests, and Chi-squared instead of F tests) and plug-in of other covariance matrices are coeftest() and waldtest() in lmtest.

A. Ohri, *R for Business Analytics*, DOI 10.1007/978-1-4614-4343-8_6,

3. Tests of more general linear hypotheses are implemented in linear.hypothesis() in car. HC and HAC covariance matrices that can be plugged into these functions are available in sandwich.
4. Diagnost checking: The packages car and lmtest provide a large collection of regression diagonstics [sic] and diagnostic tests.
5. Instrumental variables regression (two-stage least squares) is provided by ivreg() in AER, another implementation is tsls() in package sem.

6.2 Logistic Regression

Regression models remain the most commonly used modeling technique within business analytics, and, though often criticized for being overtly simplistic, logistic regression modeling will continue to be an important part of the toolkit of business analysts. This is because businesses deal overtly with many types of binary outcomes—sale/no sale, click/no click, payment/no payment. A brief exposition of the basics of logistic regression can be perused at http://en.wikipedia.org/wiki/Logistic_regression.

6.3 Risk Models

Modeling risk is both more complicated and more sensitive than modeling marketing propensity. This is due to the impact on business profitability of the adverse value of a bad model (in risk) as compared to the loss in opportunity revenue due to a bad model (in marketing). It is important to review the exact usage and applicability of regression for modeling; not every binary outcome should be modeled with regression.

6.4 Scorecards

A commonly used term in the modeling field is scorecards, which are simple matrixlike representations of where individual scores (or probabilities) are bucketed into categories. Scorecards are thus useful in decision management in helping convert a linear metric (propensity or probability) into a categoric metric (decision/no decision).

Scorecards are used as follows.

6.4.1 Credit Scorecards

A guideline to building credit scoring models using R is given at http://cran.r-project.org/doc/contrib/Sharma-CreditScoring.pdf. It is an effective guide to creating scoring models for credit usage.

6.4.2 Fraud Models

Fraud models are characterized by an extremely low number of fraud cases as compared to the overall population. Thus regression models are just one of the techniques; other techniques include decision trees and artificial neural networks (ANNs). The reader is advised to use the Rattle GUI for both ANNs and decision trees.

6.4.3 Marketing Propensity Models

Marketing propensity models are a major use of regression models. They are used in multiple domains for marketing including web analytics, telecom, direct marketing, and marketing campaigns.

6.5 Useful Functions in Building Regression Models in R

Using the following commands we can build a regression model in R with much greater ease than any other comparable statistical software. The main intention for the simplified approach is for the user and reader to navigate the often confusing and intimidating variety of packages available and to ease them into the learning environment that the R Analytics platform requires from its users.

Syntax

- Model Equation

```
library(car)
  outp=lm(y~x1+x2+xn,data=dataset)
```

- Model Summary

```
summary(outp)
  par(mfrow=c(2,2)) +
  plot(outp) #Model graphs
```

- MultiCollinearity

What is multicollinearity?

Multicollinearity refers to the correlation between explanatory variables. VIF stands for variance inflation factor. The bigger the measure of VIF for a particular explanatory variable, the more standard error is introduced in the model. Multicollinearity is generally dealt with in regression models by replacing variables with a high VIF.

vif(outp)

- Heteroscedasticity using GVLMA package

What is heteroscedasticity?

Heteroscedasticity refers to the problem of varying variances within variables. A variable is heteroscedastic if there are subpopulations that have different variabilities than others. This can be found out using the *gvlma* package.

library(gvlma)
gvlma(outp)

Taking logarithms of the data can remove the heteroscedasticity.

Note *hccm()* from the CAR package can also help deal with heteroscedasticity.

- Outliers for model

outlierTest (outp)

- Scoring dataset with scores

outp$fitted

6.6 Using R Cmdr to Build a Regression Model

- *Step 1:* Loading data

This was covered previously. Here we load the data from the dataset Anscombe from the car package.

- *Step 2:* Data exploration

Graphical Analysis

A simple graphical analysis entails using the Graphs>Scatterplot Matrix

There are multiple options here.

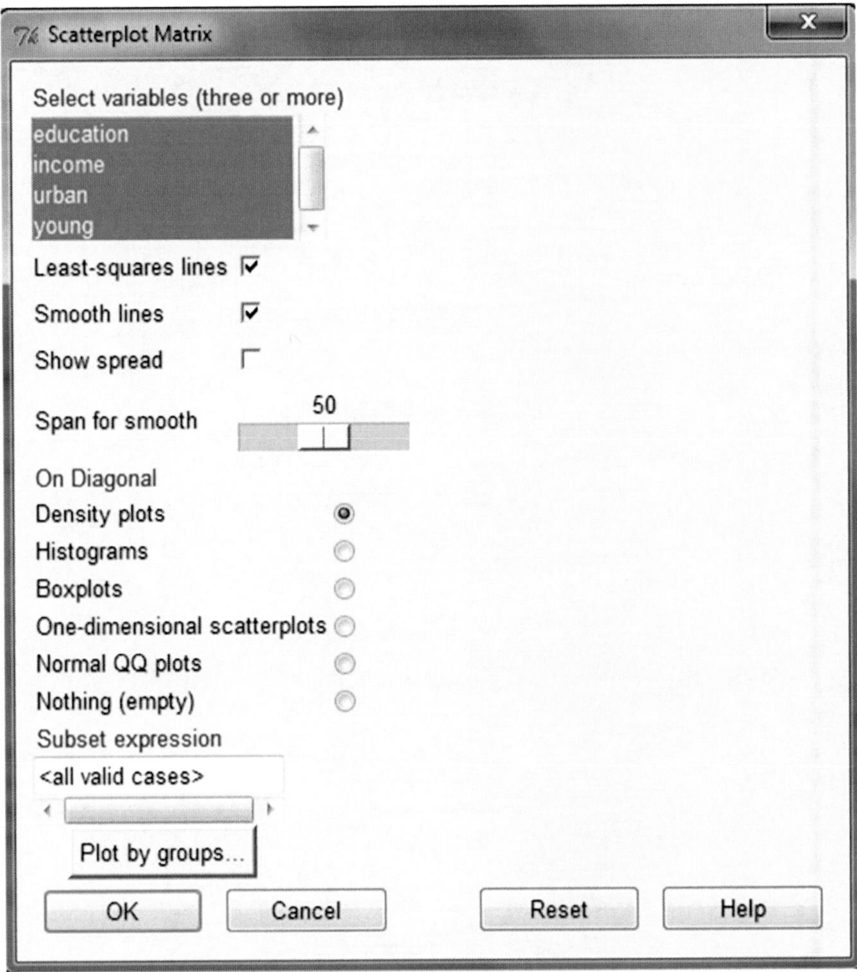

The output gives a good overview of all the data.

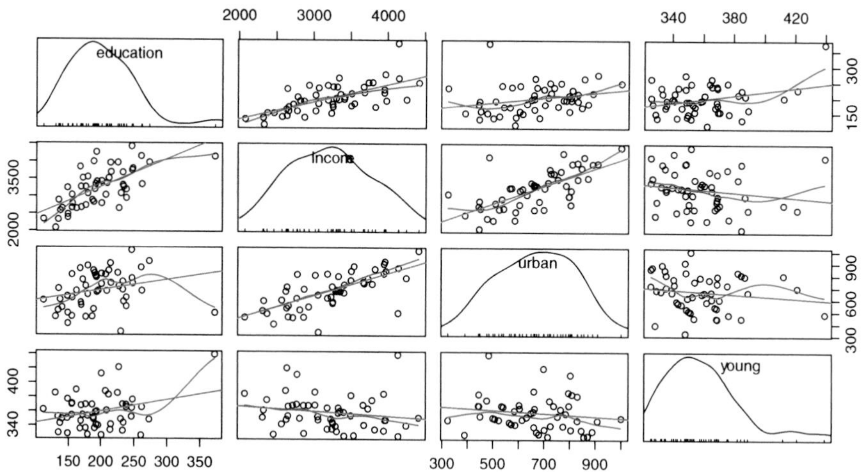
education
income
urban
young
2000
3000
4000
340
380
420
300
150
3500
2000
900
600
300
400
340
150 200 250 300 350
300 500 700 900

For more detailed graphical options, refer to Chap. 5 and to the corresponding Graphs menu in R Commander.

Data Statistics

Evaluate numerical summaries (and missing values as a quality check) from the menu in Statistics>Summaries.

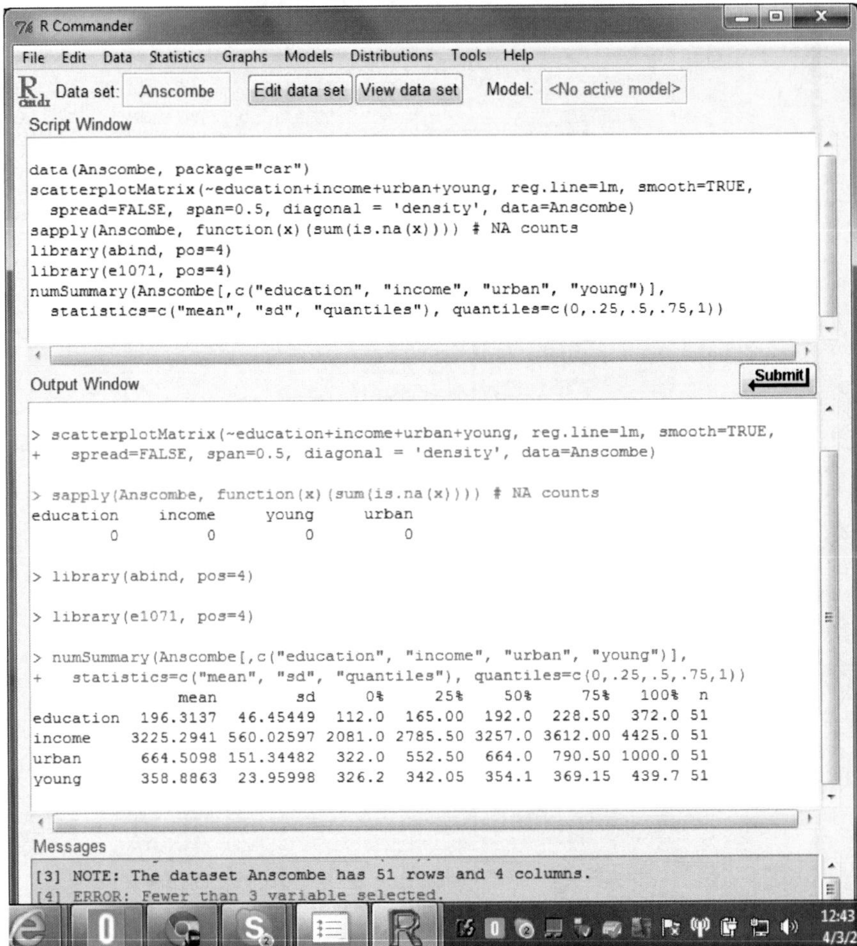

Correlation between variables.

1. Navigate to the Statistics>Summaries>Correlation Matrix menu

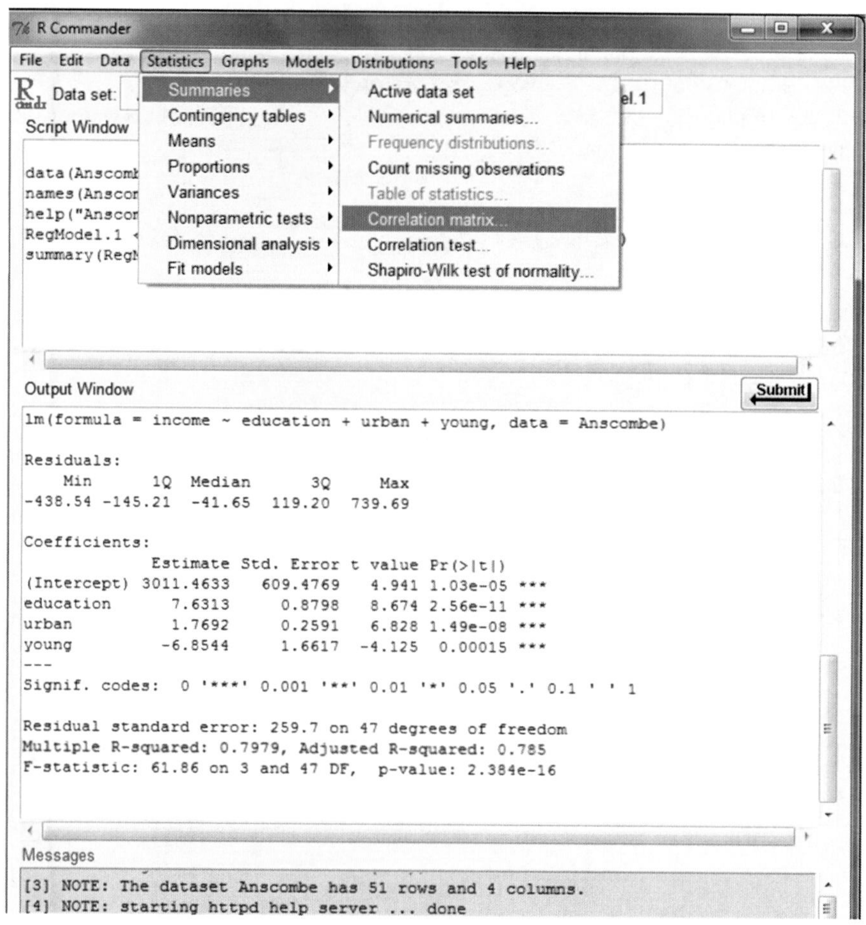

2. Choose variables for correlation analysis

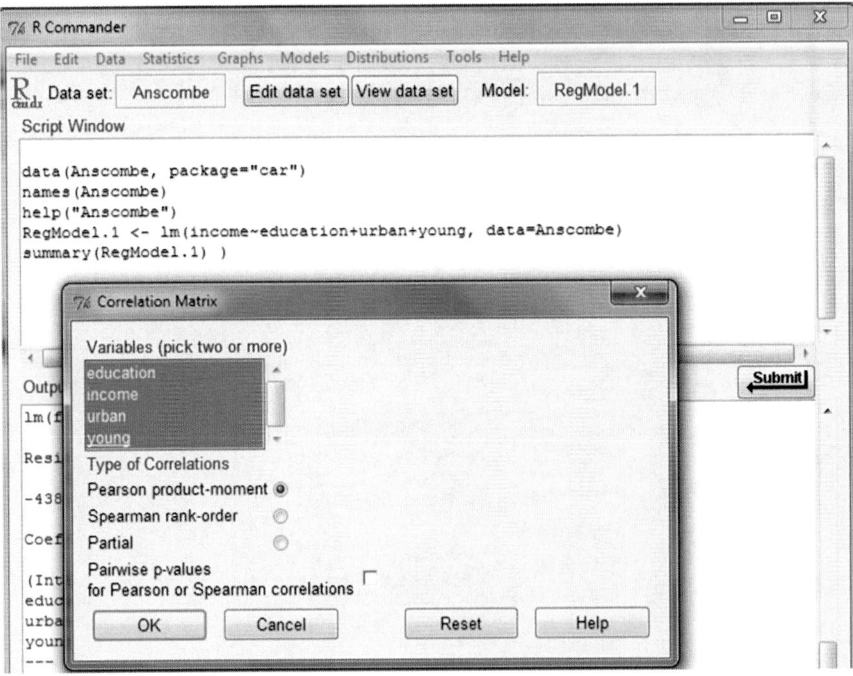

3. Evaluate the correlation matrix

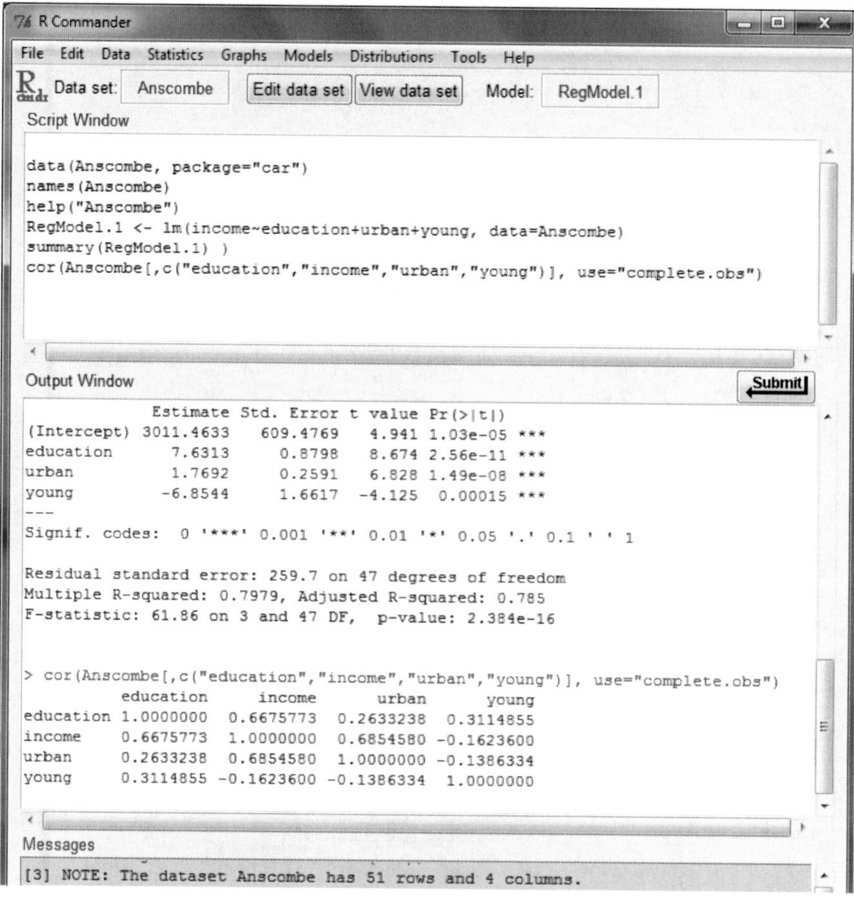

- *Step 3:* Running a regression model and model summary

1. Navigate to Statistics>Fit Models>Linear Regression.

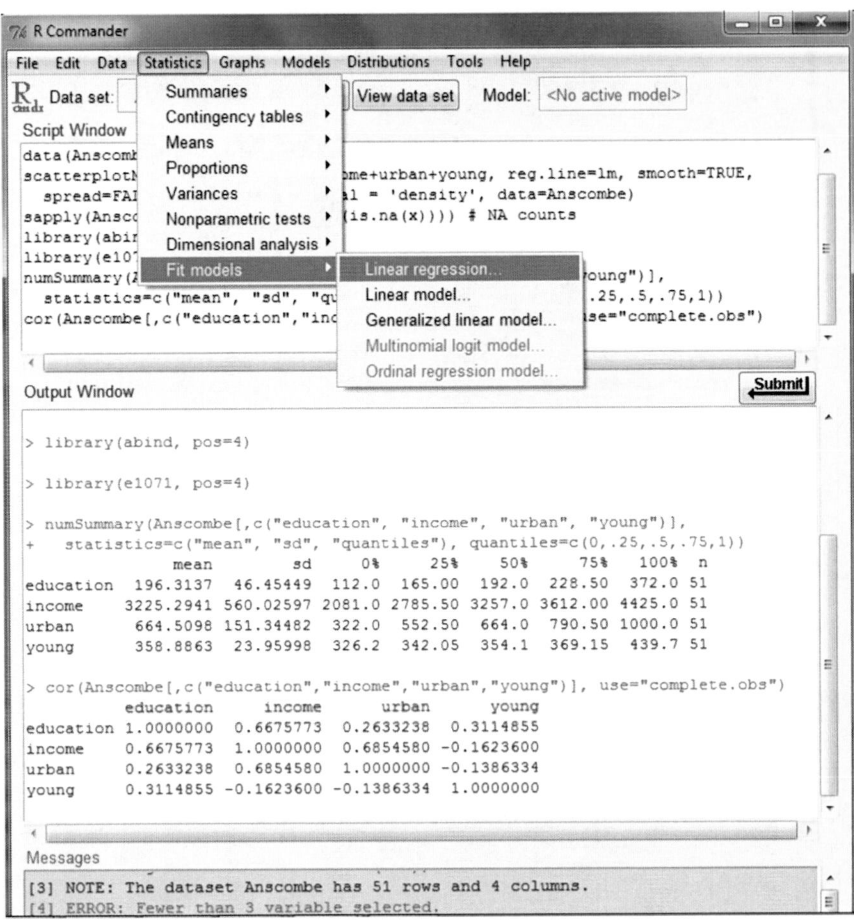

2. Choose the response variable (the value that changes—also known as the dependent variable) and explanatory variables (also known as independent variables).

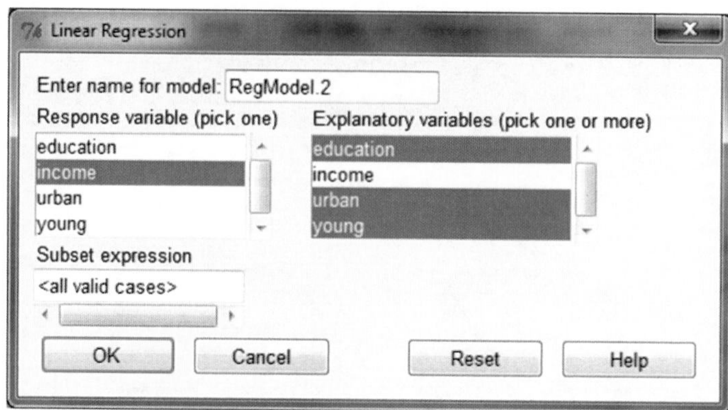

You can also choose linear model (which is the next menu item from linear regression) options.

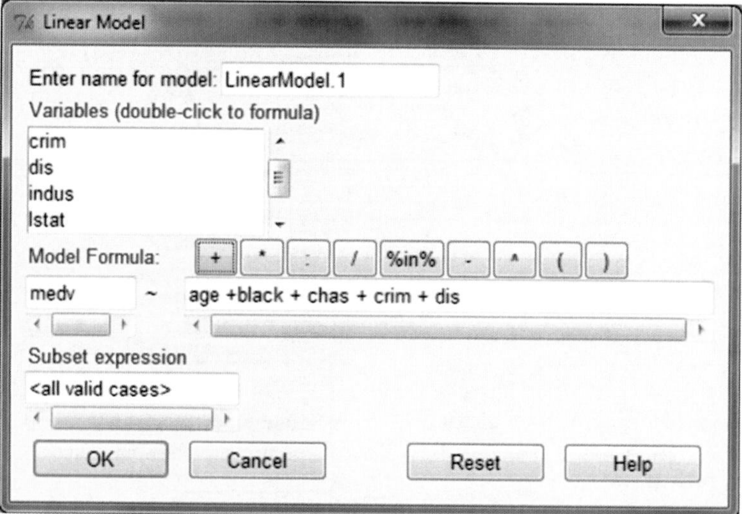

3. The regression output and model summary is displayed in the output window of R Commander. Note the script window of R Commander automatically generates the syntax relevant to the GUI clicks.

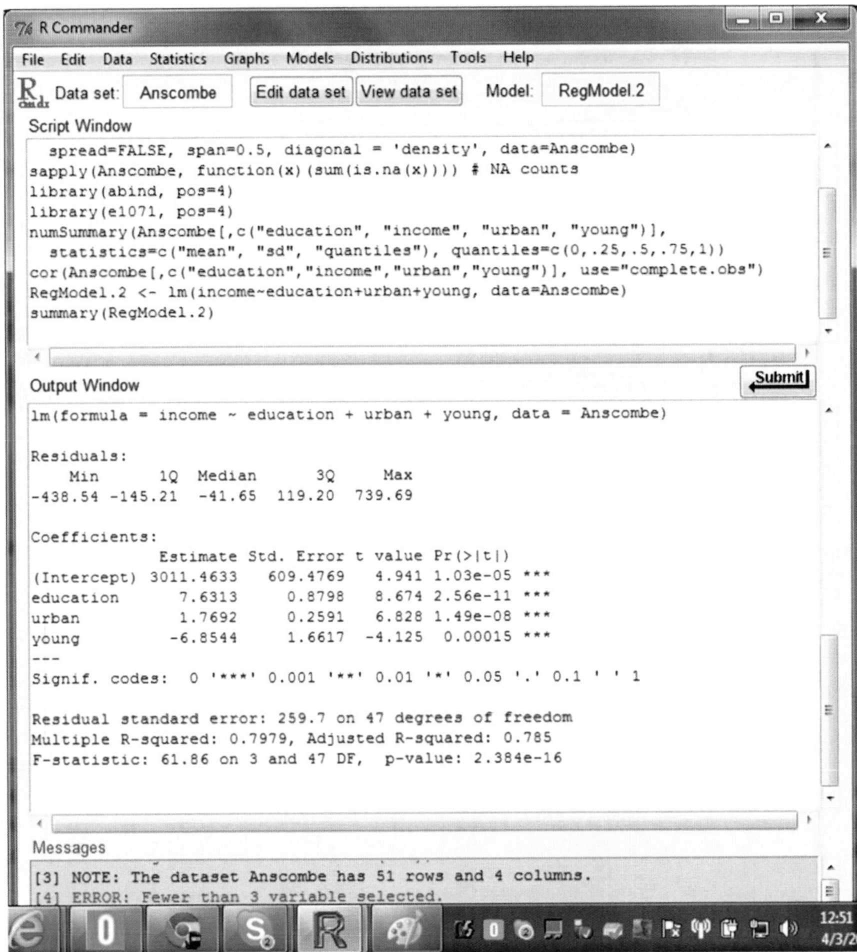

- *Step 4:* Model analysis

Navigate to the Models tab.

We want to know the variance inflation and outliers. These are in the Numerical Diagnostics tab.

Note a VIF above 2.5–3 is considered bad for adding that particular variable to the model. Remove the variables with high VIF and rerun the preceding model equation.

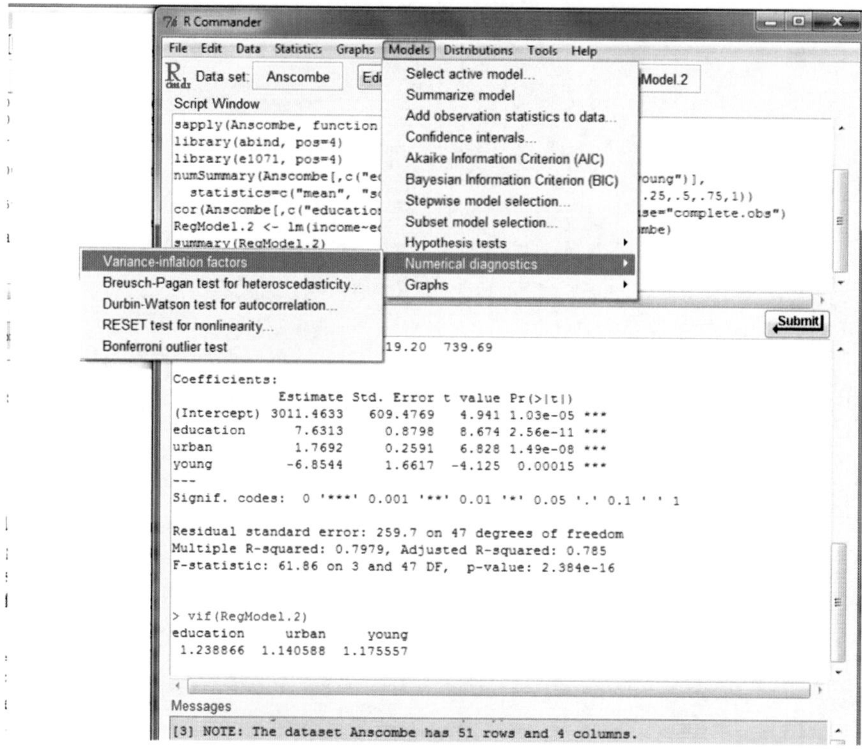

Lastly, go to Models>Graphs>Basic Diagnostic Graphs

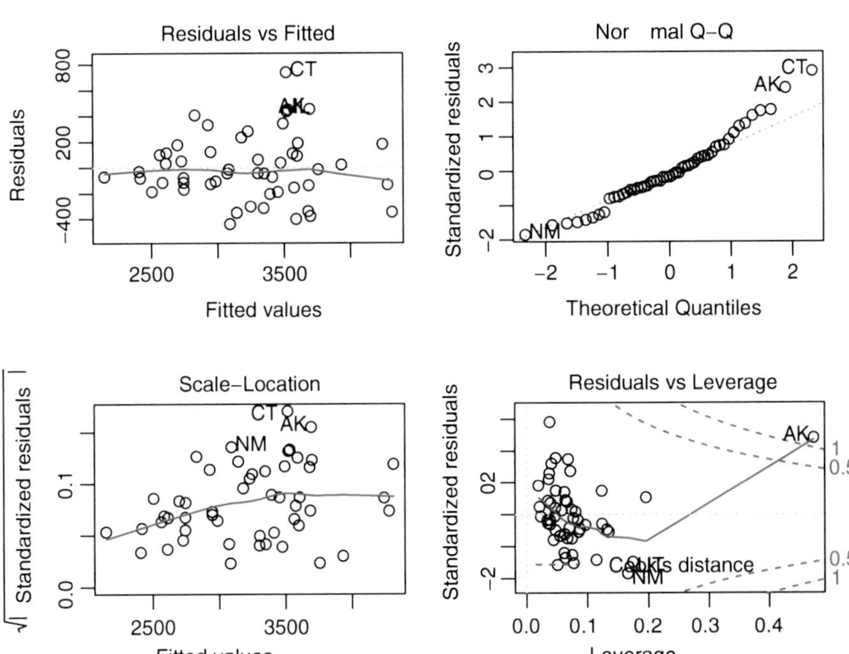

These will help you to find out how the model behaves in response to actual data. You can also explore other graphs here.

Note: "Residuals" means the difference between actual values and estimated values.

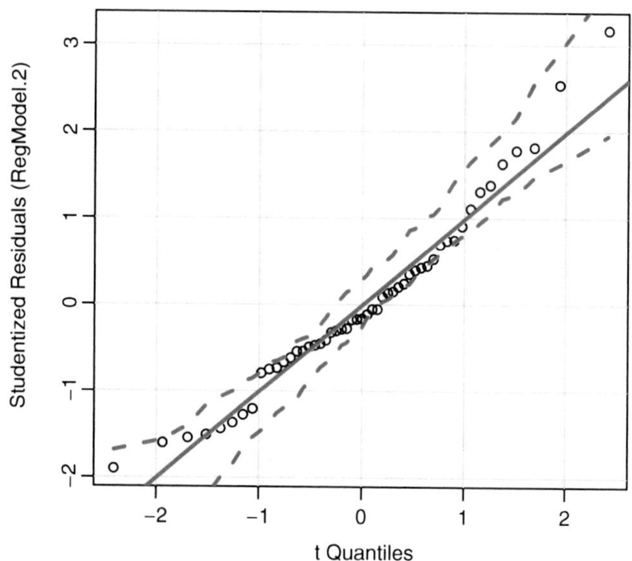

If you have a large number of explanatory variables, then you can select Stepwise Model.

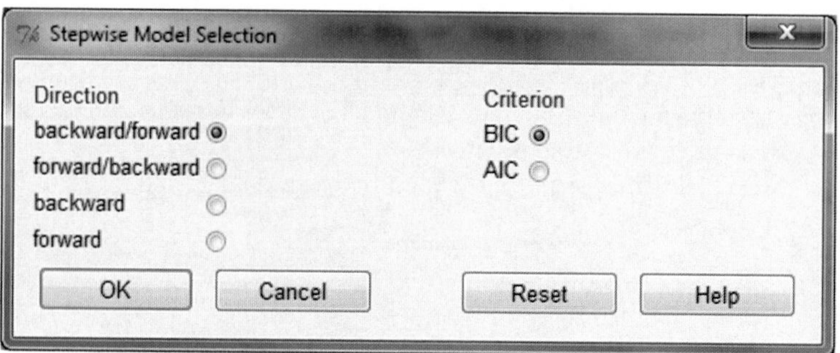

Scoring the model

You can add prediction using the third menu in the Models tab in R Commander. Add observation statistics to data.

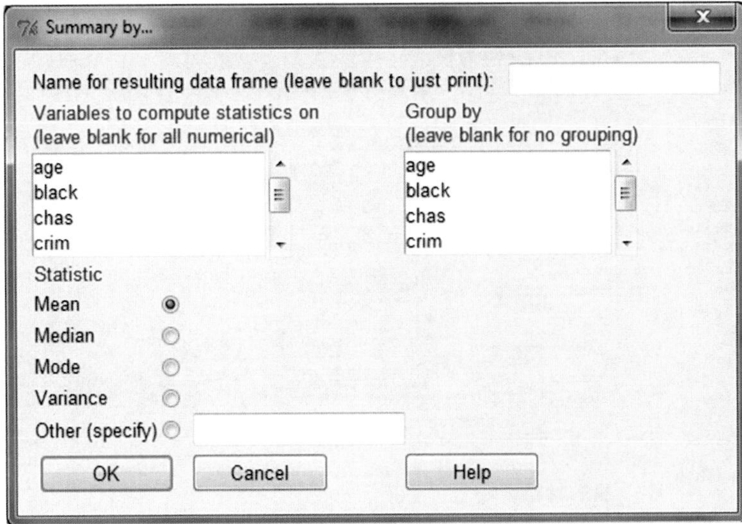

You can also shift between models by clicking the Models tab. This is especially useful when we have huge numbers of models to build and deploy.

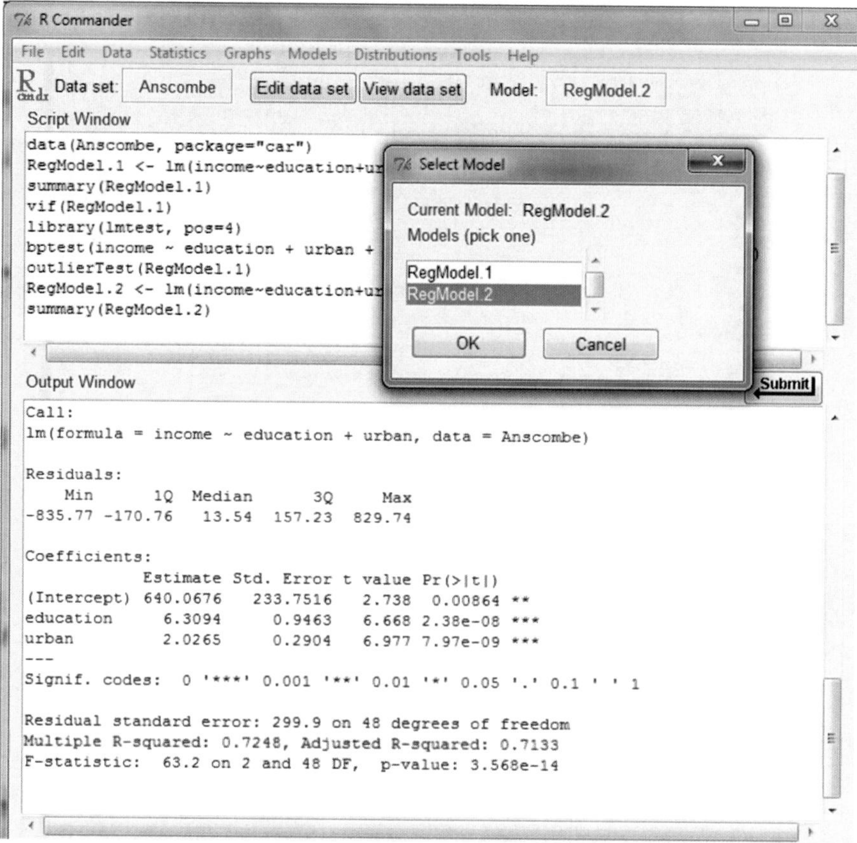

6.7 Other Packages for Regression Models

6.7.1 *ROCR for Performance Curves*

We can use the R package ROCR to plot the performance curves for logistic regression models. Despite theoretical objections to their validity or robustness, ROC curves continue to be widely used in regression models.

6.7.2 *rms Package*

Professor Frank Harell's *rms* package does regression modeling, testing, estimation, validation, graphics, prediction, and typesetting by storing enhanced model design attributes in the fit. It is statistically more complex for the average business analyst to understand, but it is quite comprehensive in its details on regression models.

An example of an OLS model using the RMS package is given below.

```
data(Boston, package="MASS")
library(rms)
ddBoston=datadist(Boston)
summary(ddBoston)
options(datadist="ddBoston")
LinearModel.2=ols(medv ~ age +black + chas + crim + dis, data=Boston)
summary(LinearModel.2)
p=Predict(LinearModel.2)
plot(p)
```

6.8 PMML

After you have created your regression model, you will want it to score new data as they comes in. This is typically done for propensity and risk models. You can use the predictive modeling markup language (PMML) to export models.

PMML provides an open standard for representing data mining models.

PMML is a language for representing models using XML in an application-independent way. Such models can then be shared with other applications that support PMML (http://www.dmg.org/products.html). The generic pmml() function (from the pmml package) takes an R model as its argument and returns the corresponding PMML. Currently supported models for export include linear regression (lm and glm), support vector machines (ksvm), decision trees (rpart), neural networks (nnet, multinom), association rules (arules), survival models (coxph, survreg), random survival forests (randomSurvivalForest), and clusters (kmeans, hclust).

For more on PMML see http://www.dmg.org/v4-1/GeneralStructure.html.

6.8.1 Zementis: Amazon EC2 for Scoring

The Zementis company helped create the PMML package for exporting R data mining models (including regression and other models) to PMML formats so that they could be used for data in other analytical packages as well. In addition, the company has an Amazon EC2 hosted solution called ADAPA, which can be used for scoring models with a large amount of data.

What follows is an extract from an interview with Dr. Michael Zeller, CEO of Zementis.

Ajay: How do Zementis and ADAPA and PMML fit?

Mike: Zementis has been an avid supporter of the PMML standard and is very active in the development of the standard. We contributed to the PMML package for the open source R Project. Furthermore, we created a free PMML Converter tool that helps users to validate and correct PMML files from various vendors and convert legacy PMML files to the latest version of the standard. Most prominently with ADAPA, Zementis launched the first cloud-computing scoring engine on the Amazon EC2 cloud. ADAPA is a highly scalable deployment, integration, and execution platform for PMML-based predictive models. Not only does it give you all the benefits of being fully standards-based using PMML and Web services, but it also leverages the cloud for scalability and cost-effectiveness. By being a Software as a Service (SaaS) application on Amazon EC2, ADAPA provides extreme flexibility, from casual usage that only costs a few dollars a month all the way to high-volume mission-critical enterprise decision management that users can seamlessly launch in the United States or in European data centers.

Ajay: What are some examples of where PMML helped companies save money?

Mike: For any consulting company focused on developing predictive analytics models for clients, PMML provides tremendous benefits, both for clients and service providers. In standardizing on PMML, it defines a clear deliverable—a PMML model—that clients can deploy instantly. No fixed requirements on which specific tools to choose for development or deployment, it is only important that the model adhere to the PMML standard, which becomes the common interface between the business partners. This eliminates miscommunication and lowers the overall project cost. Another example is where a company has taken advantage of the capability to move models instantly from development to operational deployment. It allows them to quickly update models based on market conditions, say in the area of risk management and fraud detection, or to roll out new marketing campaigns. Personally, I think the biggest opportunities are still ahead of us as more and more businesses embrace operational predictive analytics. The true value of PMML is to facilitate a real-time decision environment where we leverage predictive models in every business process, at every customer touch point and on demand, to maximize value.

6.9 Summary of Commands Used in This Chapter

6.9.1 Packages

- rms
- car
- ROCR
- RCommander
- gvlma

6.9.2 *Functions*

- ols(rms): Build an OLS model using RMS
- plot(predict(rms))-
- fitted(car)
- lm
- vif(car)
- outlierTest(car)
- gvlma(gvlma)
- hccm(car)

Citations and References

- Edsel A. Pena and Elizabeth H. Slate (2010). gvlma: Global Validation of Linear Models Assumptions. R package version 1.0.0.1. http://CRAN.R-project.org/package=gvlma
- John Fox and Sanford Weisberg (2011). An {R} Companion to Applied Regression, Second Edition. Thousand Oaks, CA: Sage. URL: http://socserv.socsci.mcmaster.ca/jfox/Books/Companion
- Graham Williams, Michael Hahsler, Zementis Inc., Hemant Ishwaran, Udaya B. Kogalur and Rajarshi Guha (2012). pmml: Generate PMML for various models. R package version 1.2.29. http://CRAN.R-project.org/package=pmml
- PMML: An Open Standard for Sharing Models by Alex Guazzelli, Michael Zeller, Wen-Ching Lin and Graham Williams. R Journal http://journal.r-project.org/2009-1/RJournal_2009-1_Guazzelli+et+al.pdf
- Tobias Sing, Oliver Sander, Niko Beerenwinkel, Thomas Lengauer. ROCR: visualizing classifier performance in R. Bioinformatics 21(20):3940–3941 (2005).
- Frank E Harrell Jr (2012). rms: Regression Modeling Strategies. R package version 3.5-0. http://CRAN.R-project.org/package=rms

Chapter 7
Data Mining Using R

Data mining is a commonly used term that is interchangeably used with business analytics, but it is not exactly the same.

7.1 Definition

One definition of data mining is as follows:

The practice of examining large databases in order to generate new information.

While much of statistical theory and literature until the twentieth century dealt with use cases where data were either insufficient or sparse, in the twenty-first century we now have cases where data collection is often automated by logging software on machines, and database sizes are huge compared to earlier days. Chapter 6 deals with regression models, which are covered separately because they are frequently used in the business world. Clustering, covered in Chap. 8, is an integral part of data mining, though it is covered separately because it is a very popular technique. Accordingly, we will cover a small set of remaining data mining procedures here.

The various statistics involved in data mining are also beautifully illustrated in an online data map at http://chem-eng.utoronto.ca/~datamining/dmc/data_mining_map.htm, which has relevant sections that can be clicked for further study.

As defined by the U.S. Government Accountability Office (formerly the U.S. General Accounting Office) (GAO), data mining is "the application of database technology and techniques—such as statistical analysis and modeling—to uncover hidden patterns and subtle relationships in data and to infer rules that allow for the prediction of future results."

A. Ohri, *R for Business Analytics*, DOI 10.1007/978-1-4614-4343-8_7,

7.1.1 Information Ladder

The information ladder below shows the various stages in the conversion of data to information. It was made by Norman Longworth.

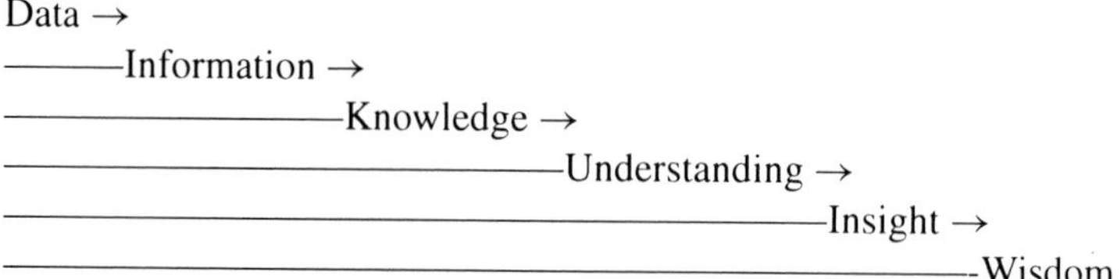

Data →

———Information →

——————————Knowledge →

———————————————————Understanding →

————————————————————————————Insight →

—————————————————————————————————-Wisdom

Whereas the first two steps can be scientifically exactly defined, the latter parts belong to the domain of psychology and philosophy.

The process and practical methodology of using data mining has been defined using three primary industry classifications—knowledge discovery in databases (KDD), cross-industry standard process for data mining (CRISP-DM), and SEMMA.

7.1.2 KDD

The KDD process is commonly defined by the following stages:

- Selection
- Preprocessing
- Transformation
- Data mining
- Interpretation/evaluation

KDD:

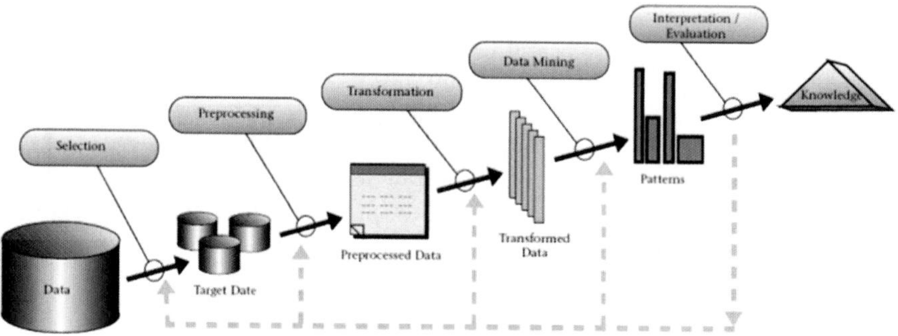

7.1.3 CRISP-DM

CRISP-DM has six subdivisions

- Business understanding
- Data understanding
- Data preparation
- Modeling
- Evaluation
- Deployment

CRISP-DM:

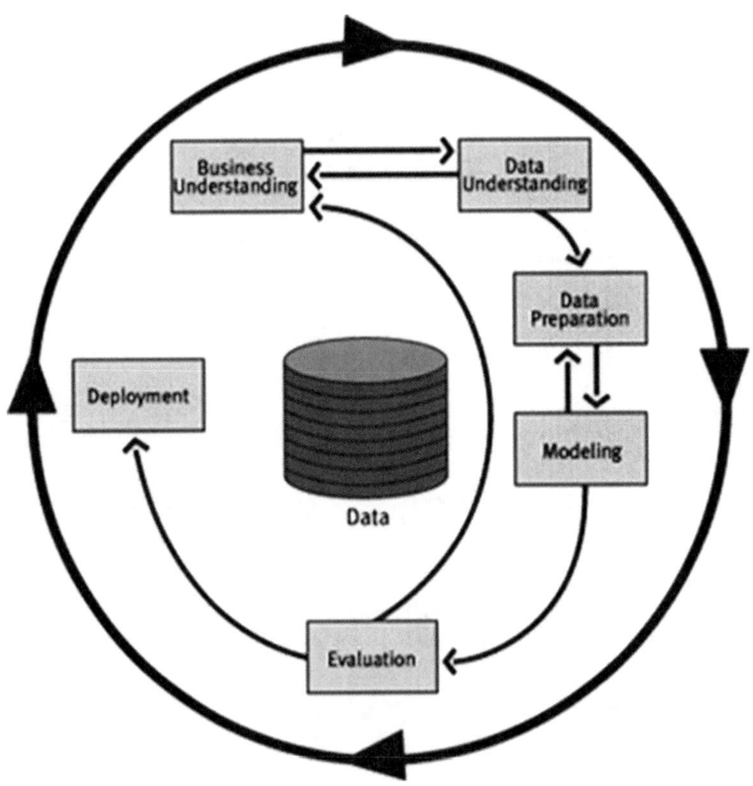

7.1.4 SEMMA

SEMMA is a process for data mining and has five subparts:

- Sample
- Explore
- Modify
- Model
- Assess

Visual representation of KDD, CRISP-DM, and SEMMA:
http://www.decisionstats.com/visual-guides-to-crisp-dm-kdd-and-semma/. SEMMA

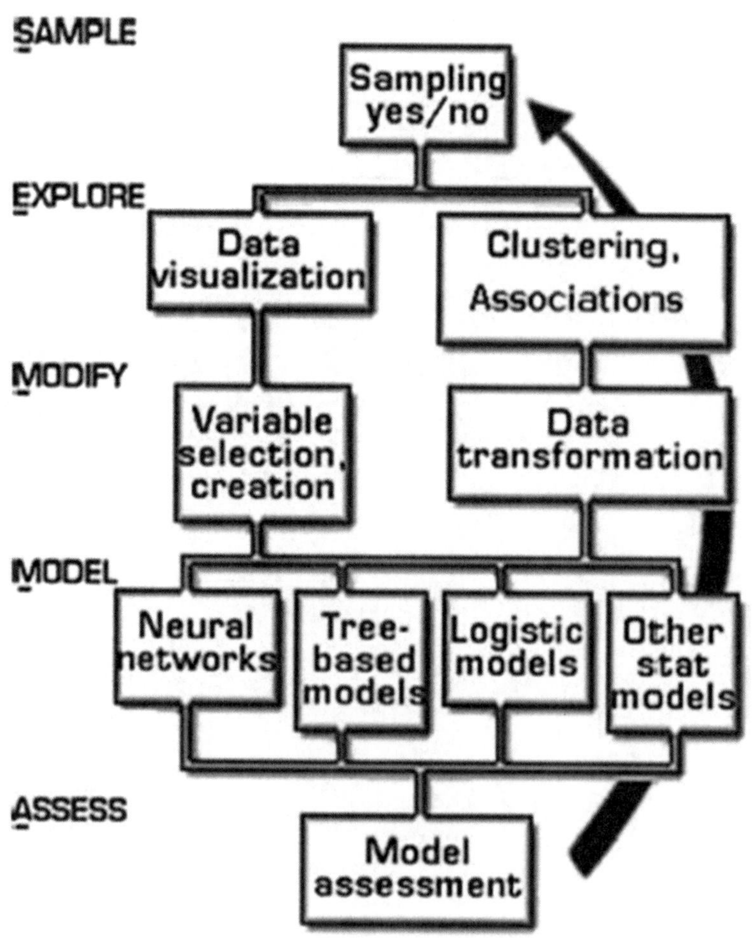

7.1.5 *Four Phases of Data-Driven Projects*

According to the author, data-driven projects fall into these four broad phases from a business or managerial point of view.

7.1.5.1 *Business Problem Phase*: What Needs to Be Done?

- Increase revenues
- Cut costs
- Investigate unusual events
- Project timelines

7.1.5.2 *Technical Problem Phase*: Technical Problems in Project Execution

- Data availability/data quality/data augmentation costs
- Statistical (technique-based approach), hypothesis formulation, sampling, iterations
- Programming (tool-based approach), analytics platform coding (input, formats, processing)

7.1.5.3 *Technical Solution Phase*: Problem Solving Using the Tools and Skills Available

- Data cleaning/outlier treatment/missing value imputation
- Statistical (technique-based approach), error minimization, model validation, confidence levels
- Programming (tool-based approach), analytics platform coding (output, display, graphs)

7.1.5.4 Business Solution Phase: Put It All Together in a Word Document, Presentation, or Spreadsheet

- Finalized forecasts, models and data strategy improvements in existing processes
- Control and monitoring of analytical results post implementation
- Legal and compliance guidelines to execution (internal or external)
- Client satisfaction and expectation management audience
- Feedback based on presenting final deliverable to broader audience

7.1.5.5 Visual Representation of Four Phases of a Data-Driven Project

Business Problem - **Increase Revenues /Cut Costs/Investigate Unusual Events**
Technical Problem - **Data Availability /Data Quality/** **Statistical -(Technique based approach) , Hypothesis Formulation,Sampling, Iterations** **Programming-(Tool based approach) Analytics Platform Coding (Input, Formats,Processing)**
Technical Solution - **Data Cleaning /Outlier Treatment/Missing Value Imputation** **Statistical -(Technique based approach)** **Error Minimization, Model Validation, Confidence Levels** **Programming-(Tool based approach) Analytics Platform Coding (Output, Display,Graphs)**
Business Solution- **Forecasts/ Improvements / Control and Monitoring of Analytical Results Implementation/ Legal and Compliance**

7.1.6 Data Mining Methods

The primary data mining methods are as follows:

- Classification
- Regression
- Clustering
- Summarization
- Dependency modeling
- Change and deviation detection

Source: Usama Fayyad, Gregory Piatetsky-Shapiro, and Padhraic Smyth
From Data Mining to KDD: http://www.kdnuggets.com/gpspubs/aimag-kdd-overview-1996-Fayyad.pdf

7.2 Rattle: A GUI for Data Mining in R

Rattle, or R Analytical Tool To Learn Easily, is a more advanced user interface than R Commander, though not as popular in academia. This may be due to hypothesis testing as a more traditional option for statistical thinking rather than data mining within academia. It was designed explicitly for data mining and also has a commercial version for sale.

Rattle has a Tab and Radio button/checkbox rather than a menu–dropdown approach to graphical interface design for the user. Also, the Execute button must be clicked after checking certain options, just like the Submit button is clicked after writing code. This is different from clicking on a dropdown menu and thus minimizes the risk of junk analysis by inadvertent clicking. Rattle can be downloaded from http://rattle.togaware.com/.

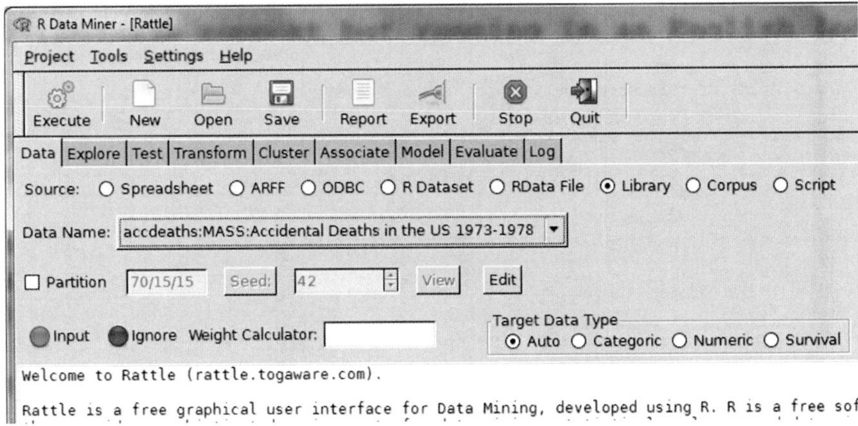

7.2.1 *Advantages of Rattle*

- Useful for beginner in R language for building models, clustering, and data mining.
- Has separate tabs for data entry, summary, visualization, model building, clustering, association, and evaluation. The design is intuitive and easy to understand, even for those without a background in statistics, as the help is

conveniently explained as each tab or button is clicked. Also, the tabs are placed in a very sequential and logical order. It uses a lot of other R packages to build a complete analytical platform. Very good for correlation graphs, clustering, and decision trees.

- Easy-to-understand interface even for first-time user.
- Log for R code is auto-generated and timestamp is placed.
- Complete solution for model building from partitioning datasets randomly for testing, validation to building model, evaluating lift and ROC curve, and exporting PMML output of model for scoring.
- Has a well-documented online help as well as in-software documentation. The help explains terms even to nonstatistical users and is highly useful for business users.

Example: Hypothesis testing in Rattle with easy-to-understand descriptions.

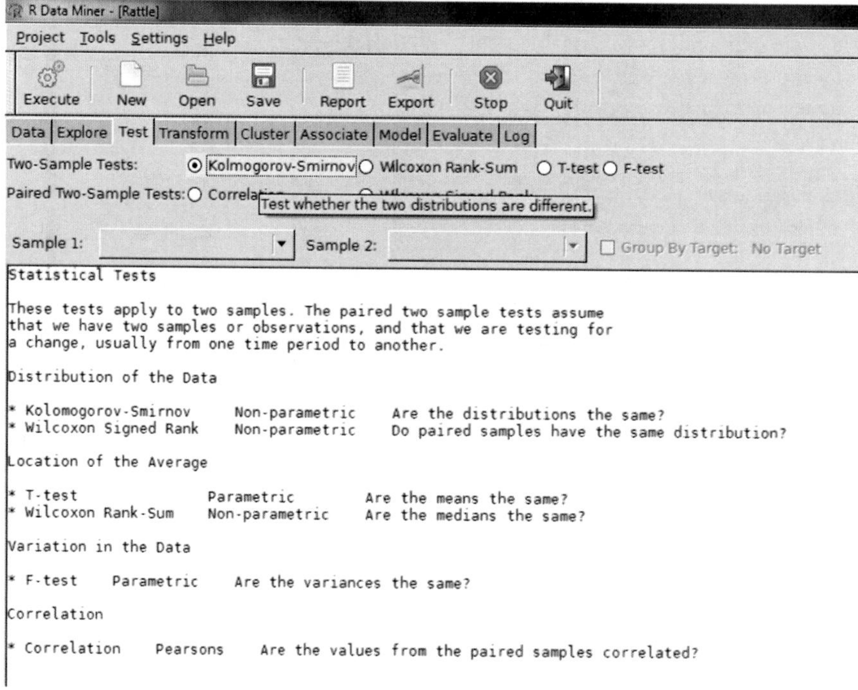

While R Commander may be preferred by a pure statistician, Rattle shows easier help and is thus better designed for business users.

7.2.2 Comparative Disadvantages of Using Rattle

- It is basically aimed at a data miner. Hence it is more of a data mining GUI rather than an analytics GUI. It thus has difficulty in loading data apart from a data frame format.
- It has limited ability to create different types of graphs or visualizations from a business analyst's perspective. Numeric variables can be made into a boxplot, histogram, cumulative, and Benford graphs. While interactivity using ggobi and latticist is involved, the number of graphical options is still less than in other GUIs.
- It is not suited for projects that involve multiple graphical analyses and that do not have model building or data mining. For example, a data plot is given in the Clustering tab but not in the general Explore tab.
- Despite the fact that it is meant for data miners, no support for biglm packages or parallel programming is enabled in the GUI for bigger datasets, though these can be done by the R command line in conjunction with the Rattle GUI. Data mining is typically done on bigger datasets.
- There may be some problems installing it as it is dependent on GTK and has a lot of packages as dependencies.

7.2.3 Description of Rattle

Top –This has the Execute button (shown as two gears) and has keyboard shortcut F2. It is used to execute the options in tabs and is the equivalent of the Submit Code button. Other buttons include new Projects, Save, and Load projects, Projects are files with extensions to .rattle and which store all related information from Rattle. It also has a button for exporting information in the current tab as an open office document and buttons for interrupting the current process as well as for exiting Rattle.

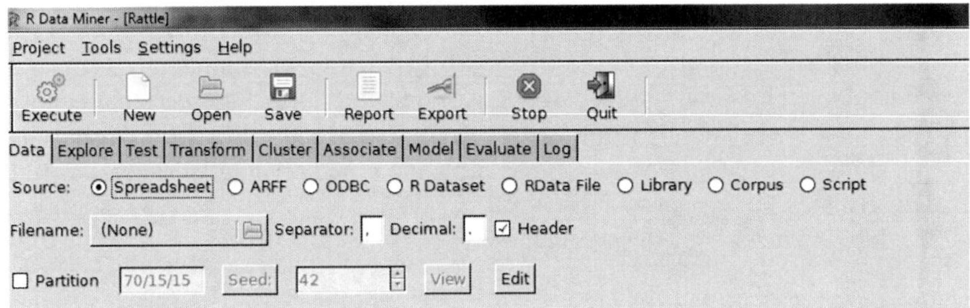

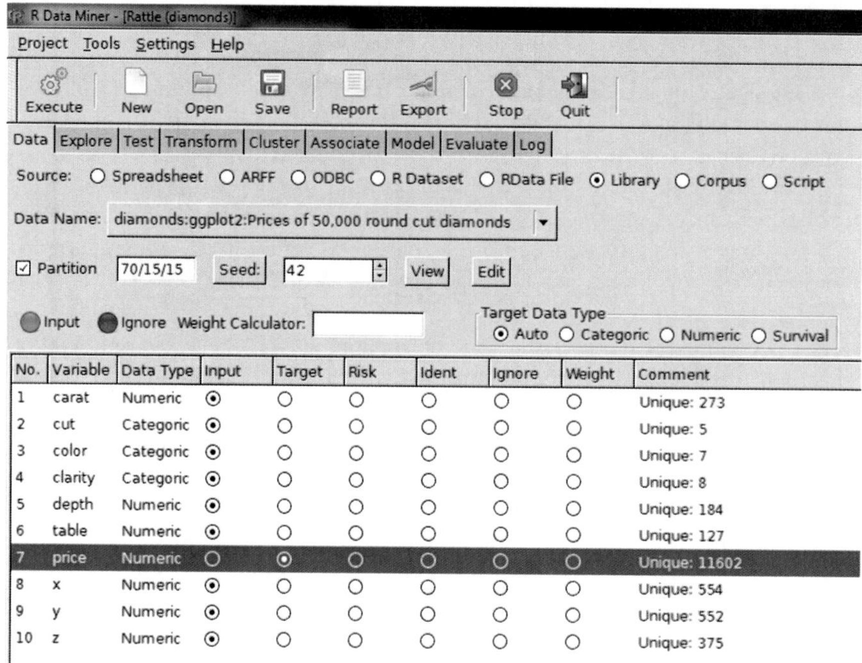

The data menu has the following options:

Data type: These are radio buttons between Spreadsheet (and comma-delimited values), ARFF files (Weka), ODBC (for database connections), Library (for datasets from packages), R Dataset or R datafile, Corpus (for text mining), and Script for generating the data by code.

The **second row** in the Data tab in Rattle is Detail on Data Type, and its appearance shifts as per the radio button selection of data type in the previous step. For Spreadsheet, it will show Path of File, Delimiters, and Header Row, while for ODBC it will show DSN, Tables, and Rows, and for Library it will show you a dropdown of all datasets in all R packages installed locally.

The **third row** is a partition field for splitting datasets into training, testing, and validation, and it shows percentage. It also specifies a random seed that can be customized for random partitions that can be replicated. This is very useful as model building requires a model to be built and tested on random subsets of full datasets.

The **fourth row** is used to specify the variable type of input data.

The variable types are as follows:

Input: Used for modeling as independent variables

Target: Output for modeling or the dependent variable. The target is a categoric variable for classification, numeric for regression, and for survival analysis both time and status need to be defined

Risk: A variable used in the risk chart

Ident: An identifier for unique observations in a dataset like AccountId or Customer Id

Ignore: Variables that are to be ignored

In addition, the weight calculator can be used to perform mathematical operations on certain variables and identify certain variables as more important than others

Explore tab, Summary subtab has Summary for a brief summary of variables, Describe for a detailed summary, and Kurtosis and Skewness for comparing them across numeric variables.

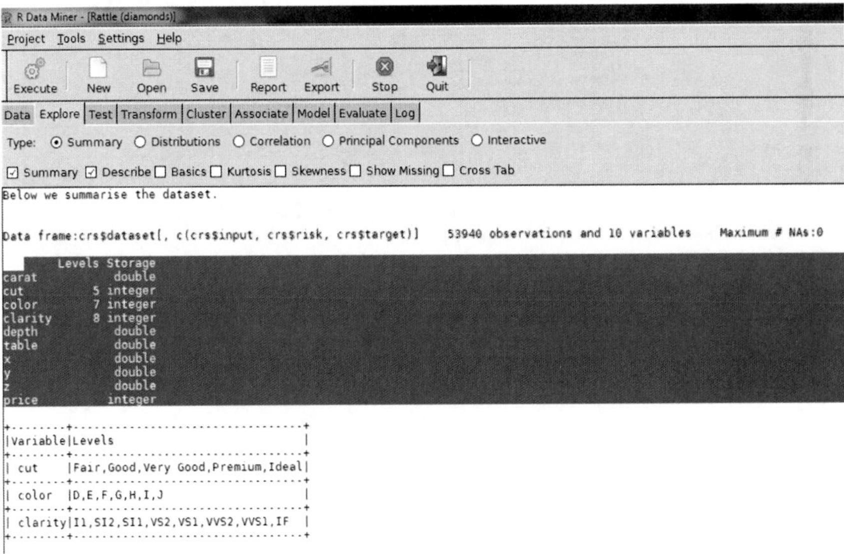

- The Distributions subtab allows for the plotting of histograms, boxplots, and cumulative plots for numeric variables and for categorical variables barplot and dotplot. It has a Benford plot for Benford's law on the probability distribution of digits.

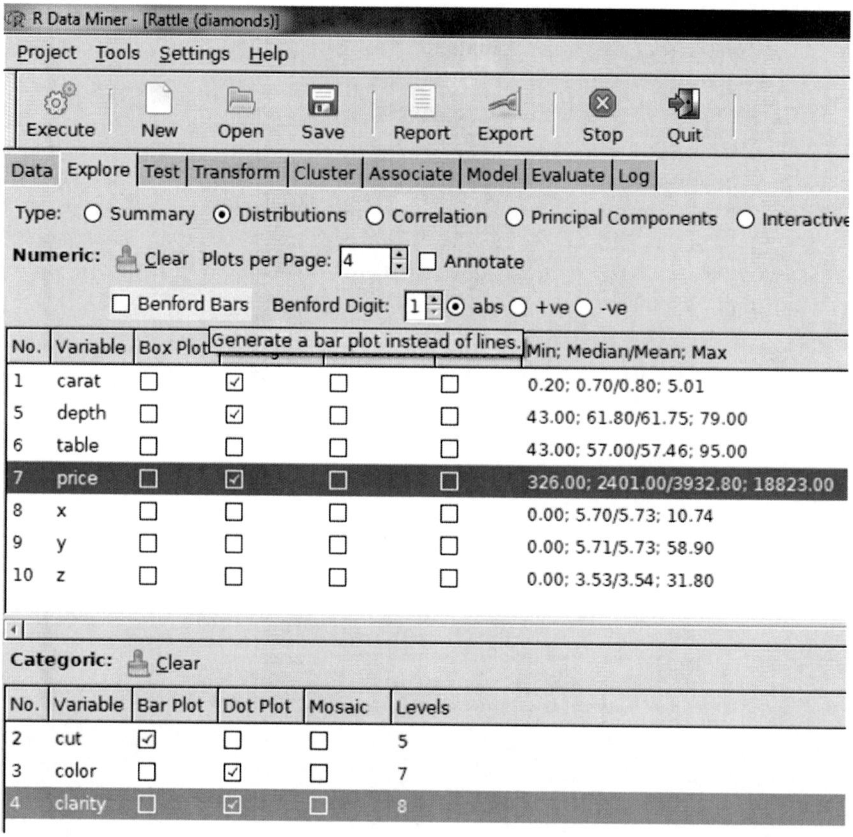

- The Correlation subtab displays the correlation between variables as a table and also as a nicely designed plot.

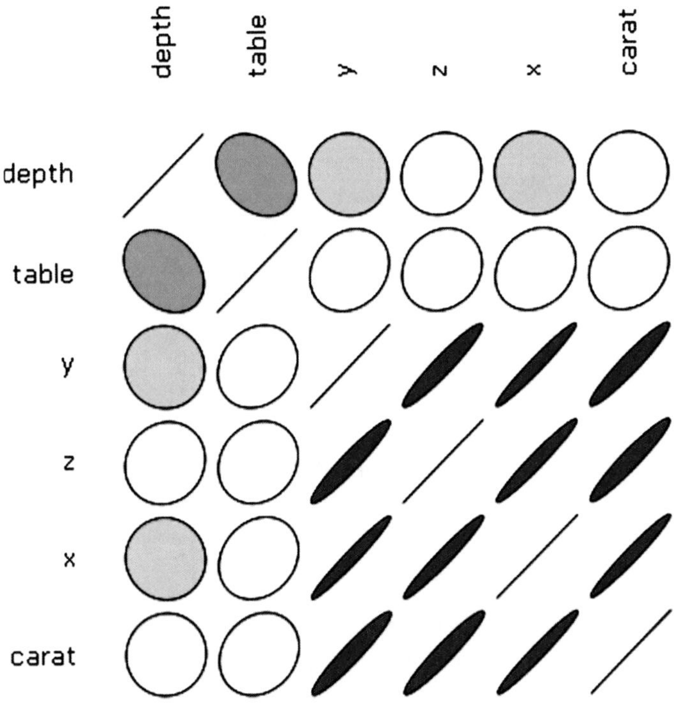

- The Principal Components subtab is for use with principal component analysis including the singular value decomposition (SVD) and eigen methods.
- The Interactive subtab allows interactive data exploration using GGobi and Lattice software. It is a powerful visual tool.

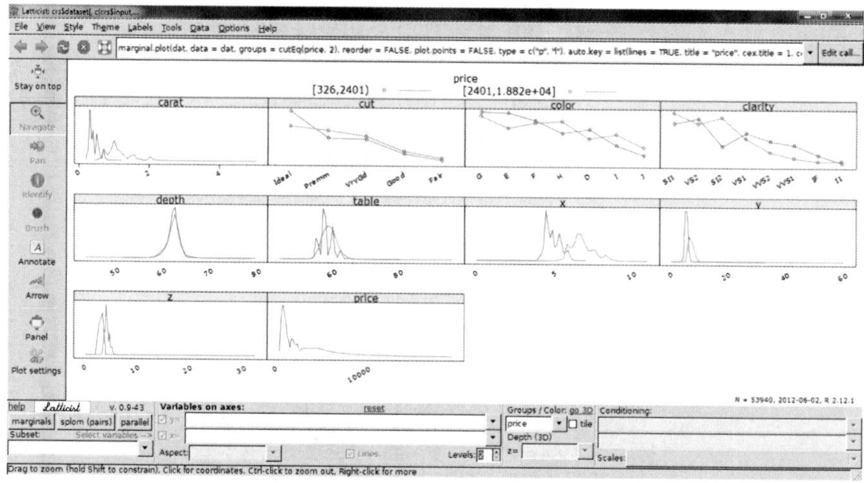

- The Test tab has options for hypothesis testing of data for two sample tests (see above in Advantages of Using Rattle).
- The Transform tab has options for rescaling data, missing values treatment, and deleting invalid or missing values.
- The Cluster tab gives an option to k-means, hierarchical, and bicluster clustering methods with automated graphs, plots (including dendrogram, discriminant plots, and data plots), and cluster results available.

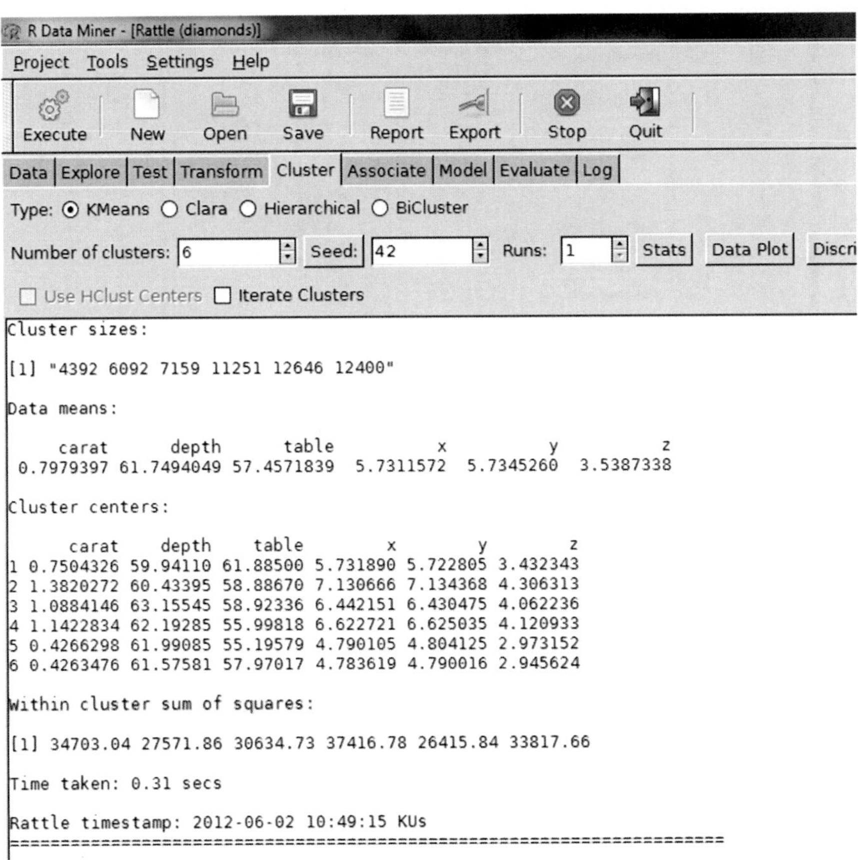

The Cluster tab is highly recommended for clustering projects especially for people who are proficient in clustering but not in R.

- The Associate tab helps in building association rules between categorical variables, which are in the form of "if then"statements.

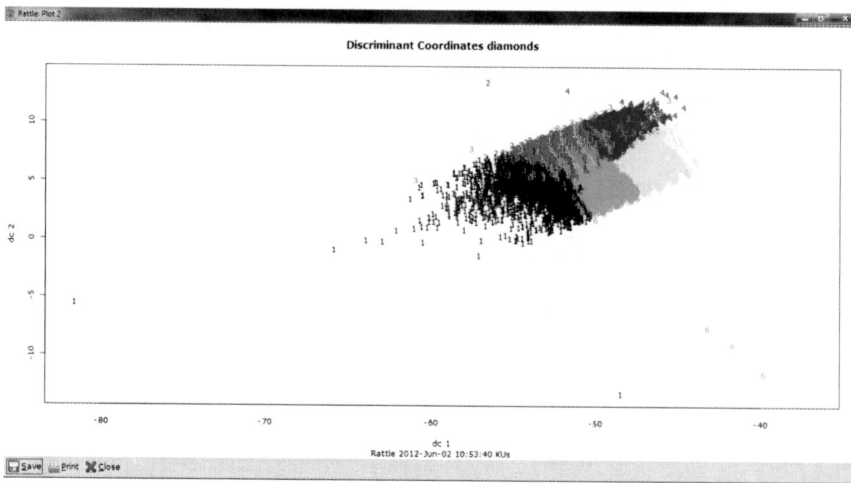

Example: If the day is Thursday and someone buys milk, there is an 80 % chance she will buy diapers. These probabilities are generated from observed frequencies.

Here is an example of the code generated by Rattle when you run association rules (on the diamond dataset):

```
# Rattle timestamp: 2011-02-05 20:06:32 i386-pc-mingw32
# Load an R dataset.
data(list = "Adult", package = "arules")
crs$dataset <- Adult
names(crs$dataset) <- gsub("-", ".", names(crs$dataset))
#==========================================
# Rattle timestamp: 2011-02-05 20:07:16 i386-pc-mingw32
# Load an R dataset.
data(list = "Agrawal", package = "arulesNBMiner")
crs$dataset <- Agrawal.pat
names(crs$dataset) <- gsub("-", ".", names(crs$dataset))
#==========================================
# Rattle timestamp: 2011-02-05 20:09:02 i386-pc-mingw32
# Load an R dataset.
data(list = "diamonds", package = "ggplot2")
crs$dataset <- diamonds
names(crs$dataset) <- gsub("-", ".", names(crs$dataset))
#==========================================
# Rattle timestamp: 2011-02-05 20:09:12 i386-pc-mingw32
# Note the user selections.
# Build the training/validate/test datasets.
set.seed(crv$seed)
crs$nobs <- nrow(crs$dataset)
# 53940 observations
```

```
crs$sample <- crs$train <- sample(nrow(crs$dataset), 0.7*crs$nobs)
# 37758 observations
crs$validate <- sample(setdiff(seq_len(nrow(crs$dataset)),
crs$train), 0.15*crs$nobs)
# 8091 observations
crs$test <- setdiff(setdiff(seq_len(nrow(crs$dataset)), crs$train), crs$validate)
# 8091 observations
# The following variable selections have been noted.
crs$input <- c("carat", "cut", "color", "clarity", "depth", "table", "price", "x",
"y", "z")
crs$numeric <- c("carat", "depth", "table", "price", "x", "y", "z")
crs$categoric <- c("cut", "color", "clarity")
crs$target <- NULL crs$risk <- NULL crs$ident <- NULL crs$ignore
 <- NULL crs$weights <- NULL
#=============================================
# Rattle timestamp: 2011-02-05 20:09:23 i386-pc-mingw32
# Note the user selections.
# The following variable selections have been noted.
crs$input <- c("carat", "cut", "color", "clarity", "depth", "table", "x", "y", "z")
crs$numeric <- c("carat", "depth", "table", "x", "y", "z")
crs$categoric <- c("cut", "color", "clarity")
crs$target <- "price"
crs$risk <- NULL
crs$ident <- NULL
crs$ignore <- NULL
crs$weights <- NULL
#=============================================
# Rattle timestamp: 2011-02-05 20:09:36 i386-pc-mingw32
# Association Rules
# The 'arules' package provides the 'arules' function.
require(arules, quietly=TRUE)
# Generate a transaction dataset.
crs$transactions <- as(crs$dataset[, crs$categoric], "transactions")
# Generate the association rules.
crs$apriori <- apriori(crs$transactions, parameter = list(support=0.100,
confidence=0.100))
# Summarize the resulting rule set.
generateAprioriSummary(crs$apriori)
# Time taken: 1.58 secs
# List rules.
inspect(SORT(crs$apriori, by = "confidence"))
#=============================================
# Rattle timestamp: 2011-02-05 20:10:26 i386-pc-mingw32
# Relative frequencies plot
# Association rules are implemented in the 'arules' package.
```

```
require(arules, quietly=TRUE)
# Generate a transaction dataset.
crs$transactions <- as(crs$dataset[,c(2:4)], "transactions")
# Plot the relative frequencies.
itemFrequencyPlot(crs$transactions, support=0.1, cex=0.8)
```

The preceding code may seem a bit much even to an experienced data miner just starting out in R, but actually it took just five mouse clicks to generate the results—and the code was auto-generated!

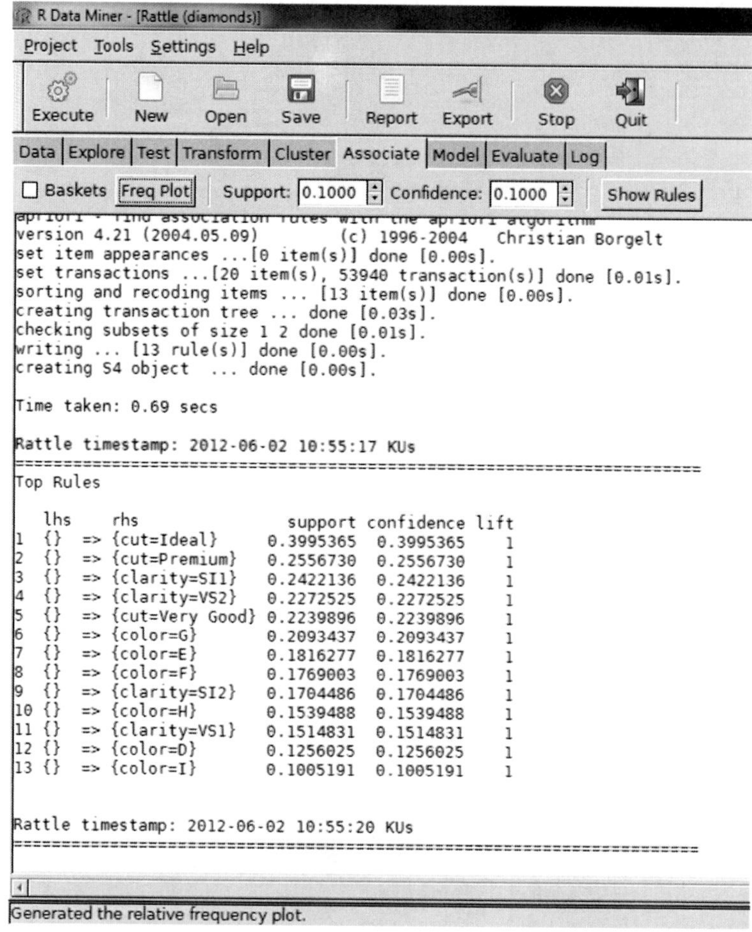

And the frequency tab was generated by a single mouse click!

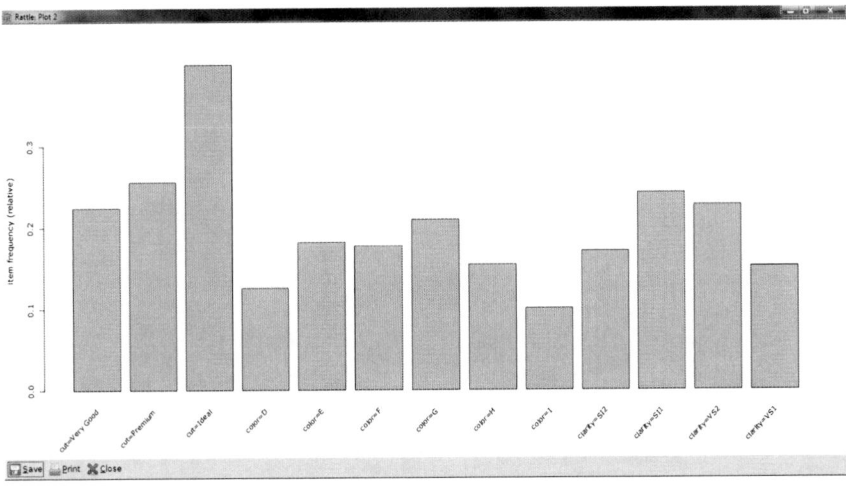

- The Model tab makes Rattle one of the most advanced data mining tools as it incorporates decision trees (including boosted models and the forest method), linear and logistic regression, SVM, neural nets, and survival models.

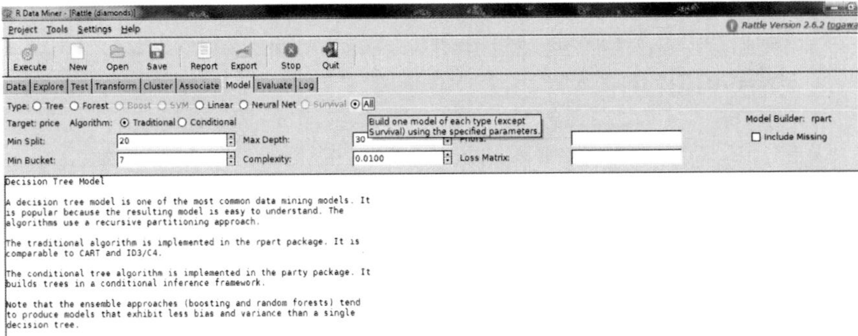

- The Evaluate tab has functionality for evaluating models including lift, ROC, confusion matrix, cost curve, risk chart, precision, specificity, and sensitivity as well as scoring datasets with a built model or models. *Example:* a ROC curve generated by Rattle for surviving passengers on the *Titanic* (as a function of age, class, sex).

  This shows a comparison of various models built.

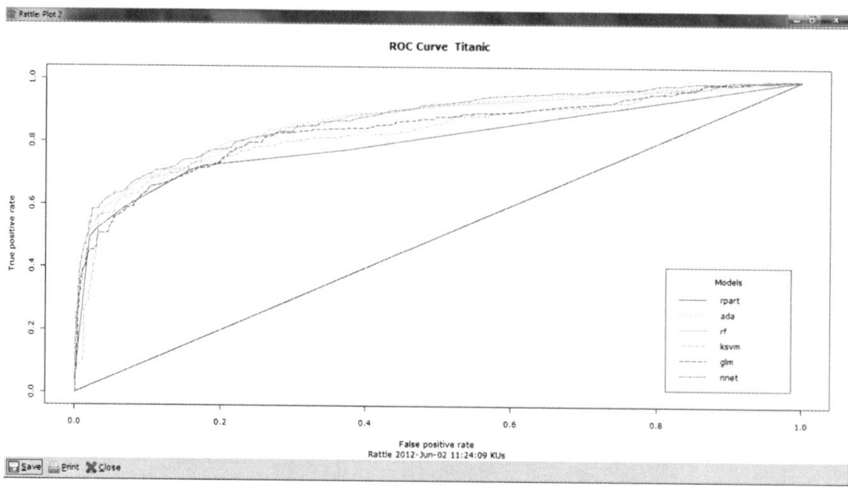

- The Log tab: R code is automatically generated by Rattle as the respective operation is executed. Also, a timestamp is made so it helps in reviewing errors as well as evaluating the speed for code optimization.

7.3 Interview: Graham Williams, Author of *Data Mining with Rattle and R*

What follows is an extract from an interview with the creator of Rattle, Dr. Graham Williams.

Ajay: What made you get involved with R? What is the advantage of using Rattle versus normal R?

Graham: I have used Clementine and SAS Enterprise Miner for many years (and IBM's original Intelligent Miner and Thinking Machines' Darwin, and many other tools that emerged early on with data mining). Commercial vendors come and go (even large ones like IBM, in terms of the products they support). Lock-in is one problem with commercial tools. Another is that many vendors, understandably, won't put resources into new algorithms until they are well accepted. Because it is open source, R is robust, reliable, and provides access to the most advanced statistics. Many research statisticians publish their new algorithms in R. But what is most important is that the source code is always going to be available. Not everyone has the skill to delve into that source code, but at least we have a chance to do so. We also know that there is a team of highly qualified developers whose work is openly peer reviewed. I can monitor their coding changes, if I so desire. This helps ensure quality and integrity. Rolling out R to a community of data analysts, though, does present challenges. Since it is primarily a language for statistics, we need to learn to speak that language. That is, we need to communicate through language rather than through pictures (or a GUI). It is, of course, easier to draw pictures, but pictures can be limiting. I believe a written language allows us to express and communicate ideas better and more formally. But it needs to be with the philosophy that we are communicating those ideas to our fellow humans, not just writing code to be executed by the computer. Nonetheless, GUIs are great as memory aids, for doing simple tasks, and for learning how to perform particular tasks. Rattle aims to do the standard data mining steps but to also expose everything that is done as R commands in the log. In fact, the log is designed to be run as an R script and to teach the user the R commands.

Ajay: What are the advantages of using Rattle instead of SAS or SPSS? What are the disadvantages of using Rattle instead of SAS or SPSS?

Graham: Because it is free and open source, Rattle (and R) can be readily used in teaching data mining. In business it is, initially, useful for people who want to experiment with data mining without the sometimes quite significant up-front costs of the commercial offerings. For serious data mining, Rattle and R offer all of the data mining algorithms offered by the commercial vendors, plus many more. Rattle provides a simple, tab-based user interface that is not as graphically sophisticated as Clementine in SPSS and SAS Enterprise Miner.

BUT WITH JUST 4 BUTTON CLICKS YOU CAN BUILD YOUR FIRST DATA MINING MODEL.

The usual disadvantage cited for R (and, thus, Rattle) is in the handling of large datasets—SAS and SPSS can handle datasets out of memory, although they do slow down when doing so. R is memory based, so going to a 64-bit platform is often necessary for the larger datasets. A very rough rule of thumb has been that the 2–3 GB limit of the common 32-bit processors can handle datasets of up to about 50,000 rows with 100 columns (or 100,000 rows and 10 columns, etc.), depending on the algorithms you deploy. I generally recommend, as quite a powerful yet inexpensive data mining machine, one running on an AMD 64 processor, running the Debian GNU/Linux operating system, with as much memory as you can afford (e.g., 4–32 GB, although some machines today can go up to 128 GB, but memory gets expensive at that end of the scale).

Ajay: Rattle is free to download and use, yet it must have taken you some time to build it. What are your revenue streams to support your time and efforts?

Graham: Yes, Rattle is free software: free for anyone to use, free to review the code, free to extend the code, free to use it for whatever purpose. I have been developing Rattle for a few years now, with a number of contributions from other users. Rattle, of course, gets its full power from R. The R community works together to help each other, and others, for the benefit of all. Rattle and R can be the basic toolkit for knowledge workers providing analyses. I know of a number of data mining consultants around the world who are using Rattle to support their day-to-day consultancy work. As a company, Togaware provides user support, does installations of R and Rattle, and runs training in using Rattle and in doing data mining. It also delivers data mining projects to clients. Togaware also provides support for incorporating Rattle (and R) into other products (e.g., as RStat for Information Builders).

7.4 Text Mining Analytics Using R

A corpus is a set of documents in text mining. A term document matrix simply gives a frequency distribution of all words/terms in that document. We use the R package *tm* for text mining.

7.4.1 Text Mining a Local Document

In this example, the folder C:/Users/KUs/Desktop/test contains all the text documents. In this case I manually copied and pasted words from Chap. 1 into one Notepad file to give a simple example.

```
library(tm)
getReaders() #Gives list of readers
getSources() #Gives list of sources
txt2 = "C:/Users/KUs/Desktop/test" #specifying the file path
b = Corpus(DirSource(txt2), readerControl = list(language = "eng"))
#Note DirSource is used as a parameter to Corpus
#inspect(b) will print out the entire text document
# This is useful for quality control if you want to test out your code of text mining
onto a small sample document
summary(b) #This gives information about the new corpus we have created
b<- tm_map(b, tolower) #Changes case to lowercase
b<- tm_map(b, stripWhitespace) #Strips white space
b <- tm_map(b, removeWords, stopwords("english")) #Removes stop words
b <- tm_map(b, removePunctuation) #Removes punctuation
tdm <- TermDocumentMatrix(b)
#or dtm <- DocumentTermMatrix(b), the difference is terms are in rows or columns
m1 <- as.matrix(tdm)
v1<- sort(rowSums(m1),decreasing = TRUE)
d1<- data.frame(word = names(v1),freq = v1)
> d3 = subset(d1,freq>20)
```

and plotting d3 we obtain the frequency of the most common words in Chap. 1 of this book!

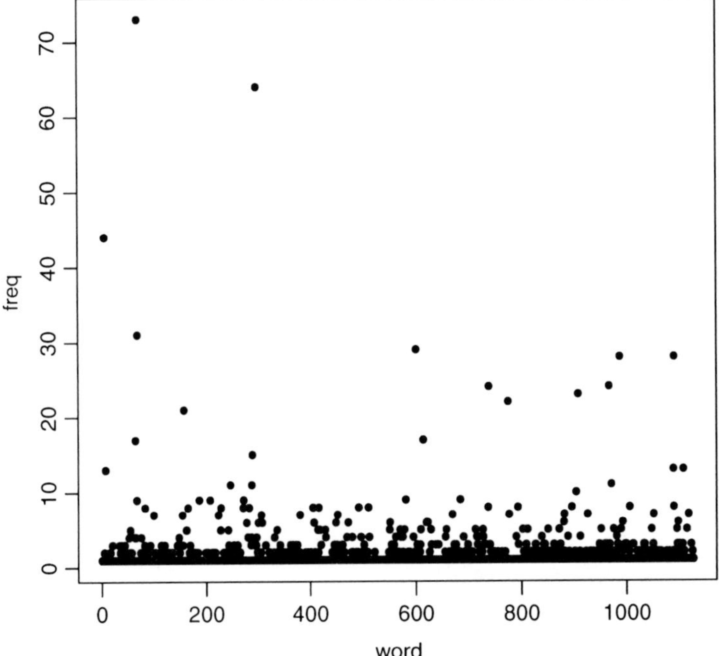

Exercise 1: Try to replicate this case study if you have an online e-book and guess which word comes out most often in Chap. 1.

Exercise 2: Also try to figure out which command or type of graph replicates this graph above.

Answer 1: We use the word cloud plot in Chap. 5 to find the most frequently occurring word. Note the size of words in a word cloud is related to their frequency of occurrence.

```
> library(wordcloud)
>wordcloud(d1$word,d1$freq)
```

7.4.2 Text Mining from the Web and Cleaning Text Data

In this case we download data from the Internet and clean the data.
Note if we need to retrieve just the links on a Web page, we can use the string package.

```
url=("http://www.nytimes.com")
html <- paste(readLines(url), collapse = "\n")
library(stringr)
matched <- str_match_all(html, "<a href = \"(.*?)\"")
links <- matched[[1]][, 2]
```

Let us say that we want to read the front page of *The New York Times* in R.

Downloading in R

```
con <- url("http://www.nytimes.com", "r")
x <- readLines(con) #We can do this using the readLines function to read the html
# or we can use the function getURL in the R package RCurl
url = ("http://www.nytimes.com")
library(RCurl)
ans <- getURL(url)
ans<-as.data.frame(ans)
library(tm)
getReaders() #gives a list of readers
#[1] "readDOC" "readGmane" "readPDF" "readReut21578XML" "readReut21578XMLasPlain" "readPlain" #[7] "readRCV1" "readRCV1asPlain" "readTabular" "readXML"
getSources() #gives a list of sources
#[1] "DataframeSource" "DirSource" "GmaneSource" "ReutersSource" "URISource" "VectorSource"
b = Corpus(DataframeSource(ans), readerControl = list(language = "eng"))
#Note DataFrameSource is used as a parameter to Corpus because we refer to the data frame created here.
```

If we change the sources option to URISource in the Corpus function, then we can avoid the earlier use of readLines.

```
library(tm)
url=("http://www.nytimes.com")
b = Corpus(URISource(url), readerControl = list(language = "eng"))
```

Now we move to the text mining stage.

```
#inspect(b) will print out the entire text document
getTransformations() #gives a list of transformations for tm_map
#1] "as.PlainTextDocument" "removeNumbers" "removePunctuation"
#[4] "removeWords" "stemDocument" "stripWhitespace"
b<- tm_map(b, stripWhitespace) #Strips white space
b<- tm_map(b, tolower) #Changes case to lowercase
b <- tm_map(b, removeWords, stopwords("english")) #Removes stop words
b <- tm_map(b, removePunctuation) #Removes punctuation
tdm <- TermDocumentMatrix(b)
#or dtm <- DocumentTermMatrix(b), the difference is terms are in rows or columns
m1 <- as.matrix(tdm)
v1<- sort(rowSums(m1),decreasing = TRUE)
```

```
d1<- data.frame(word = names(v1),freq = v1)
d2<- subset(d1,d1$freq<6)
```

Visualizing a word cloud using the wordcloud function from the wordcloud package. To obtain the most frequently used code, we can run the word cloud on data frame d1 and not d2, but we would see a lot of junk HTML values.

```
library(wordcloud)
wordcloud(d2$word,d2$freq)
```

This is not what the *The New York Times* says! We need to clean up the html so we can see only English words and none of the html.

We do that by modifying the stop words.

```
# modifying stopwords
# keep "r" by removing it from stop words and delete href by adding it to the stop words
myStopwords <- c(stopwords('english'), "available", "via", "href")
idx <- which(myStopwords == "r") #Note the new data created based on a condition
myStopwords <- myStopwords[-idx]
#Note the negative operator to delete a particular data based on a condition
b <- tm_map(b, removeWords, myStopwords)
```

After this step we redo the analysis reusing the preceding code.

A much simpler approach would be to save the Web page as a text document and then read it by changing the sources option to DirSource in the Corpus function in the tm package.

However, this brings an error message:

In readLines(y, encoding = x$Encoding) : incomplete final line found on 'C:/Users/KUs/Desktop/test/nytimes2.txt'

So rather inelegantly, we copy the Web page using select all (Ctrl+A) and copy (Ctrl+C) and paste into a text file (Ctrl+V).

Then we read the pasted dataset using read.table:

```
ans <- read.table("C:/Users/KUs/Desktop/nytimes2.txt",header = F, sep="\t",
na.strings="NA", dec=".", strip.white=TRUE)
library(tm)
b=Corpus(DataframeSource(ans), readerControl = list(language = "eng"))
b<- tm_map(b, stripWhitespace) #Strips white space
b<- tm_map(b, tolower) #Changes case to lowercase
b <- tm_map(b, removeWords, stopwords("english")) #Removes stop words
b <- tm_map(b, removePunctuation) #Removes punctuation
tdm <- TermDocumentMatrix(b)
#or dtm <- DocumentTermMatrix(b), the difference is that terms are in rows or
columns
m1 <- as.matrix(tdm)
v1<- sort(rowSums(m1),decreasing=TRUE)
d1<- data.frame(word = names(v1),freq=v1)
head(d1)
library(wordcloud)
wordcloud(d1$word,d1$freq)
```

review
city brother romantic
april time real ago times
movies caring style 2012 movie video education
york dissident national served
fashion chinese sunday life garden estate
music physicist town dies bits
lizhi china trying technology nyt theater jobs
cars subscriptions
television obama cash search issue shot
china' colleges american arts eat
news fang boys caribbean home oped
words health travel fight world
science day carseminal
masters editorial sports measure
opinion business

But using color palettes we can make the words stand out more.
wordcloud(d1$word,d1$freq,col=terrain.colors(8))

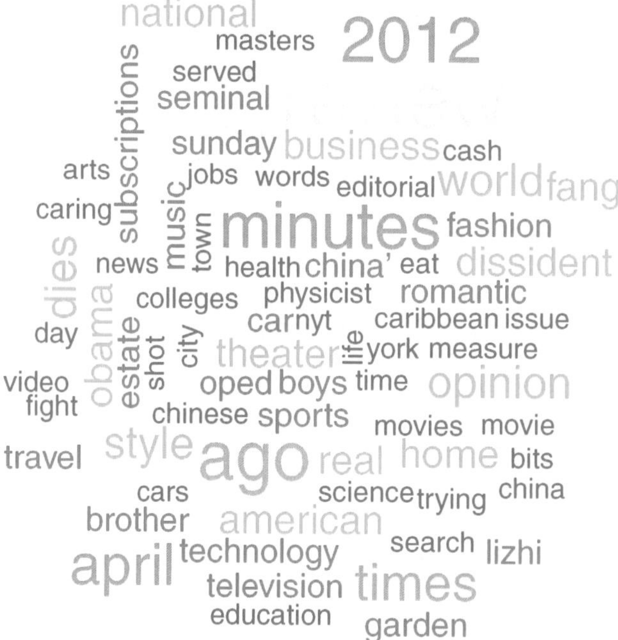

We notice that the RColorBrewer package is attached to the wordcloud package. So we can also try some new color palettes using the examples in Chap. 5.

Note an example of using the R package RJSONIO to do analysis on the *New York Times* API is given at http://www.stanford.edu/~cengel/cgi-bin/anthrospace/scraping-new-york-times-articles-with-r.

We can also explore the readHTMLTable and XMLTreeParse functions in XML library to read tables.

7.5 Google Prediction API

The Google Prediction API helps users use Google's machine learning libraries to build their prediction models using the cloud. The cloud-based machine learning tools can be accessed using the R packages for Google Prediction API. Note both these packages and functions are best used in the Linux environment.

- R package implementing Google Prediction API v1.2-Modification and some extension of "R client library for the Google Prediction API" (Markko Ko, Google Inc., 2010) for Google Prediction API v1.2 http://code.google.com/p/r-google-prediction-api-v12/.
- R client library for Google Prediction API: http://code.google.com/p/google-prediction-api-r-client/.

An older (and easier version) of the Google Prediction API is available at http://onertipaday.blogspot.in/2010/11/r-wrapper-for-google-prediction-api.html. Download the wrapper code from https://github.com/onertipaday/predictionapirwrapper. Source: Paolo Sonego.

Note we need to enable both Google Storage API and Google Prediction API from the API console at https://code.google.com/apis/console/. For Google Storagegotohttps://sandbox.google.com/storage/. (*Note:* if you are using Internet Explorer as a browser, then you can use Google Frame.) Note the name of your bucket and upload your file here. *Here we upload the language_id file from https://developers.google.com/prediction/docs/language_id.txt to a bucket in Google Storage named saysay*. In addition, we use the source method to read in code (since we are using the Windows operating system) instead of installing packages from source code. This method can also be used for packages that are not available on CRAN.

```
library(RCurl)
library(rjson)
download.file(url="http://curl.haxx.se/ca/cacert.pem", destfile="cacert.pem")
curl <- getCurlHandle()
options(RCurlOptions = list(capath = system.file("CurlSSL", "cacert.pem", package
= "RCurl"), ssl.verifypeer = FALSE))
curlSetOpt(.opts = list(proxy = 'proxyserver:port'), curl = curl)
#
source("C:/Users/KUs/Desktop/CANADA/googlepredictionapi/GetAuthToken
.R")
source("C:/Users/KUs/Desktop/CANADA/googlepredictionapi/GooglePredict.R")
source("C:/Users/KUs/Desktop/CANADA/googlepredictionapi/GoogleTrainCheck
.R")
source("C:/Users/KUs/Desktop/CANADA/googlepredictionapi/GoogleTrain.R")
source("C:/Users/KUs/Desktop/CANADA/googlepredictionapi/DeleteTrained
Model.R")
token=GetAuthToken(email="user@gmail.com", passwd="mypassword")
GoogleTrain(auth_token=token$Auth, mybucket="saysay", mydata=" lan-
guage_id.txt")
GoogleTrainCheck(auth_token=token$Auth, mybucket="saysay", mydata=" lan-
guage_id.txt")
GooglePredict(auth_token=token$Auth, mybucket="saysay", mydata= "lan-
guage_id.txt", myinput="J'm appelle Ajay")
> GooglePredict(auth_token=token$Auth, mybucket="saysay", mydata=" lan-
guage_id.txt", myinput="Vous etes")
$data $data$kind
[1] "prediction#output"
$data$outputLabel
[1] "English"
$data$outputMulti
$data$outputMulti[[1]]
```

```
$data$outputMulti[[1]]$label
[1] "English"
$data$outputMulti[[1]]$score
[1] 0.3333333
$data$outputMulti[[2]]
$data$outputMulti[[2]]$label
[1] "French"
$data$outputMulti[[2]]$score
[1] 0.3333333
$data$outputMulti[[3]]
$data$outputMulti[[3]]$label
[1] "Spanish"
$data$outputMulti[[3]]$score
[1] 0.3333333
```

7.6 Data Privacy for Data Miners

Individuals should be know the following information before data are collected: the purpose of the data collection and any data mining projects, how the data will be used, who will be able to mine the data and use them, the security surrounding access to the data, and in addition, how collected data can be updated.

Data miners want the maximum data that are available at a reasonable price, while data privacy advocates want a minimum of data to be collected and used. Proper safeguards in data collection, data storage with proper encryption, and adequate communication regarding the guidelines that are in place for data privacy can help this serious issue from derailing the growth of data mining in a business.

7.7 Summary of Commands Used in This Chapter

7.7.1 Packages

- rattle GUI and its dependencies
- Google prediction API package
- tm
- RCurl
- rjson
- wordcloud

7.7.2 *Functions*

- readLines()
- as.matrix()
- rattle()
- tdm(tm)
- Corpus(tm)
- tm_map(tm)
- wordcloud(wordcloud)

Citations and Sources

- http://www.nascio.org/publications/documents/NASCIO-dataMining.pdf
- Data Mining Reference Card: http://cran.r-project.org/doc/contrib/YanchangZhao-refcard-data-mining.pdf
- Longworth, Norman and Davies, W. Keith. Lifelong Learning. London: Kogan Page, 1996; p. 93
- Tutorial: Data Mining in R: http://www2.isye.gatech.edu/~shan/ISyE7406/Introduction_to_Data_Mining.pdf
- http://cran.cnr.berkeley.edu/web/views/MachineLearning.html
- Williams, G. J. (2011) Data Mining with Rattle and R: The Art of Excavating Data for Knowledge Discovery, Use R! Springer, Berlin Heidelberg New York.
- Ingo Feinerer (2012) tm: Text Mining Package. R package version 0.5-7.1. Ingo Feinerer, Kurt Hornik, and David Meyer (2008). Text Mining Infrastructure in R. Journal of Statistical Software 25/5. URL: http://www.jstatsoft.org/v25/i05/
- Duncan Temple Lang (2012) RCurl: General network (HTTP/FTP/...) client interface for R. R package version 1.91-1.1. http://CRAN.R-project.org/\penalty-\@Mpackage=RCurl
- Alex Couture-Beil (2012) rjson: JSON for R. R package version 0.2.8. http://CRAN.R-project.org/package=rjson
- Ian Fellows (2012) wordcloud: Word Clouds. R package version 2.0. http://CRAN.R-project.org/package=wordcloud
- Michael Hahsler, Bettina Gruen and Kurt Hornik (2011) arules: Mining Association Rules and Frequent Itemsets. R package version 1.0-7. Michael Hahsler, Bettina Gruen and Kurt Hornik (2005) arules—A Computational Environment for Mining Association Rules and Frequent Item Sets. Journal of Statistical Software 14/15. http://www.jstatsoft.org/v14/i15/

Chapter 8
Clustering and Data Segmentation

8.1 When to Use Data Segmentation and Clustering

Cluster analysis is basically a data reduction technique to reduce a large number of objects in groups or clusters in such a manner that objects belonging to one group or cluster are more similar to each other and more different from objects in another group or cluster. Clustering is used in business analytics to identify groups of customers that can be targeted with similar products, to understand products and markets, and basically to reduce data for an actionable strategy especially in cases where data are not sufficiently clean or exhaustive to create predictive models.

8.2 R Support for Clustering

R has extensive packages and literature in clustering. There are almost 67 packages in R for clustering, and packages like **mclust** have an annual license fee and require completion of a license agreement except for academic use. The huge number of packages can often be confusing to the business analytics user who is more concerned with business problems and may not be technologically savvy enough to understand the difference between all the algorithms. (There are almost 30 algorithms available at Wikipedia for those new to them at http://en.wikipedia.org/wiki/Category:Data_clustering_algorithms.)

8.2.1 Clustering View

A dedicated view to Clustering is available in R. It distinguishes and classifies the extensive number of clustering packages available in R to five different sections:

- Hierarchical clustering
- Partition clustering

A. Ohri, *R for Business Analytics*, DOI 10.1007/978-1-4614-4343-8_8,

- Model-based clustering
  - ML estimation
  - Bayesian estimation
- Other clustering algorithms
- Clusterwise regression

You can read more at http://cran.r-project.org/web/views/Cluster.html.

8.2.2 GUI-Based Method for Clustering

The R GUI for data mining is Rattle. The latest version of Rattle supports five clustering methods.

- K-means: aims to partition points into k groups such that the sum of squares from points to the assigned cluster centers is minimized. At the minimum, all cluster centers are at the mean of their Voronoi sets (the set of data points that are nearest to the cluster center).
- Clara: compared to other partitioning methods, it can deal with much larger datasets. Internally, this is achieved by considering subdatasets of fixed size (sampsize) such that the time and storage requirements become linear in n rather than quadratic.
- Ewkm: an entropy weighted subspace k-means clustering method.
- Hierarchical: uses a set of dissimilarities for the n objects being clustered. Initially, each object is assigned to its own cluster and then the algorithm proceeds iteratively, at each stage joining the two most similar clusters, continuing until there is just a single cluster.
- Bicluster: consider a two-way dataset. The goal of biclustering is to find subgroups of rows and columns that are as similar as possible to each other and as different as possible from others.

A good tutorial on using Rattle is given at http://eric.univ-lyon2.fr/~ricco/tanagra/fichiers/en_Tanagra_Rattle_Package_for_R.pdf. As a business analyst dealing with data mining, you should especially read the article in the R Journal on Rattle at http://journal.r-project.org/archive/2009-2/RJournal_2009-2_Williams.pdf.

8.3 Using RevoScaleR for Revolution Analytics

RevoScaleR from Revolution Analytics is a package that supports the analysis of big datasets using the xdf file format. You can use the snippets and auto-suggest features present in Revolution Analytics software to easily create clusters. RevolScaleR supports k-means clustering for bigger datasets. It claims to find two clusters (out of seven variables) on a 123 million plus row airline dataset in just under 6 min. For a detailed example of k-means clustering in Revolution Analytics, refer to the case study at http://blog.revolutionanalytics.com/2011/06/kmeans-big-data.html.

The k-means function, rxKmeans, is implemented as an external memory algorithm that works on a chunk of data at a time. Since the k-means algorithm is "embarrassingly parallel," rxKmeans reads chunks of data (rows/observations) at a time and iterates the Lloyd algorithm on each chunk (in parallel if multiple processors are available). Once all of the chunks have been processed, the means are updated one last time to produce the final result.

As you can see, using the Insert Snippet feature means that doing your own cluster analysis in Revolution Analytics on big data is just three clicks away.

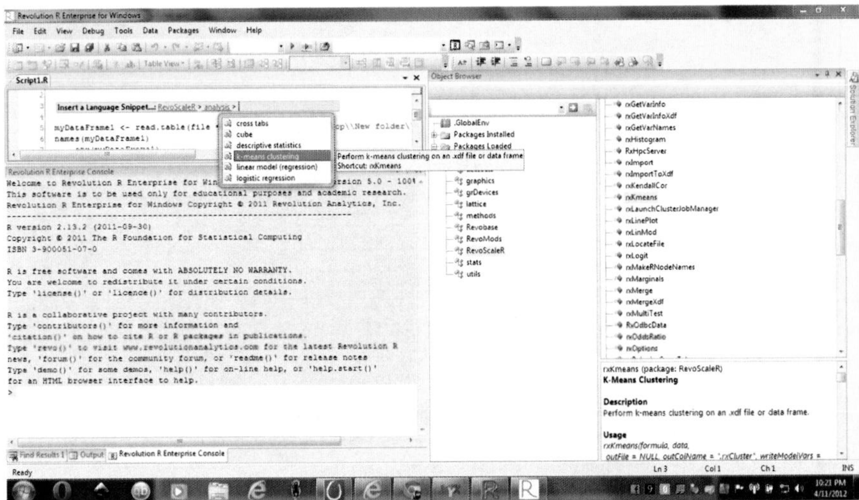

8.4 A GUI Called Playwith

Playwith is another GUI that can be used for interactive data visualization. It has a clustering application, too, that can be used for simple clusters.

Let us test a clustering app called clusterApp in playwith. This gives a nice interactive way to play with clustering.

clusterApp: A mini application (a playwith toolbar) for clustering. The clustering and distance methods can be chosen. The "Cut Tree" button allows clusters to be defined. Buttons open linked plots (marginal distributions, parallel plot, or MDS plot) showing clusters.

http://code.google.com/p/playwith/wiki/ClusterApp

```
library(playwith)
demo(clusterApp)
clusterApp(iris)
```

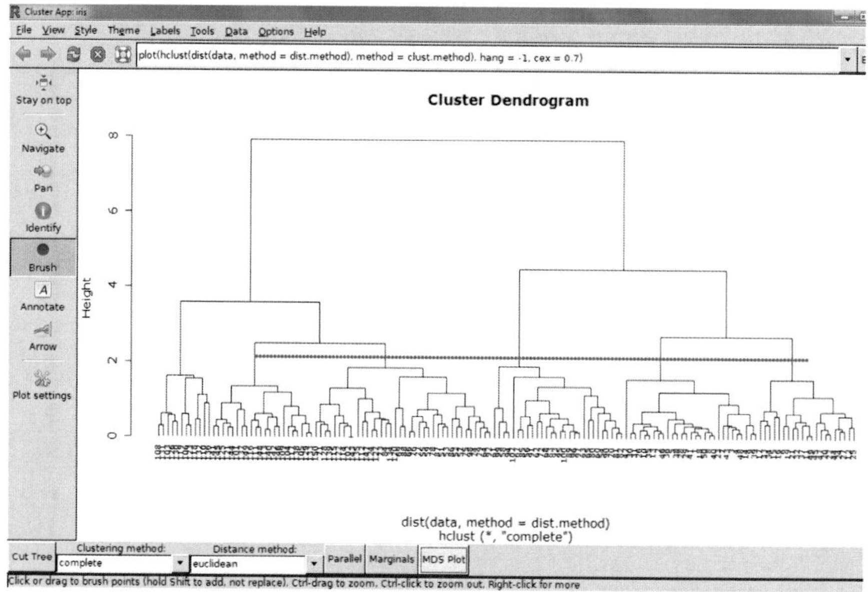

and the output can be saved as a PDF

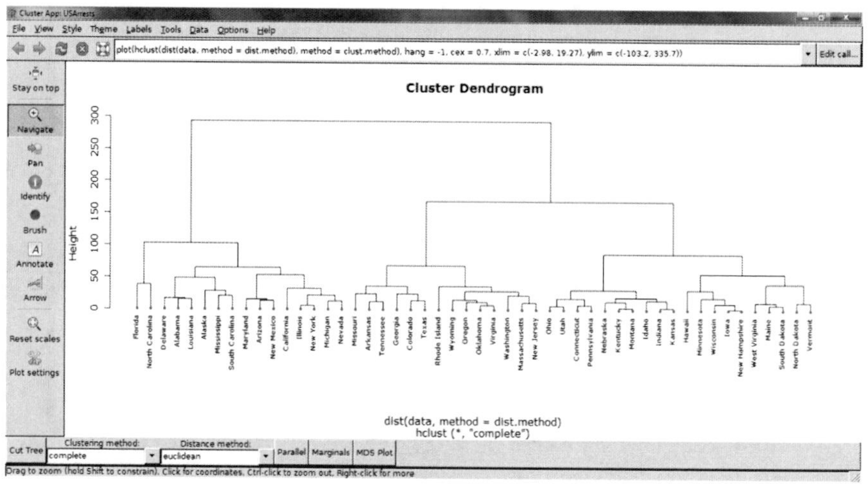

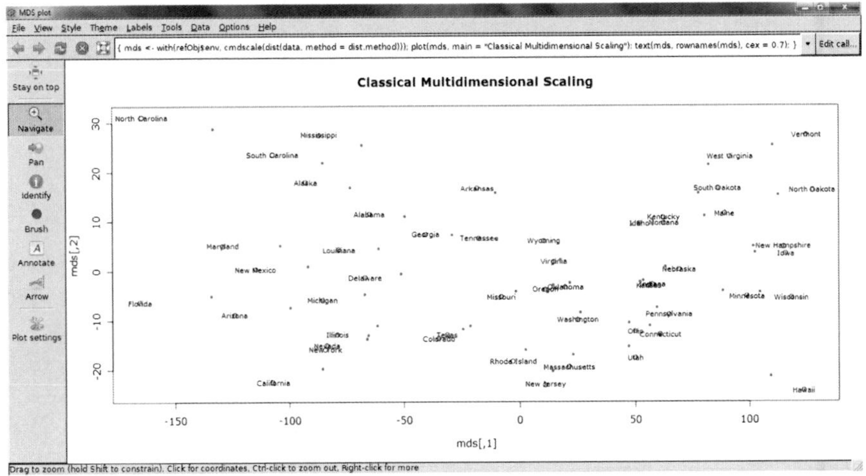

8.5 Cluster Analysis Using Rattle

Let us do a test case of a cluster analysis using Rattle. Before we go to the software, let us evaluate the process of clustering.

From http://faculty.ucr.edu/~hanneman/soc203b/examples/cluster.htm

The numerous variations of cluster analysis differ primarily in:

1. *Whether the observations are to be standardized or not.*
2. *Whether the analysis is "agglomerative" or "divisive" (does one begin with each case as its own cluster and join them or in all cases a single cluster and split cases off?).*
3. *The measure of similarity or distance used to measure the association between cases (or between variables).*
4. *Whether cases are allowed to be members of more than one class or not (hierarchical or overlapping clusters).*
5. *How we decide which cases are to be joined (or divided) in the next step.*

Because of all of the choices, the family of "cluster analyses" is a large one. And it usually does matter what choices one makes. It is usually wise to

1. *Choose a method that can be justified on theoretical grounds for the problem at hand and*
2. *See if the results are robust against modest changes in technique.*

The basic process of cluster analysis can be simply listed here:

- Formulate the business and statistical problem—why do you need clustering?

- Select a clustering procedure based on data availability
- Decide on the number of clusters (this can be iterative)
- Map and interpret clusters
- Check stability of results by redoing analysis

A succinct explanation between the various kinds of clustering methods is also given at

http://stat.ethz.ch/R-manual/R-patched/library/cluster/html/agnes.html

Cluster analysis divides a dataset into groups (clusters) of observations that are similar to each other.

Hierarchical methods like agnes, diana, and mona construct a hierarchy of clusterings, with the number of clusters ranging from one to the number of observations.

Partitioning methods like pam, clara, and fanny require that the number of clusters be given by the user.

One must also choose how to evaluate how many clusters to finally make. The ways to do this are with dendrograms and perceptual maps.

Here is a brief explanation of doing a cluster analysis using the Rattle GUI in R. Note we will not be coding but simply using the GUI, assuming that the user knows basics of R and clustering basics.

1. We use the Davis dataset within the car package to demonstrate and input the data using the Data tab.

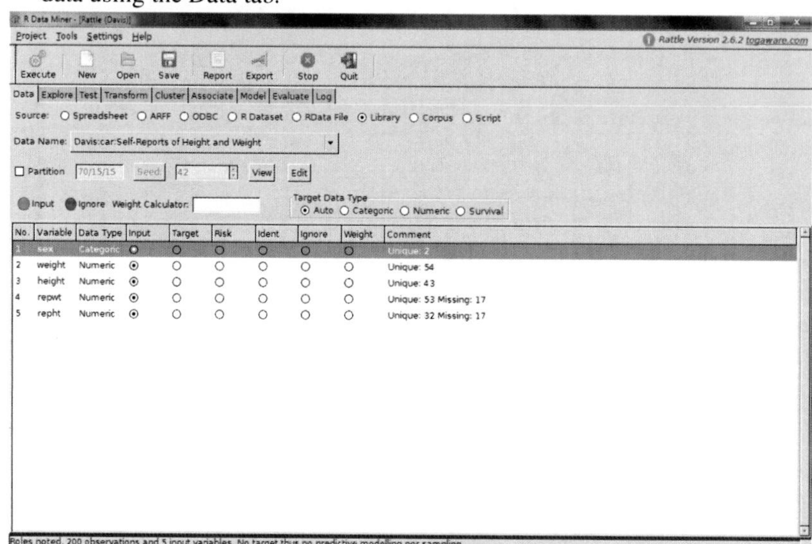

2. We use the Cluster tab and build the cluster using the k-means algorithm as shown. Note we have not shown the ewkm tab in the cluster as we are using version 2.6.2 of Rattle. Note the cluster statistics generated including cluster sizes and cluster centers.

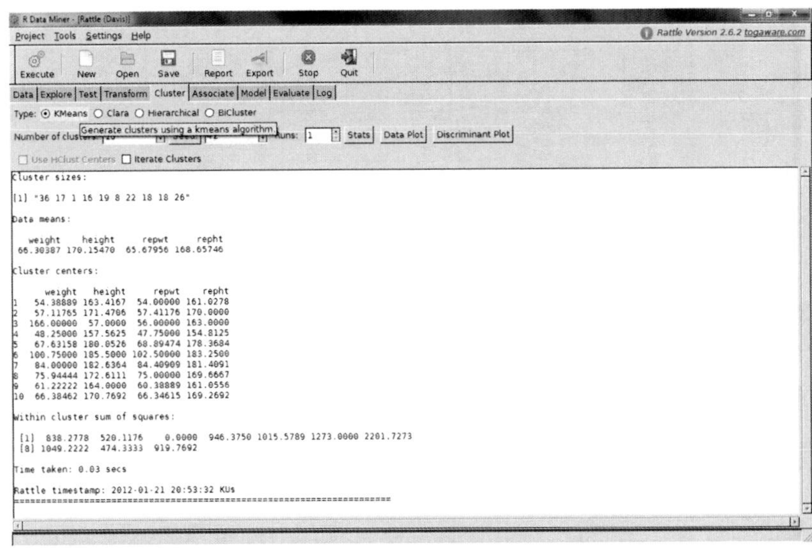

3. Using the Data Plot feature we see the graphical plot of the dataset:

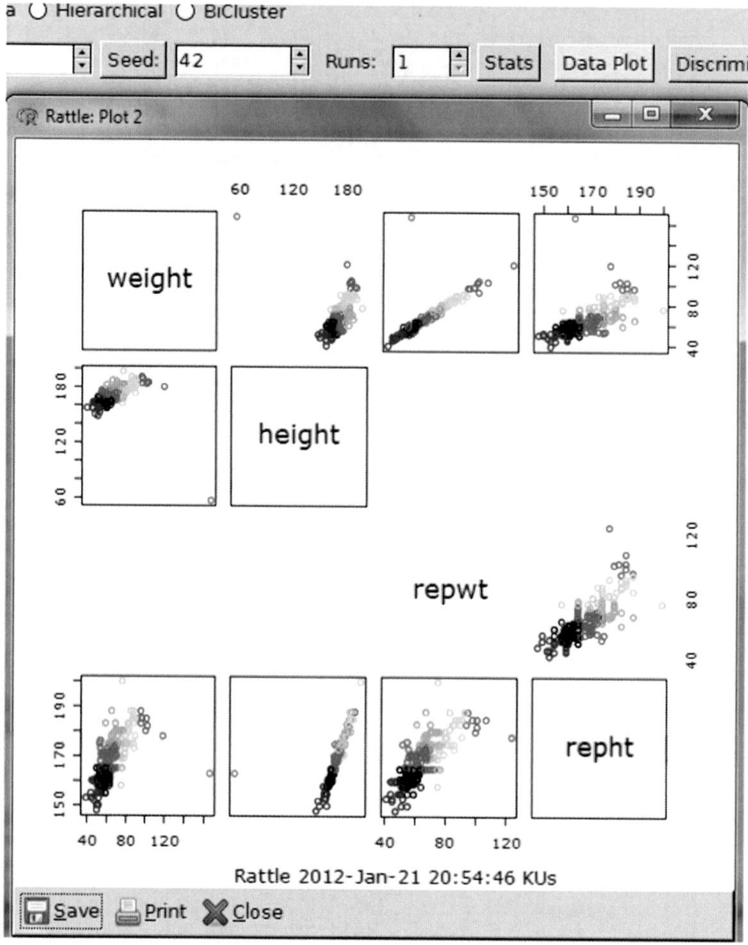

4. Note we can use the Iterate feature within a cluster in Rattle so as to iterate and find the number of clusters:

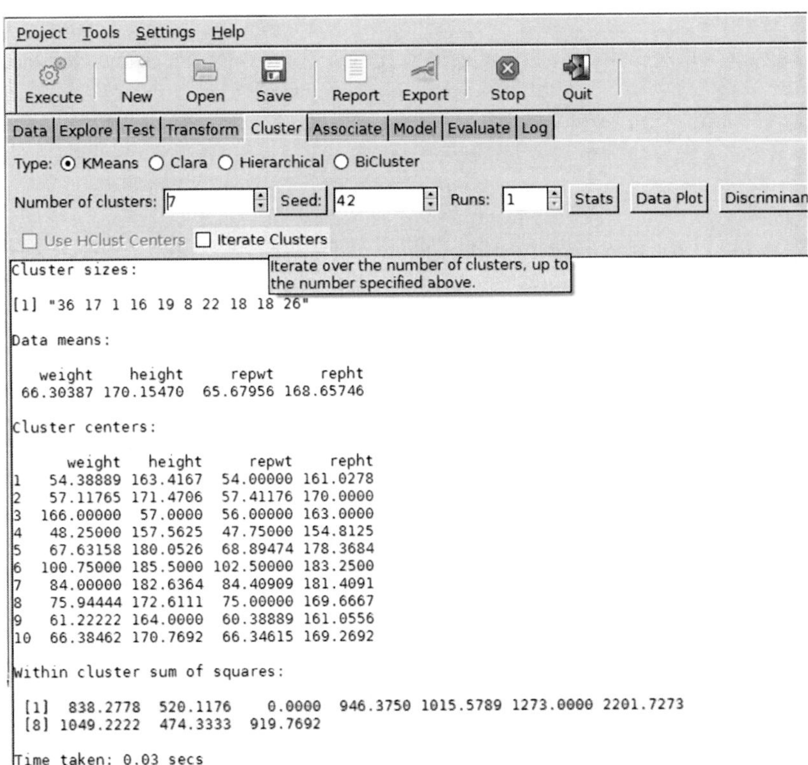

5. The Iterate feature shows the appropriate number of clusters to be decided based on the difference in the sum of squares. Note the big difference in the sum of squares between four and five clusters:

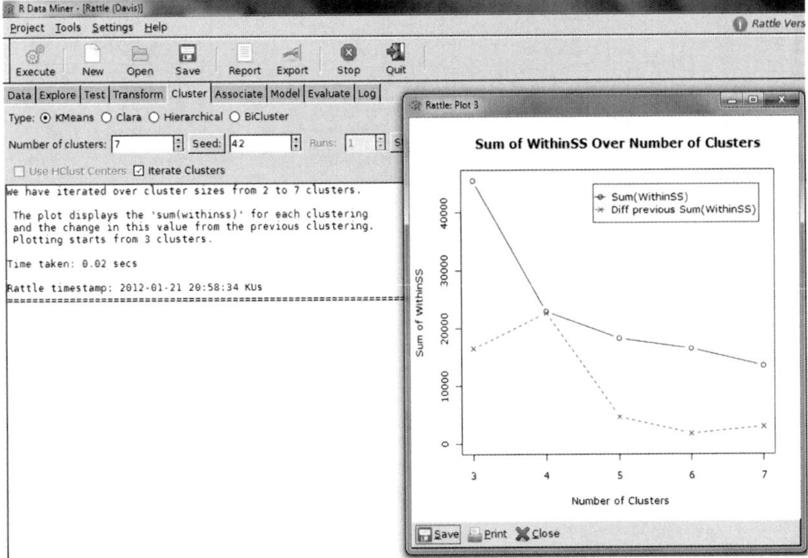

6. We move to hierarchical cluster algorithms. Rattle supports seven types of agglomerative methods; a detailed description of the differences between these methods is available at http://stat.ethz.ch/R-manual/R-devel/library/stats/html/hclust.html: *a hierarchical cluster analysis using a set of dissimilarities for the n objects being clustered. Initially, each object is assigned to its own cluster and then the algorithm proceeds iteratively, at each stage joining the two most similar clusters, continuing until there is just a single cluster. At each stage distances between clusters are recomputed by the Lance–Williams dissimilarity update formula according to the particular clustering method being used. Ward's minimum variance method aims at finding compact, spherical clusters. The complete linkage method finds similar clusters. The single-linkage method (which is closely related to the minimal spanning tree) adopts a 'friends of friends' clustering strategy. The other methods can be regarded as aiming at clusters with characteristics somewhere between the single and complete link methods. Note, however, that the "median" and "centroid" methods do not lead to a monotone distance measure, or, equivalently, the resulting dendrograms can have so-called inversions (which are hard to interpret).*

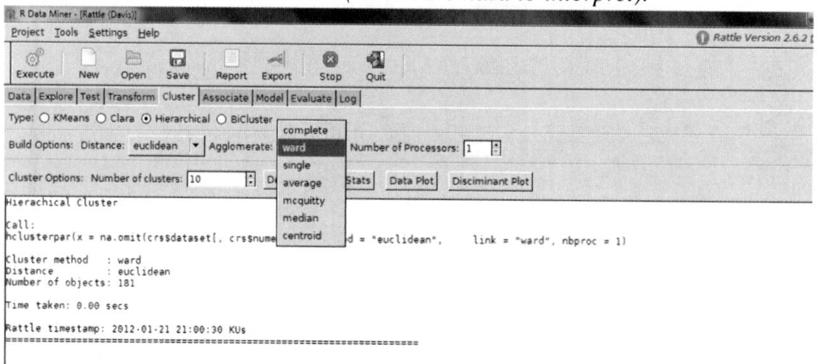

7. There are eight kinds of distance metrics for determining the distance between cluster centroids. An example is given here: *in a 2D space, the distance between the point (1,0) and the origin (0,0) is always 1 according to the usual norms, but the distance between the point (1,1) and the origin (0,0) can be 2, Sq.root(2), or 1 under Manhattan distance, Euclidean distance, or maximum distance respectively.*

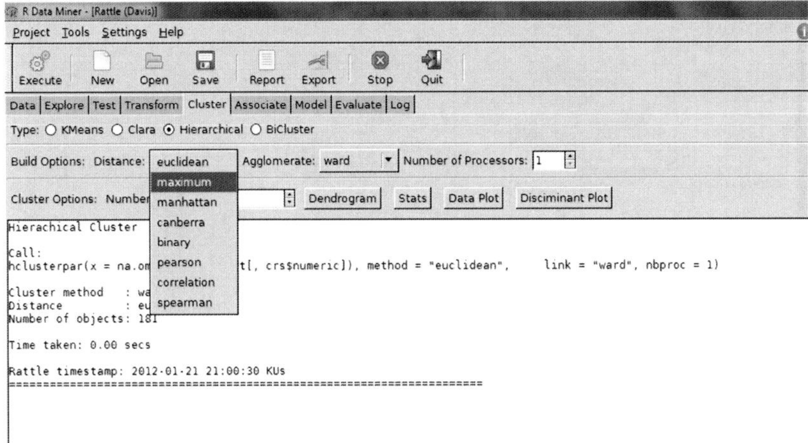

8. A dendrogram is produced at the end of this cluster analysis and it can be shown by clicking the Dendrogram tab:

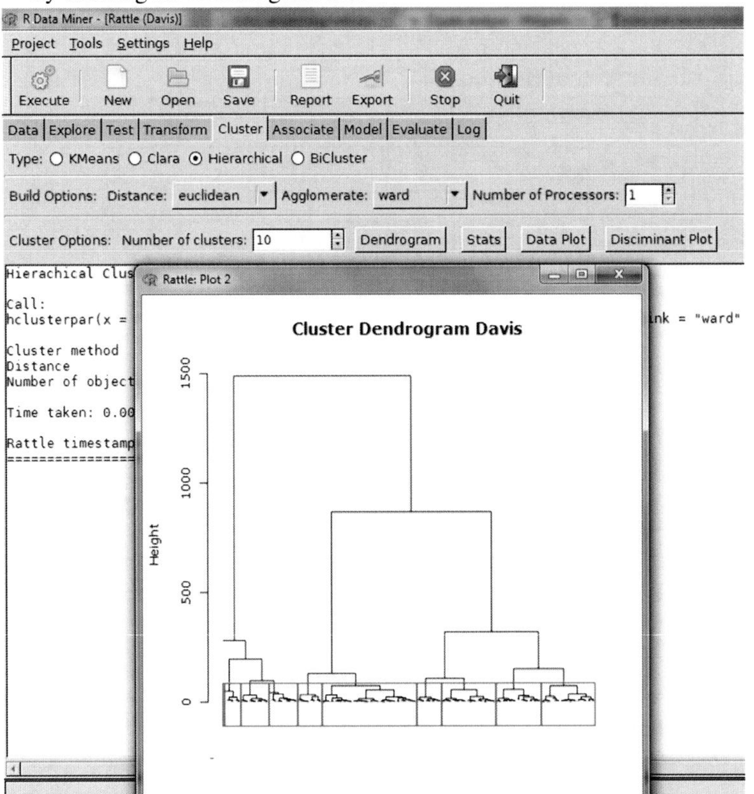

9. The Stats tab shows the statistics associated with hierarchical clusters:

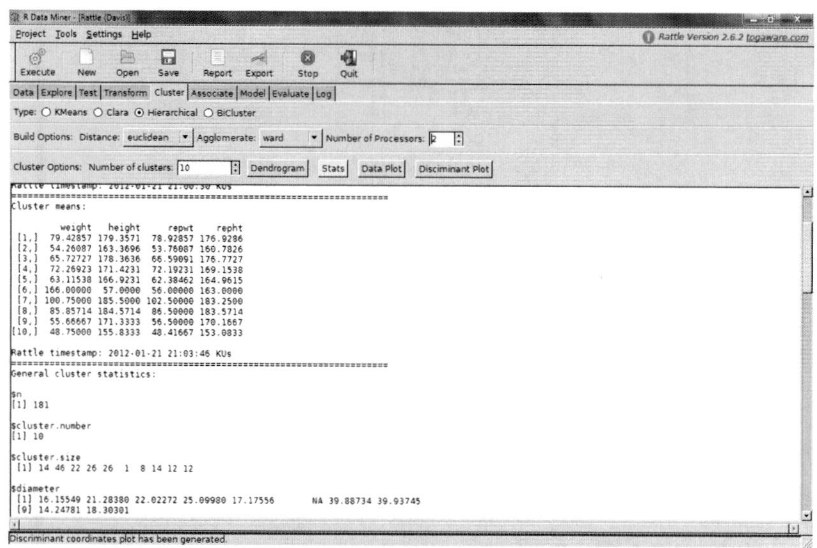

10. We move to the bicluster method. An explanation of this method was given previously in the section *GUI-based Method of Clustering*

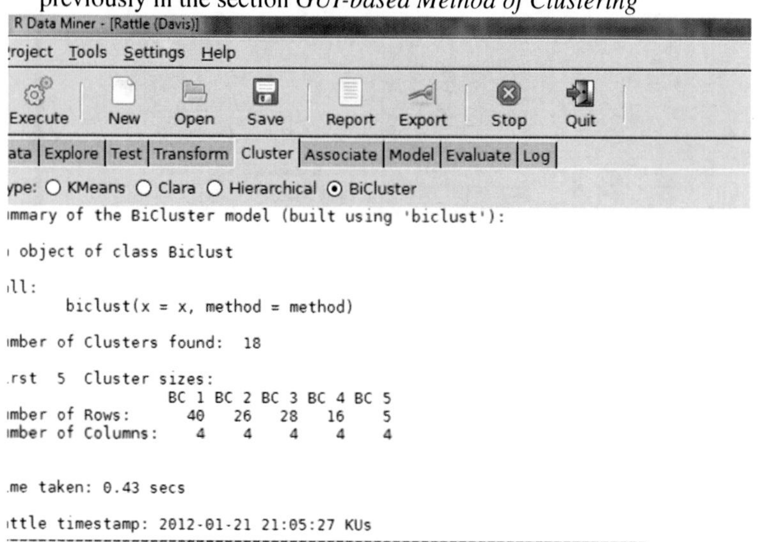

11. Version 2.6.7 of Rattle GUI introduced the Ewkm method of clustering (note I am using Ubuntu Linux as the operating system on a virtual VMWare partition as sometimes GTK2 can create installation issues for Rattle in Windows):

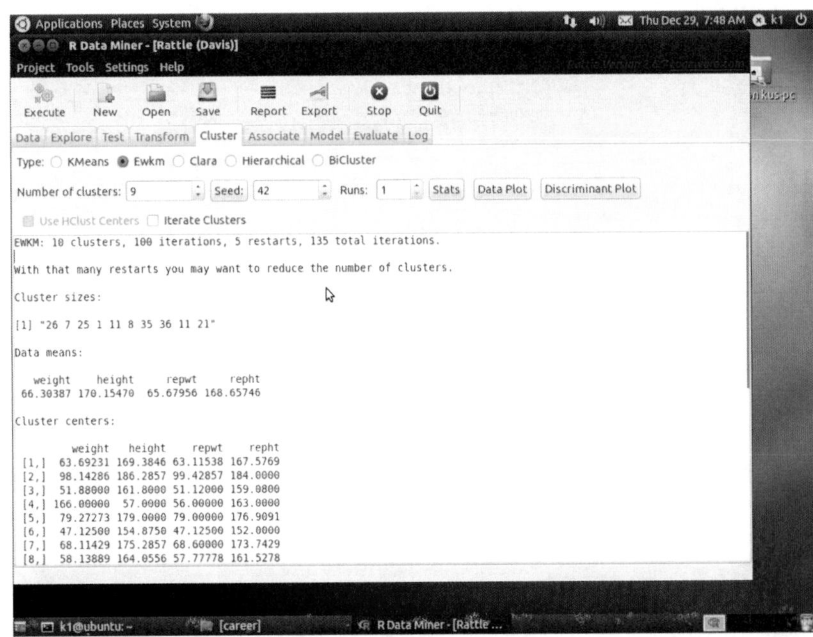

12. A nice feature in Rattle is that it allows you to easily use multiple methods of clustering analysis, like creating hierarchial clusters and using them as inputs for k-means clustering:

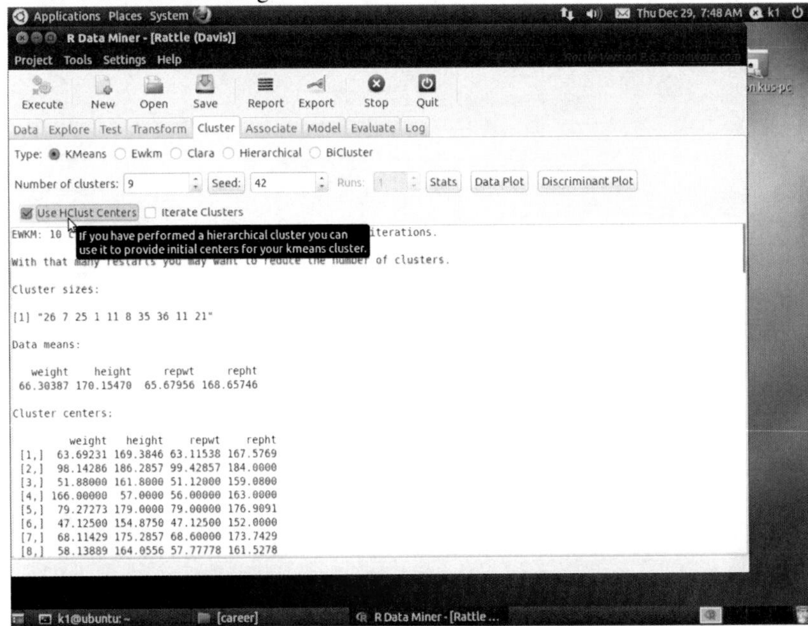

13. We again use the iterate cluster method to fine-tune the number of clusters:

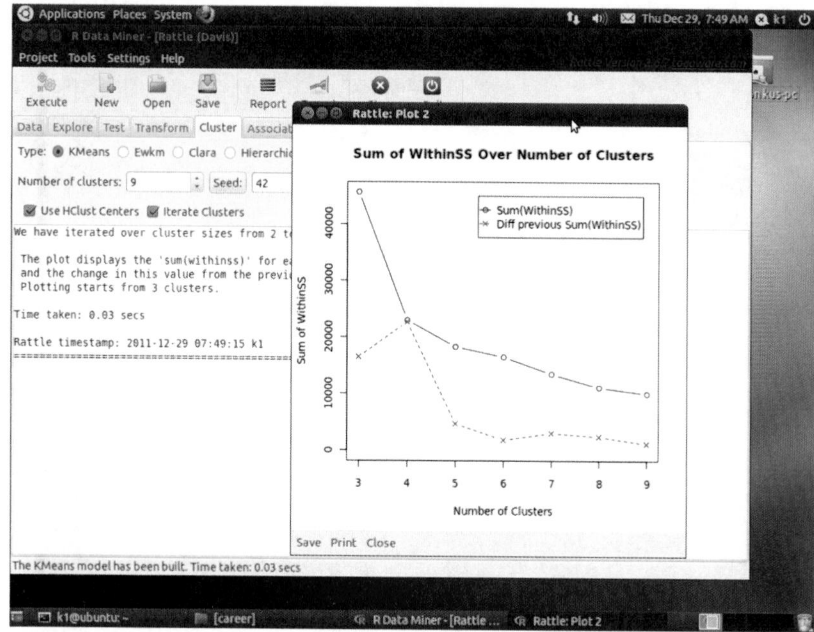

14. Note the number of processors can be changed using Rattle to increase the speed of processing:

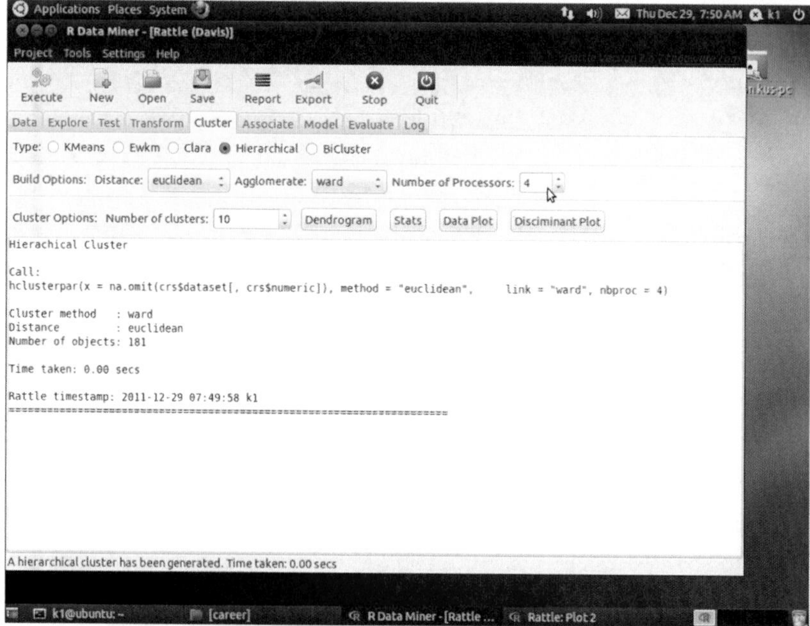

8.6 Summary of Commands Used in This Chapter

8.6.1 Packages

- Rattle
- playwith
- RevoScaleR

8.6.2 Functions

- ClusterApp(playwith)
- rattle()
- rxKmeans(RevoScaleR)

Citations and Sources

- Graham Williams (2011) Data Mining with R and Rattle: The Art of Excavating Data for Knowledge Discovery, Springer
- Kaiser, Sebastian and Leisch, Friedrich (16. April 2008): A Toolbox for Bicluster Analysis in R. Department of Statistics, University of Munich: Technical Reports. http://epub.ub.uni-muenchen.de/3293/1/S_Kaiser_biclust.pdf
- Revolution Analytics (2011) RevoScaleR: Scalable, distributable, fast, and extensible data analysis in R. R package version 2.0-0.
- Partitioning (clustering) of data into k clusters "around medoids", a more robust version of k-means: http://stat.ethz.ch/R-manual/R-patched/library/cluster/html/pam.html
- Felix Andrews (2010). playwith: A GUI for interactive plots using GTK+. R package version 0.9-53. http://CRAN.R-project.org/package=playwith
- K-Means Documentation: http://stat.ethz.ch/R-manual/R-patched/library/stats/html/kmeans.html
- Hierarchical Clustering Documentation: http://stat.ethz.ch/R-manual/R-devel/library/stats/html/hclust.html
- Clara: Clustering Large Applications: http://stat.ethz.ch/R-manual/R-patched/library/cluster/html/clara.html
- CRAN View on Clustering: http://cran.r-project.org/web/views/Cluster.html
- Rattle Tutorial: http://eric.univ-lyon2.fr/~ricco/tanagra/fichiers/en_Tanagra_Rattle_Package_for_R.pdf
- Graham Williams, Togaware, Journal of Statistical Software, Rattle Article: http://journal.r-project.org/archive/2009-2/RJournal_2009-2_Williams.pdf

- Graham Williams (2011) Data Mining with R and Rattle: The Art of Excavating Data for Knowledge Discovery, Springer
- David Smith, Revolution Analytics Case Study on Using rxKmeans: http://blog.revolutionanalytics.com/2011/06/kmeans-big-data.html
- Robert A. Hanneman, Department of Sociology, University of California, Riverside: http://faculty.ucr.edu/~hanneman/soc203b/examples/cluster.htm
- Wikipedia: http://en.wikipedia.org/wiki/Category:Data_clustering_algorithms and http://en.wikipedia.org/wiki/Cluster_analysis
- Wikipedia: http://en.wikipedia.org/wiki/Hierarchical_clustering
- R Online Documentation HClust: http://stat.ethz.ch/R-manual/R-devel/library/stats/html/hclust.html
- James L. Schmidhammer, University of Tennessee: http://bus.utk.edu/stat/stat579/Hierarchical%20Clustering%20Methods.pdf
- Cosma Shalizi, Carnegie Mellon University: http://www.stat.cmu.edu/~cshalizi/350/lectures/08/lecture-08.pdf

Chapter 9
Forecasting and Time Series Models

9.1 Introduction to Time Series

Time series are series in which some quantity or variable varies with respect to time intervals (in the form of months, weeks, days, hours, etc.). This basically implies that the future value of a particular variable is in some way related to its present value as well as to the time interval difference.

Naive forecasts assume that the future value is a function of the current value. If the time series is stable, then the naive forecast would be the same as the current value. Another version of a naive forecast would be to assume that the future value is a percentage increase or decrease from the current value.

Seasonal naive forecasts assume that the future value is a function of the previous value in the same seasonal period. Drift naive forecasts assume that the future value is a function of the present value plus the average change per time period. You can use the rwf() function to create naive forecasts with drift.

For the business analytics user, we recommend automatic forecasting methods, principally the ets and auto.arima functions for a forecast package and the R Commander E Pack plugin for a GUI solution.

9.2 Time Series and Forecasting Methodology

The first step to a time series analysis is to plot the time series and observe it. See the tutorial by Shaddick (2004) at http://people.bath.ac.uk/masgs/time%20series/TimeSeriesR2004.pdf.

View the data using the plot command (or the plot.ts command).

In each case, answer the questions:

(a) Does there appear to be any trend in the data (that is, a tendency for the values to rise or fall)?
(b) Does there appear to be any periodic behavior of the data?

A. Ohri, *R for Business Analytics*, DOI 10.1007/978-1-4614-4343-8_9,

(c) Does there appear to be any change of regime in the data?
(d) Do there appear to be any anomalous values in the data?

http://www.statistik.uni-dortmund.de/useR-2008/slides/Barbosa.pdf

1. Create time series periods appropriately using the R package lubricate or function strptime.
2. Plot the time series (plot.ts) and freq option
   (a) Use tsdisplay () in the forecast package to plot data and autocovariance and partial autocorrelations.
   (b) Use seasonplot() in the forecast package to plot seasonal variations.
   (c) Time series decomposition: A time series is assumed to have three to four components—trend, seasonality, cyclicality, and irregularity (remainder). The stl() function in the forecast package provides better visualization than the decompose() function.
3. Check for stationarity.
   (a) A time series is stationary if the mean of the series over some reasonable range does not change when different endpoints for that range are chosen. So if we have a time series with 100 periods and we sample periods 1–20, 30–50, and 70–100, the sample means should all be roughly the same.
   (b) Optional: difference and log transform the series till you see a stationary series by plotting the time series.
   (c) See acf and pacf. If the acf plot tapers down and decreases, then the series is stationary. Also, peaks of acf give periodicity.
4. Use ets, auto arima, or bulk fit to generate initial models.
5. Use the accuracy function for comparing models.
6. Use forecast and predict to build the model.
7. Plot forecasts.

In R, time series or "ts" is the basic class for regularly spaced time series.

Let us see an example.

```
library(forecast)
data(AirPassengers)
str(AirPassengers)
tsdisplay(AirPassengers)
```

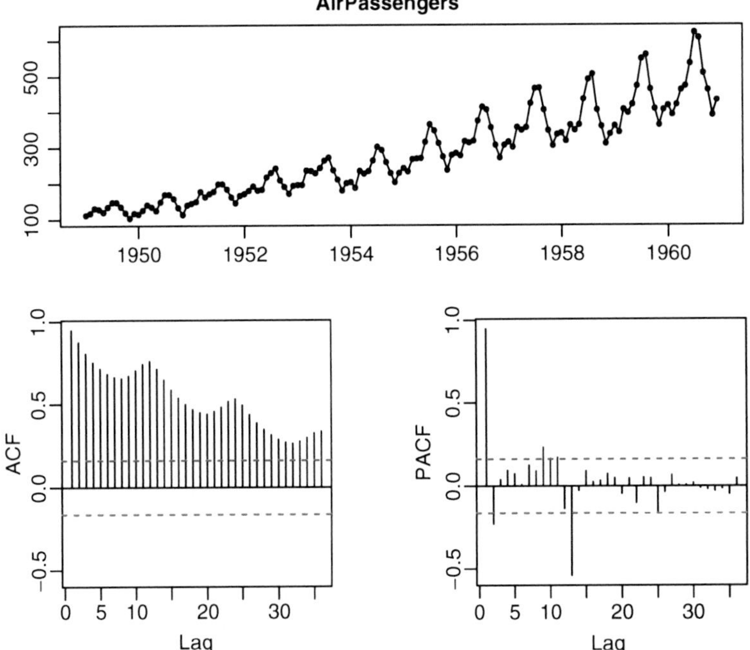

seasonplot(AirPassengers)

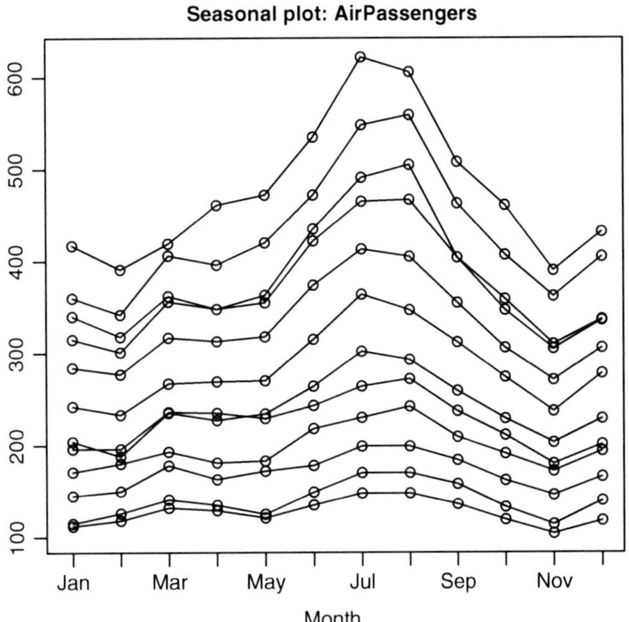

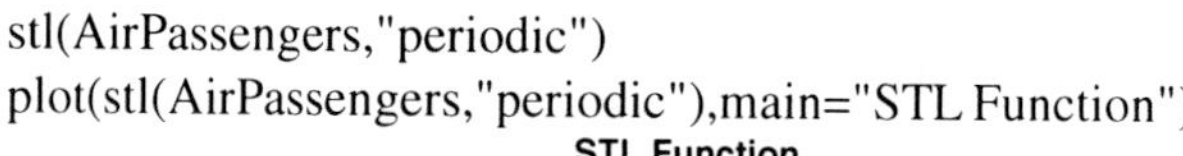

```
stl(AirPassengers,"periodic")
plot(stl(AirPassengers,"periodic"),main="STL Function")
```

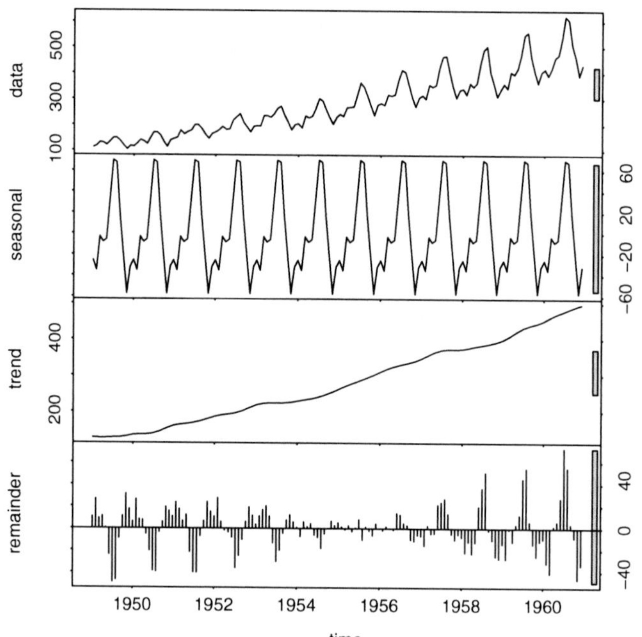


```
model1 = ets(AirPassengers)
plot(model1)
```

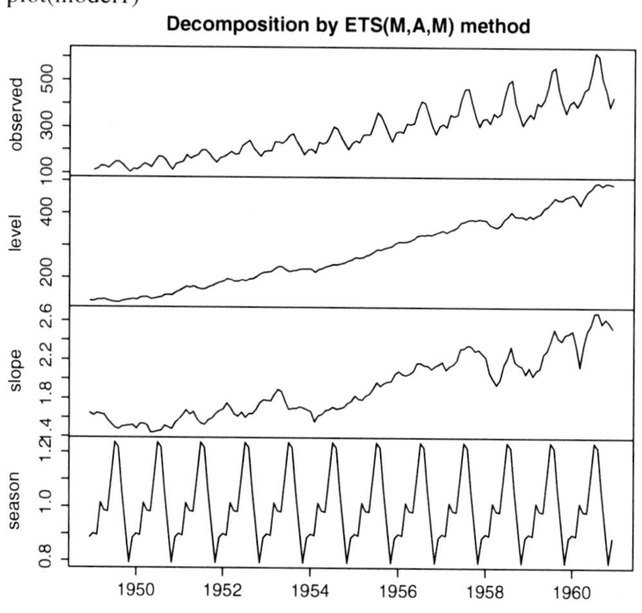

```
model2 = auto.arima(AirPassengers)
par(mfrow = c(2,1))
forecast(model1,10)
plot(forecast(model1,10))
forecast(model2,10)
plot(forecast(model2,10))
```

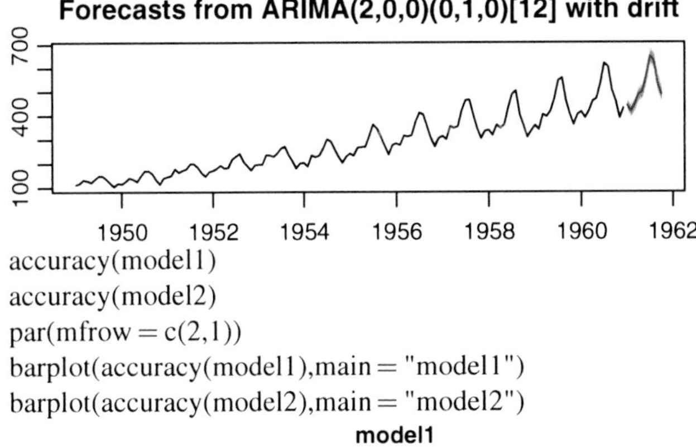


```
accuracy(model1)
accuracy(model2)
par(mfrow = c(2,1))
barplot(accuracy(model1),main = "model1")
barplot(accuracy(model2),main = "model2")
```

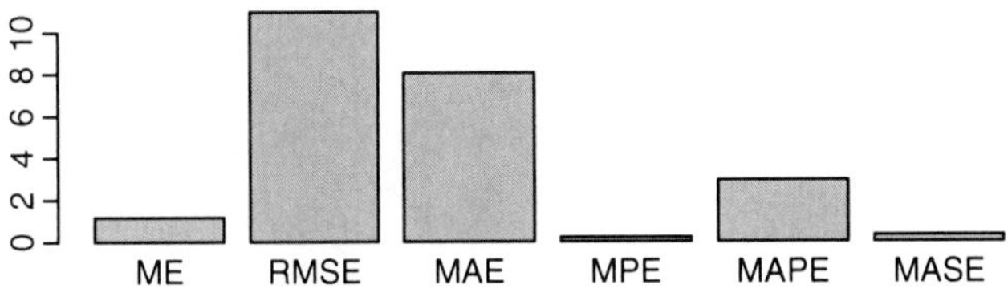

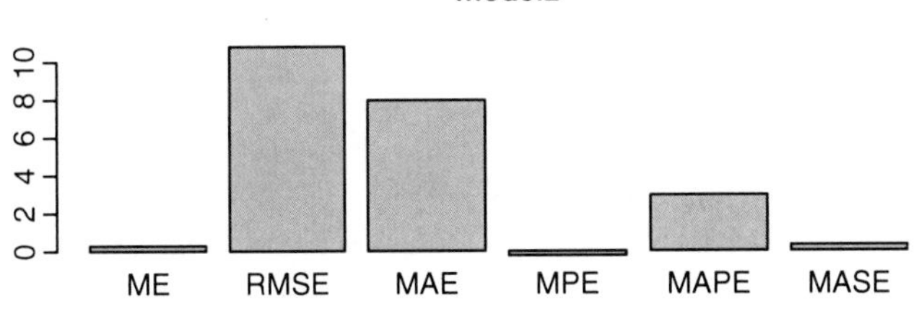

9.3 Time Series Model Types

The following are the principal model types for time series forecasting. A discussion on theoretical differences is beyond the scope of this book.

- ARCH: Autoregressive Conditional Heteroscedasticity
- GARCH: Generalized Autoregressive Conditional Heteroscedastic GARCH (p, q) time series model; found in tseries package
- Spectrum: for spectral density use the spectrum() function
- Exponential: used in ets method of forecast package
  - Winters
  - Holt
- ARIMA: Auto Regressive Integrated Moving Average
- ARIMAX: ARIMA with a regressor variable (X)
- **Time Series Models Error Metrics**

Error = Forecast value −Actual Value

If forecast value > Actual value, then error is positive (or higher forecast)
If forecast value < Actual value, then error is negative (or lower forecast)
Square Error = Error^2
Absolute Value of Error = Absolute Value (Error)
Mean Absolute Error, MAE = Average of All Absolute Value of Error

- Average error
  - MAE: Mean Absolute Error
  - MSE: Mean Square Error
  - RMSE: Root Mean Square Error
- MAPE: Mean Absolute Percentage Error
- MASE: Mean Absolute

Information Value Metrics

- AIC

R Time Series View

A very comprehensive documentation of time series within R is given using the view: http://cran.cnr.berkeley.edu/web/views/TimeSeries.html. To quote a summary line:

Methods for analyzing and modeling time series include ARIMA models in arima(), AR(p) and VAR(p) models in ar(), structural models in StructTS(), visualization via plot(), (partial) autocorrelation functions in acf() and pacf(), classical decomposition in decompose(), STL decomposition in stl(), moving average and autoregressive linear filters in filter(), and basic Holt–Winters forecasting in HoltWinters().

Exponential smoothing: HoltWinters() in stats provides some basic models with partial optimization, ets() from the forecast package provides a larger set of models and facilities with full optimization.

ARIMA models: arima() in stats is the basic function for ARIMA, SARIMA, ARIMAX, and subset ARIMA models. It is enhanced in the forecast package along with auto.arima() for automatic order selection.

9.4 Handling Date-Time Data

Use the strptime(dataset, format) function to convert character variables into strings.

For example, if the variable dob is ("*01/04/1977*") then the following will convert into a date object:

z=strptime(dob, "%d/%m/%Y")

and if the same date is ("*01Apr1977*"), then

z=strptime(dob,"%d%b%Y")

does the same.

For troubleshooting help with date and time, remember to enclose the formats *%d,%b,%m*, and *%Y* in the same exact order as the original string, and if there are any delimiters like "–" or "/", then these delimiters are entered in exactly the same order in the format statement of the *strptime.*

Sys.time() gives you the current date-time, while the function *difftime(time1, time2)* gives you the time intervals (if you have two columns as date-time variables). The various formats for inputs in date-time are given below. These will help in the conversion of string variables into date-time variables.

- %a Abbreviated weekday name in current locale (also matches full name on input)
- %A Full weekday name in current locale (also matches abbreviated name on input)
- %b Abbreviated month name in current locale (also matches full name on input)
- %B Full month name in current locale (also matches abbreviated name on input)
- %c Date and time; locale-specific on output, "%a %b %e %H:%M:%S %Y" on input
- %d Day of month as decimal number (01–31)
- %H Hour as decimal number (00–23)
- %I Hour as decimal number (01–12)
- %j Day of year as decimal number (001–366)
- %m Month as decimal number (01–12)
- %M Minute as decimal number (00–59)
- %p AM/PM indicator in locale; used in conjunction with %I and not with %H; an empty string in some locales
- %S Second as decimal number (00–61), allowing for up to two leap-seconds (but POSIX-compliant implementations will ignore leap seconds)
- %U Week of year as decimal number (00–53) using Sunday as day 1 of the week (and typically with the first Sunday of the year as day 1 of week 1)

%w	Weekday as decimal number (0–6, Sunday is 0)
%W	Week of year as decimal number (00–53) using Monday as day 1 of week (and typically with the first Monday of the year as day 1 of week 1)
%x	Date; locale-specific on output, "%y/%m/%d" on input
%X	Time; locale-specific on output, "%H:%M:%S" on input
%y	Year without century (00–99); values 00–68 are prefixed by 20 and 69–99 by 19—, that is, the behavior specified by the 2004 POSIX standard, but it does also say "it is expected that in a future version the default century inferred from a two-digit year will change"
%Y	Year with century
%z	Signed offset in hours and minutes from UTC, so −0800 is 8 h behind UTC
%Z	(Output only) time zone as character string (empty if not available)

We can also use the "*lubridate*" package to convert date-time data easily.

9.5 Using R Commander GUI with epack Plugin

We present here a method to do time series analysis using the GUI R Commander with the epack plugin.

On loading epack, we can see two additional menus have been added to R Commander. One of them is TS-Data and the other is TS-Models. TS-Data is used for converting the input data into the appropriate form.

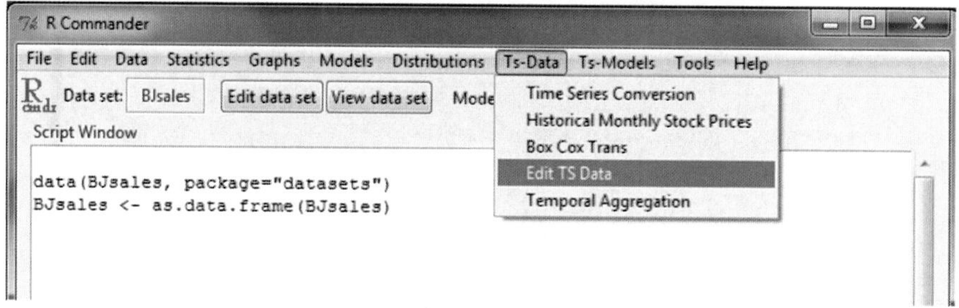

The TS-Models menu is used for creating and running ARIMA, GARCH, Holt–Winters and exponential smoothing models. It is also used for creating forecasts based on a particular model and for decomposition of time series.

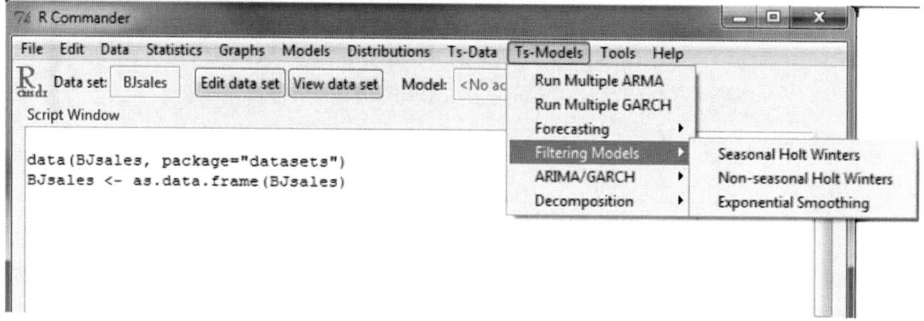

The Decomposition tab has multiplicative and additive decomposition and an option for both.

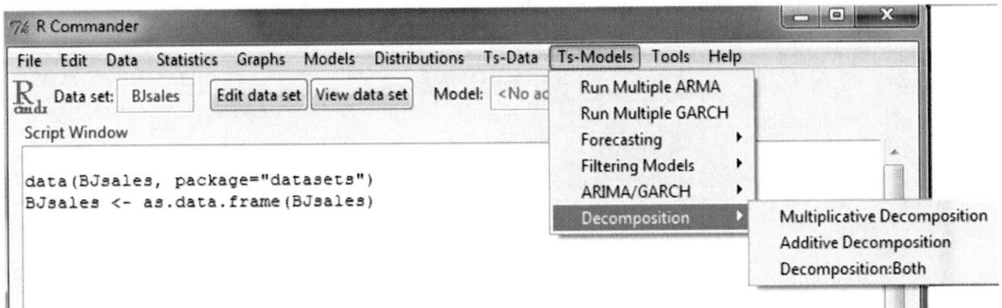

On running multiple ARMA models, this is the output we get. Note that the function called is bulkfit and R Commander automatically generates the syntax.

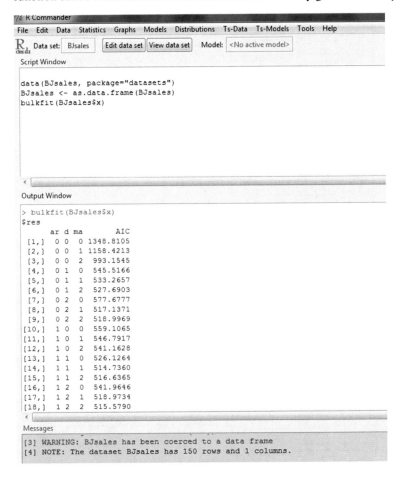

R Commander

File Edit Data Statistics Graphs Models Distributions Ts-Data Ts-Models Tools Help

Data set: BJsales | Edit data set | View data set | Model: <No active model>

Script Window

```
data(BJsales, package="datasets")
BJsales <- as.data.frame(BJsales)
bulkfit(BJsales$x)
```

Output Window

```
> bulkfit(BJsales$x)
$res
      ar d ma       AIC
 [1,]  0 0  0 1348.8105
 [2,]  0 0  1 1158.4213
 [3,]  0 0  2  993.1545
 [4,]  0 1  0  545.5166
 [5,]  0 1  1  533.2657
 [6,]  0 1  2  527.6903
 [7,]  0 2  0  577.6777
 [8,]  0 2  1  517.1371
 [9,]  0 2  2  518.9969
[10,]  1 0  0  559.1065
[11,]  1 0  1  546.7917
[12,]  1 0  2  541.1628
[13,]  1 1  0  526.1264
[14,]  1 1  1  514.7360
[15,]  1 1  2  516.6365
[16,]  1 2  0  541.9646
[17,]  1 2  1  518.9734
[18,]  1 2  2  515.5790
```

Messages

```
[3] WARNING: BJsales has been coerced to a data frame
[4] NOTE: The dataset BJsales has 150 rows and 1 columns.
```

The bulkfit function basically runs multiple ARMA models and selects the model with the lowest value of AIC.

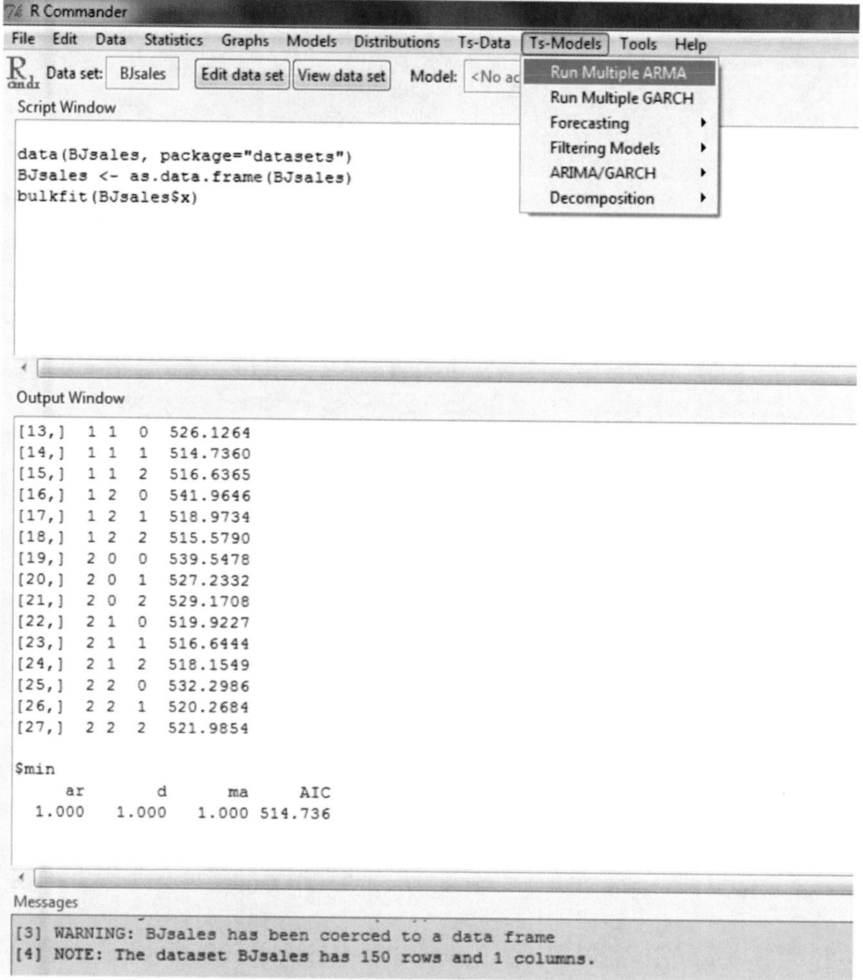

For running multiple GARCH models, the menu option is the second in TS-Models.

The command used is bulkfitg().

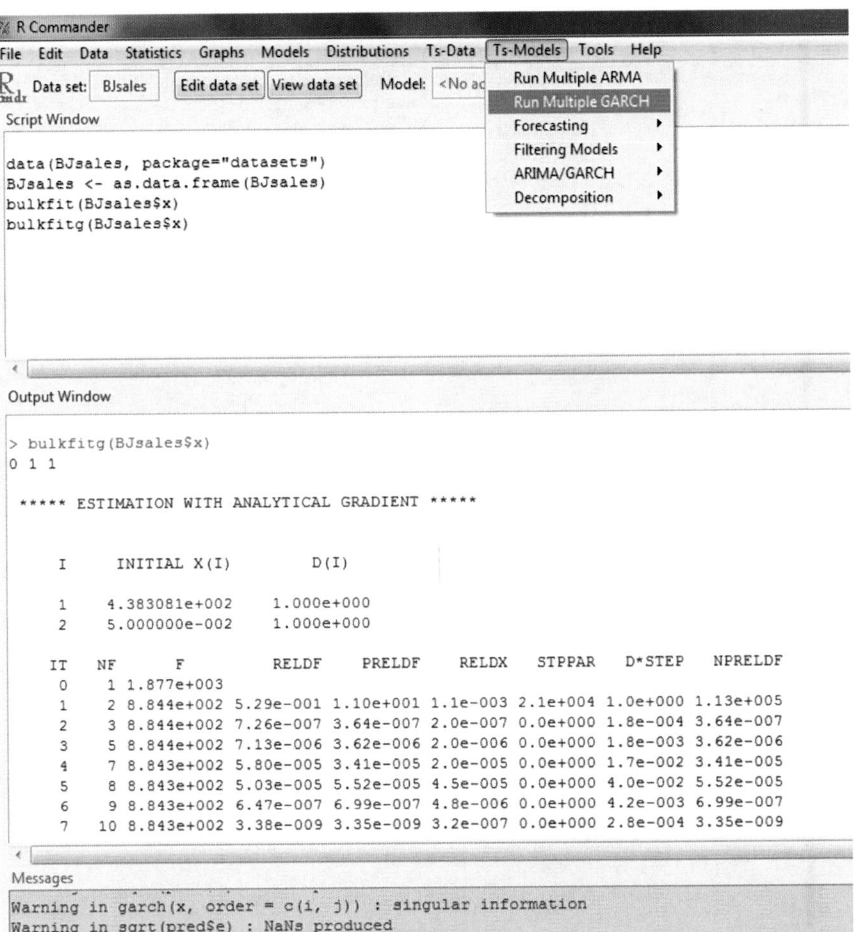
R Commander
File Edit Data Statistics Graphs Models Distributions Ts-Data Ts-Models Tools Help
Data set: BJsales
Edit data set
View data set
Model: <No ac
Run Multiple ARMA
Run Multiple GARCH
Forecasting
Filtering Models
ARIMA/GARCH
Decomposition
Script Window
data(BJsales, package="datasets")
BJsales <- as.data.frame(BJsales)
bulkfit(BJsales$x)
bulkfitg(BJsales$x)
Output Window
> bulkfitg(BJsales$x)
0 1 1
***** ESTIMATION WITH ANALYTICAL GRADIENT *****
I INITIAL X(I) D(I)
1 4.383081e+002 1.000e+000
2 5.000000e-002 1.000e+000
IT NF F RELDF PRELDF RELDX STPPAR D*STEP NPRELDF
0 1 1.877e+003
1 2 8.844e+002 5.29e-001 1.10e+001 1.1e-003 2.1e+004 1.0e+000 1.13e+005
2 3 8.844e+002 7.26e-007 3.64e-007 2.0e-007 0.0e+000 1.8e-004 3.64e-007
3 5 8.844e+002 7.13e-006 3.62e-006 2.0e-006 0.0e+000 1.8e-003 3.62e-006
4 7 8.843e+002 5.80e-005 3.41e-005 2.0e-005 0.0e+000 1.7e-002 3.41e-005
5 8 8.843e+002 5.03e-005 5.52e-005 4.5e-005 0.0e+000 4.0e-002 5.52e-005
6 9 8.843e+002 6.47e-007 6.99e-007 4.8e-006 0.0e+000 4.2e-003 6.99e-007
7 10 8.843e+002 3.38e-009 3.35e-009 3.2e-007 0.0e+000 2.8e-004 3.35e-009
Messages
Warning in garch(x, order = c(i, j)) : singular information
Warning in sqrt(pred$e) : NaNs produced

The output from running multiple GARCH models is also shown here.

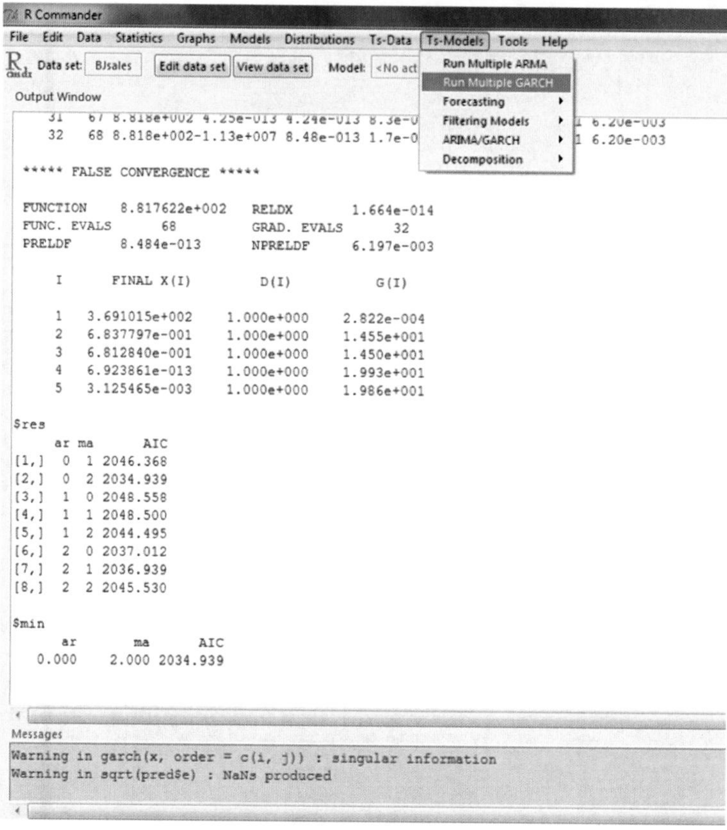

We now create a GARCH model based on the value we found in the bulkfitg() function.

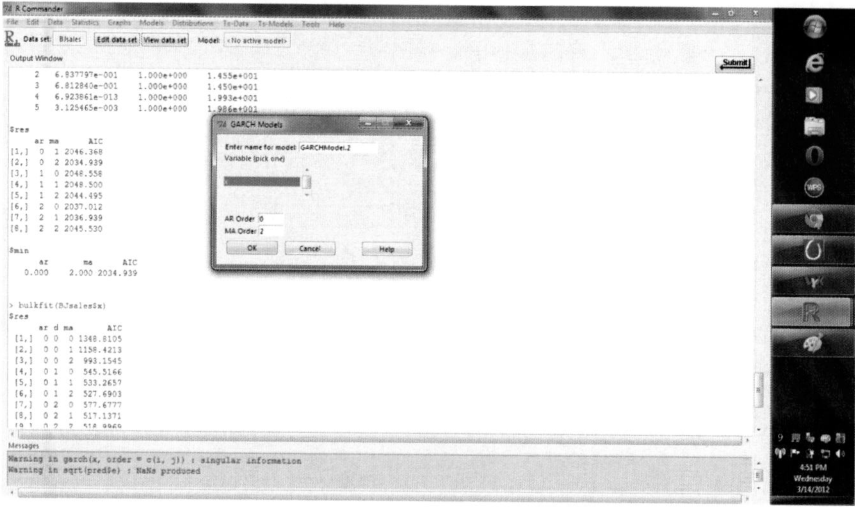

Note that the menu in R Commander shows the active model loaded.

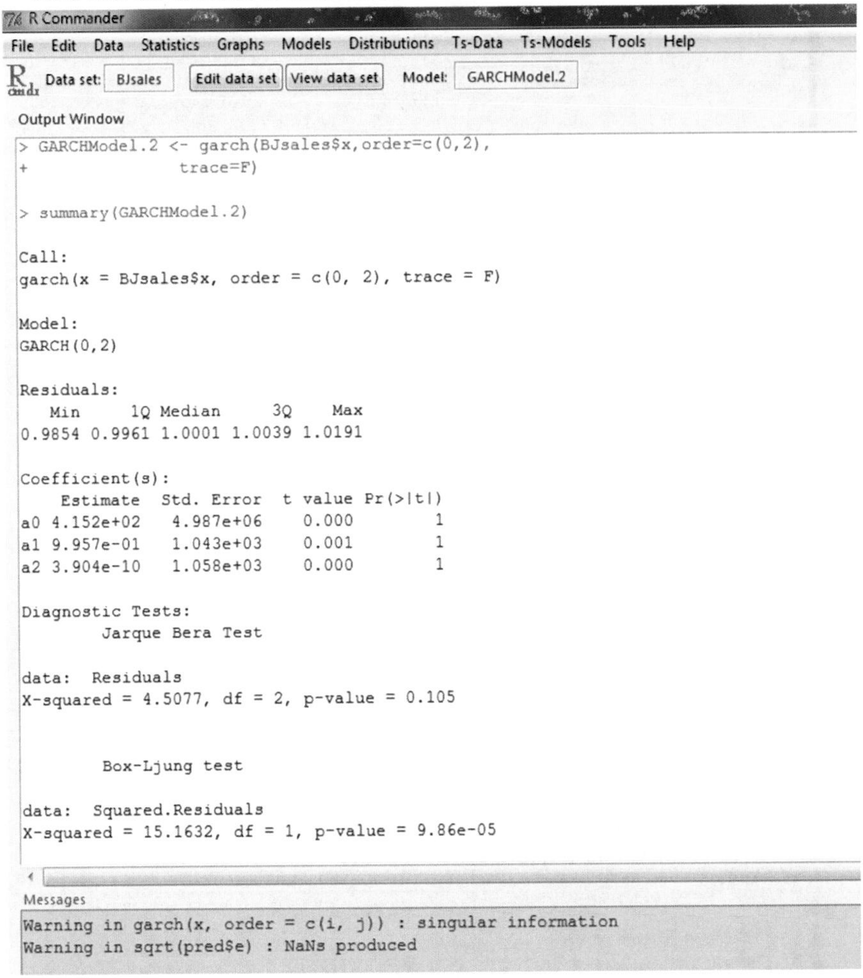

```
R Commander
File Edit Data Statistics Graphs Models Distributions Ts-Data Ts-Models Tools Help
Data set: BJsales   Edit data set   View data set   Model: GARCHModel.2
Output Window
> GARCHModel.2 <- garch(BJsales$x,order=c(0,2),
+                trace=F)

> summary(GARCHModel.2)

Call:
garch(x = BJsales$x, order = c(0, 2), trace = F)

Model:
GARCH(0,2)

Residuals:
   Min     1Q Median     3Q    Max
0.9854 0.9961 1.0001 1.0039 1.0191

Coefficient(s):
    Estimate  Std. Error  t value Pr(>|t|)
a0 4.152e+02   4.987e+06    0.000        1
a1 9.957e-01   1.043e+03    0.001        1
a2 3.904e-10   1.058e+03    0.000        1

Diagnostic Tests:
        Jarque Bera Test

data:  Residuals
X-squared = 4.5077, df = 2, p-value = 0.105


        Box-Ljung test

data:  Squared.Residuals
X-squared = 15.1632, df = 1, p-value = 9.86e-05

Messages
Warning in garch(x, order = c(i, j)) : singular information
Warning in sqrt(pred$e) : NaNs produced
```

Similarly, we create the ARIMA model based on the values we obtained in bulkfit () in the earlier step.

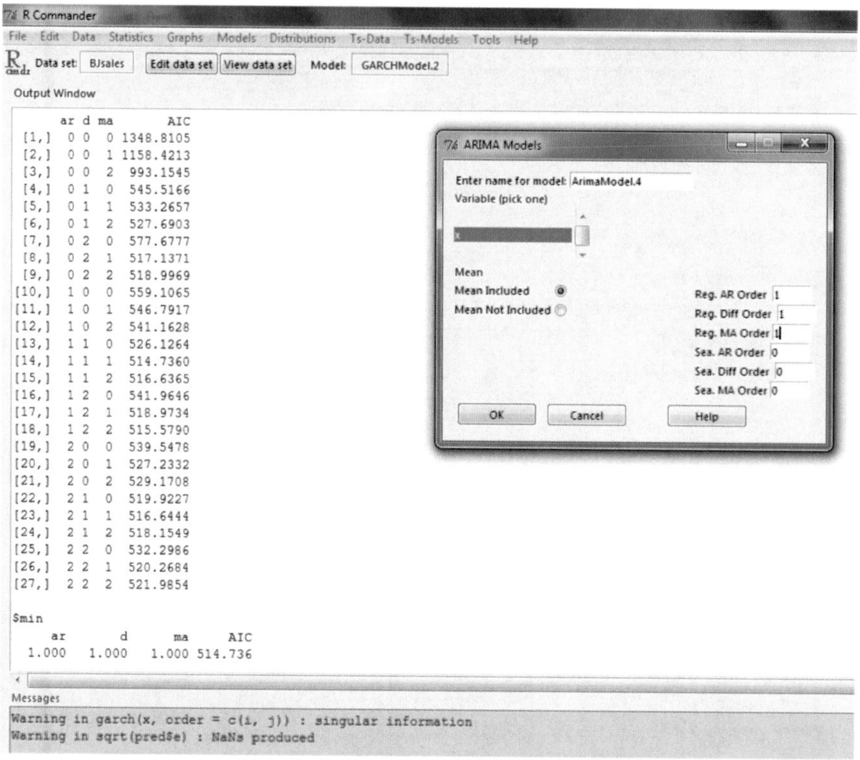

R Commander
File Edit Data Statistics Graphs Models Distributions Ts-Data Ts-Models Tools Help
Data set: BJsales Edit data set View data set Model: GARCHModel.2
Output Window
ar d ma AIC
[1,] 0 0 0 1348.8105
[2,] 0 0 1 1158.4213
[3,] 0 0 2 993.1545
[4,] 0 1 0 545.5166
[5,] 0 1 1 533.2657
[6,] 0 1 2 527.6903
[7,] 0 2 0 577.6777
[8,] 0 2 1 517.1371
[9,] 0 2 2 518.9969
[10,] 1 0 0 559.1065
[11,] 1 0 1 546.7917
[12,] 1 0 2 541.1628
[13,] 1 1 0 526.1264
[14,] 1 1 1 514.7360
[15,] 1 1 2 516.6365
[16,] 1 2 0 541.9646
[17,] 1 2 1 518.9734
[18,] 1 2 2 515.5790
[19,] 2 0 0 539.5478
[20,] 2 0 1 527.2332
[21,] 2 0 2 529.1708
[22,] 2 1 0 519.9227
[23,] 2 1 1 516.6444
[24,] 2 1 2 518.1549
[25,] 2 2 0 532.2986
[26,] 2 2 1 520.2684
[27,] 2 2 2 521.9854
$min
ar d ma AIC
1.000 1.000 1.000 514.736
ARIMA Models
Enter name for model: ArimaModel.4
Variable (pick one)
x
Mean
Mean Included
Mean Not Included
Reg. AR Order 1
Reg. Diff Order 1
Reg. MA Order 1
Sea. AR Order 0
Sea. Diff Order 0
Sea. MA Order 0
OK
Cancel
Help
Messages
Warning in garch(x, order = c(i, j)) : singular information
Warning in sqrt(pred$e) : NaNs produced

We use the forecasting tab and plot the forecasts as shown in the screenshot. We can customize the number of forecast periods and choose which model to represent. The colored values in the graph represent the confidence intervals or range of probable values of the future forecast.

R Graphics: Device 2 (ACTIVE)
Forecasts from ARIMA(1,1,1)
200
210
220
230
240
250
260
0
50
100
150
Data set: BJsales Edit data set View data set Model: ArimaModel.4
Output Window
Forecasting
Current Model: ArimaModel.4
Models (pick one)
ArimaModel.4
GARCHModel.2
Number of Forecasts: 1
Enter name for data set: Dataset
OK
Cancel
> ArimaModel.4 <- Arima(BJsales$x,order=c(1,1,1),
include.mean=1,seasonal=list(order=c(0,0,0),period=1))
> ArimaModel.4
Series: BJsales$x
ARIMA(1,1,1)
Coefficients:
ar1 ma1
0.8800 -0.6415
s.e. 0.0644 0.1035
sigma^2 estimated as 1.775: log likelihood=-254.37
AIC=514.74 AICc=514.9 BIC=523.75
> Dataset <- predar3(ArimaModel.4,fore1=2)
Messages
Warning in garch(x, order = c(i, j)) : singular information
Warning in sqrt(pred$e) : NaNs produced

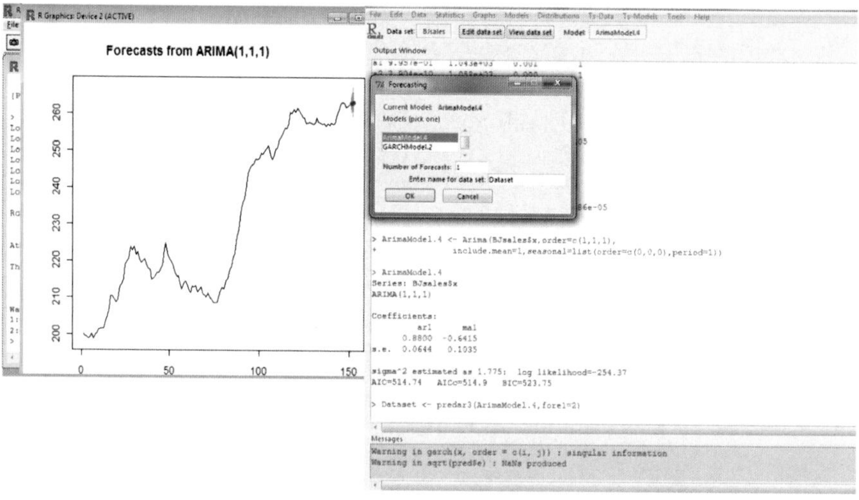

9.5.1 Syntax Generated Using R Commander GUI with epack Plugin

Source: http://www.statistik.uni-dortmund.de/useR-2008/abstracts/Hodgess+Vobach.pdf

The epack plugin provides time series functionality to R Commander.

Note the GUI helps explore various time series functionality.

Using bulkfit you can fit various ARMA models to a dataset and choose based on minimum AIC.

```
> bulkfit(AirPassengers$x)
$res ar d ma AIC
[1,] 0 0 0 1790.368
[2,] 0 0 1 1618.863
[3,] 0 0 2 1522.122
[4,] 0 1 0 1413.909
[5,] 0 1 1 1397.258
[6,] 0 1 2 1397.093
[7,] 0 2 0 1450.596
[8,] 0 2 1 1411.368
[9,] 0 2 2 1394.373
[10,] 1 0 0 1428.179
[11,] 1 0 1 1409.748
[12,] 1 0 2 1411.050
[13,] 1 1 0 1401.853
[14,] 1 1 1 1394.683
[15,] 1 1 2 1385.497
[16,] 1 2 0 1447.028
[17,] 1 2 1 1398.929
[18,] 1 2 2 1391.910
[19,] 2 0 0 1413.639
[20,] 2 0 1 1408.249
[21,] 2 0 2 1408.343
[22,] 2 1 0 1396.588
[23,] 2 1 1 1378.338
[24,] 2 1 2 1387.409
[25,] 2 2 0 1440.078
[26,] 2 2 1 1393.882
[27,] 2 2 2 1392.659
$min ar d ma AIC
2.000 1.000 1.000 1378.338
> ArimaModel.5 <- Arima(AirPassengers$x,order=c(0,1,1),
+ include.mean=1,
+ seasonal=list(order=c(0,1,1),period=12))
> ArimaModel.5
```

Series: AirPassengers$x ARIMA(0,1,1)(0,1,1)[12]
Call: Arima(x = AirPassengers$x, order = c(0, 1, 1), seasonal = list(order = c(0, 1, 1), period = 12), include.mean = 1)
Coefficients:
ma1 sma1
−0.3087 −0.1074
s.e. 0.0890 0.0828
sigma^2 estimated as 135.4: log likelihood = −507.5
AIC = 1021 AICc = 1021.19 BIC = 1029.63
> summary(ArimaModel.5, cor=FALSE)
Series: AirPassengers$x ARIMA(0,1,1)(0,1,1)[12]
Call: Arima(x = AirPassengers$x, order = c(0, 1, 1), seasonal = list (order = c(0, 1, 1), period = 12), include.mean = 1)
Coefficients:
ma1 sma1
−0.3087 −0.1074
s.e. 0.0890 0.0828
sigma^2 estimated as 135.4: log likelihood = −507.5
AIC = 1021 AICc = 1021.19 BIC = 1029.63
In-sample error measures:
ME RMSE MAE MPE MAPE MASE
0.32355285 11.09952005 8.16242469 0.04409006 2.89713514 0.31563730
Dataset79 <- predar3(ArimaModel.5,fore1=5)

9.6 Summary of Commands Used in This Chapter

9.6.1 Packages

- forecast
- lubridate
- R Commander with Plugin Epack

9.6.2 Functions

- strptime()
- difftime()
- plot()
- auto.arima(forecast)
- ets(forecast)
- accuracy()

- decompose()
- acf()
- forecast()
- pacf()
- stl(forecast)
- ts.plot
- tsdisplay(forecast)
- seasonplot(forecast)

Citations and Sources

- *Using R (with applications in Time Series Analysis) Dr. Gavin Shaddick January 2004: http://people.bath.ac.uk/masgs/time%20series/TimeSeriesR2004.pdf*
- Garrett Grolemund, Hadley Wickham (2011) Dates and Times Made Easy with lubridate. Journal of Statistical Software, 40(3), 1–25. http://www.jstatsoft.org/v40/i03/
- Rob J. Hyndman with contributions from Slava Razbash and Drew Schmidt (2012) forecast: Forecasting functions for time series and linear models. R package version 3.19. http://CRAN.R-project.org/package=forecast
- Fox, J. (2005). The R Commander: A Basic Statistics Graphical User Interface to R. Journal of Statistical Software, 14(9): 1–42. http://www.jstatsoft.org/v14/i09
- Erin Hodgess (2012) RcmdrPlugin.epack: Rcmdr plugin for time series. R package version 1.2.5. http://CRAN.R-project.org/package=RcmdrPlugin.epack
- *Time Series Analysis with R - Part I. Walter Zucchini, Oleg Nenadi ´http://www.statoek.wiso.uni-goettingen.de/veranstaltungen/zeitreihen/sommer03/ts_r_intro.pdf*
- *Econometrics in R Grant V Farnsworth 2008* http://cran.r-project.org/doc/contrib/Farnsworth-EconometricsInR.pdf
- Hyndman, R.J. (n.d.) Time Series Data Library, http://robjhyndman.com/TSDL. Accessed on 21 March 2012.
- A Reference Card for Time Series functions in R: http://cran.r-project.org/doc/contrib/Ricci-refcard-ts.pdf
- A slightly more exhaustive time series reference sheet: http://www.statistische-woche-nuernberg-2010.org/lehre/bachelor/datenanalyse/Refcard3.pdf
- Issues in Time Series Analysis in R: *http://www.stat.pitt.edu/stoffer/tsa2/Rissues.htm*
- Using GUI R Commander with Plugin Econometrics: http://user2010.org/slides/Rosadi.pdf

Documentation on time-date variables (especially for time zone level and leap year seconds and differences)

- http://stat.ethz.ch/R-manual/R-patched/library/base/html/difftime.html
- http://stat.ethz.ch/R-manual/R-patched/library/base/html/strptime.html

- http://stat.ethz.ch/R-manual/R-patched/library/base/html/Ops.Date.html
- http://stat.ethz.ch/R-manual/R-patched/library/base/html/Dates.html
- Using TimeSeries http://www.r-project.org/conferences/useR-2009/slides/Chalabi+Wuertz.pdf

Chapter 10
Data Export and Output

Data export, and saving results, graphs, and code are important to help complete the final documentation and presentation for an analytical project. What are the various formats available in R for exporting graphs? The function capabilities() can be used to obtain a list of exportable formats for graphs.

> capabilities()

jpeg png tiff tcltk **X11 aqua** http/ftp sockets libxml **fifo** cledit iconv NLS profmem cairo

TRUE TRUE TRUE TRUE **FALSE FALSE** TRUE TRUE TRUE **FALSE** TRUE TRUE TRUE TRUE TRUE

getwd()

This obtains the current working directory.

setwd("C:/New Folder")

This sets the working directory to the new folder.

1. Exporting Data

write.table(df, quote = FALSE, sep = ",")

Alternatively

write.csv(x, file = "foo.csv")

read.csv("foo.csv", row.names = 1)

or without row names

write.csv(x, file = "foo.csv", row.names = FALSE)

read.csv("foo.csv")

write.matrix in the MASS package provides a specialized interface for writing matrices.

The function write.foreign in the foreign package uses write.table to produce a text file and also writes a code file that will read this text file into another statistical package.

2. Exporting Graphs and Animation

- You can export a plot to pdf, png, jpeg, or bmp by adding pdf("filename.pdf"), png("filename.png"), jpeg("filename.jpg"), or bmp("filename.bmp") prior to plotting, and dev.off() after plotting.

A. Ohri, *R for Business Analytics*, DOI 10.1007/978-1-4614-4343-8_10,

Using png

png("symbol.png", width = 20, height = 20, bg = "transparent")

Using Dev.off

Here, the graph does not appear in a graph window in R. Rather, it just goes straight to the specified file. Close the graphics device by using the dev.off function.

postscript(file = "D:/temp/graph2.eps", onefile=FALSE, horizontal=FALSE)

plot(read, write)

Dev.off()

In this example, we save the graph as a .png file.

png("D:/temp/graph3.png")

hist(read)

In this example, we save the graph as a .pdf file.

pdf("D:/temp/graph4.pdf")

boxplot(write)

Source: http://www.ats.ucla.edu/stat/r/library/lecture_graphing_r.htm#out

- using GUI.

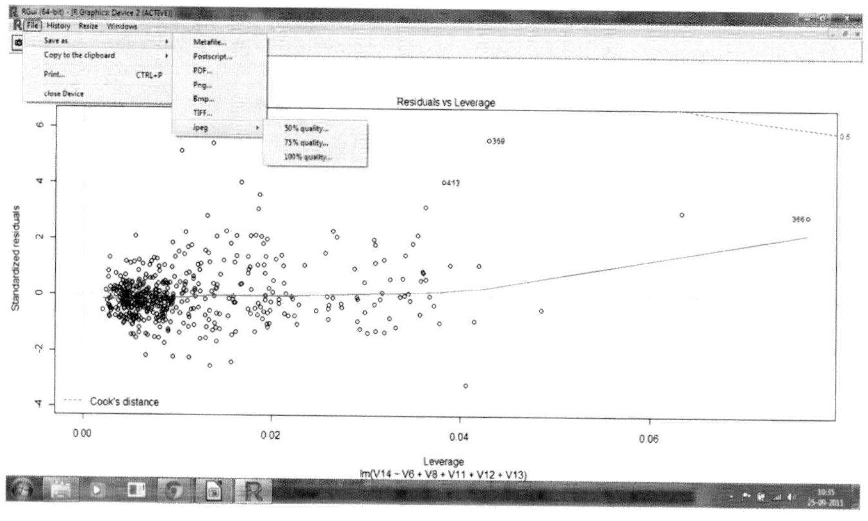

3. Exporting and Saving Models

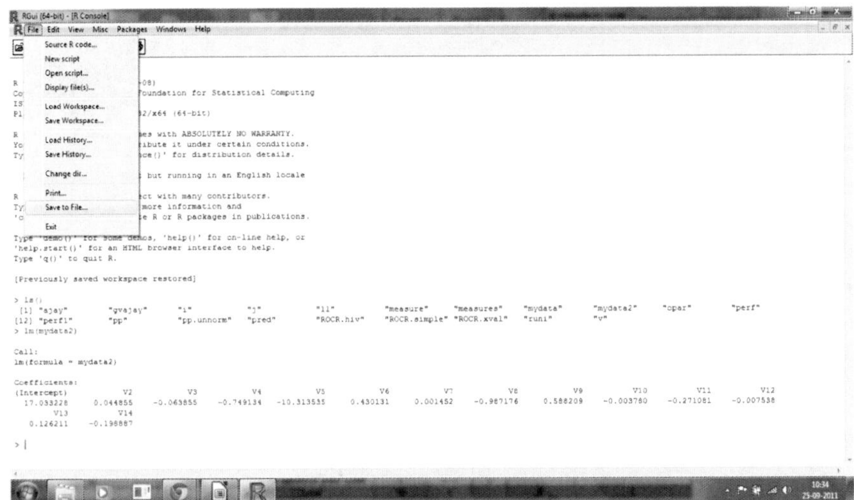

save.image() is just a shortcut for "save my current workspace",
save(list = ls(all=TRUE), file = ".RData")
It is also what happens with
q("yes")

- Using GUI

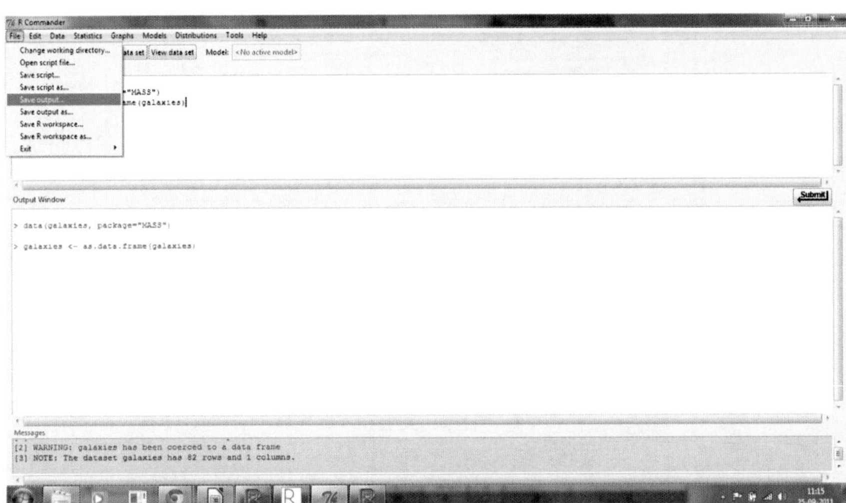

10.1 Summary of Commands Used in This Chapter

10.1.1 Packages

- graphics

10.1.2 Functions

- png
- pdf
- dev.off()

Citations and Sources

The "graphics" package is part of R. To cite R in publications use:

R Development Core Team (2012). R: A language and environment for statistical computing. R Foundation for Statistical Computing, Vienna, Austria. ISBN 3-900051-07-0, URL http://www.R-project.org/

Chapter 11
Optimizing R Code

As the previous chapters have shown, multiple techniques are available in R for powerful data-driven insights and analysis. For the average business analyst, well-designed GUI tools that are stable to use, pull data and models, and report them are essential, and all these are available within various R subcomponents and packages.

This chapter is aimed at analysts wishing to tweak their overall R experience by measuring R performance and improving it using some of the well-known and some recently introduced utilities.

11.1 Examples of Efficient Coding

This section discusses better coding practices to optimize your analytics speed and experience with R.

- Refer to parts of data.frame rather than a whole dataset.

Use square brackets to reference variable columns and rows.

The notation dataset[i,k] refers to the element in the *i*th row and *j*th column.

The notation dataset[i,] refers to all elements in the *i*th row or a record for a data.frame.

The notation dataset[,j] refers to all elements in the *j*th column or a variable for a data.frame.

For a data.frame dataset

> *n*row(dataset) #This gives the number of rows

> *n*col(dataset) #This gives the number of columns

Example of correlation between various variables in a data.frame:

A. Ohri, *R for Business Analytics*, DOI 10.1007/978-1-4614-4343-8_11,

> cor(dataset1[,4:6])
ga.visitors ga.visits ga.pageviews
ga.visitors 1.0000000 0.9936381 0.8577164
ga.visits 0.9936381 1.0000000 0.8591524
ga.pageviews 0.8577164 0.8591524 1.0000000
> cor(dataset1[,4:7])
ga.visitors ga.visits ga.pageviews ga.timeOnSite
ga.visitors 1.0000000 0.9936381 0.8577164 0.2691355
ga.visits 0.9936381 1.0000000 0.8591524 0.2701968
ga.pageviews 0.8577164 0.8591524 1.0000000 0.3453492
ga.timeOnSite 0.2691355 0.2701968 0.3453492 1.0000000
Splitting a dataset into test and control:
ts.test=dataset2[1:200] #First 200 rows
ts.control=dataset2[201:275] #Next 75 rows

- Sampling

Random sampling enables us to work on a smaller size of the whole dataset.
Use a sample to create a random permutation of vector x.
sample(x) will have the same size as length (x)/number of elements in vector x.
Sample argument size limits the size of the sample.
>sample(x, size = 5)
Assume we want a 5 % sample of a list x. Then the size will be 0.01(length(x)).
> sample(x, size = 0.01*length(x))
If you want to replace each item once it has been taken out for a sample, use the replace function.
>table(sample(x, size = 100, replace = TRUE))
Suppose we want to take a 5 % sample of a data frame with no replacement.

Let us create a dataset "ajay" of random numbers

> ajay=matrix(round(rnorm(200, 5,15)), ncol=10)

We use the round function to round off values.

> ajay=as.data.frame(ajay)

> ajay

> nrow(ajay)

[1] 20

> ncol(ajay)

[1] 10

This is a typical business data scenario where we want to select only a few records to do our analysis (or test our code) but have all the columns for those records.

Then the number of rows in the new object will be *0.05*nrow(ajay)*. That will be the size of the sample.

The new object can be referenced to choose only a sample of all rows in the original object using the size parameter.

We also use the replace = FALSE or F, to not select the same row again and again.

The new rows are thus a 5 % sample of the existing rows.

Then we use the square backets and ajay[new_rows,] to obtain

*b=ajay[sample(nrow(ajay),replace=F,size=0.05*nrow(ajay)),]*

You can change the percentage from 5 % to whatever you want accordingly.

11.2 Customizing R Software Startup

Customizing your R software startup helps you do the following.

It automatically loads packages that you use regularly (like an R GUI—Deducer, Rattle, or R Commander), sets a CRAN mirror that you mostly use or is nearest for downloading new packages, and sets some optional parameters. Instead of doing this every time—loading the same R packages, setting a CRAN mirror, setting some new functions—the user needs to do this just once by customizing the R Profile SITE file.

This is done by editing the $R_HOME/etc/Renviron file for globally setting a default or the .Renviron file that is created in your home directory for a shared system.

There are two special functions you can customize in these files.

.First( ) will be run at the start of the R session, and .Last( ) will be run when the R session is shutting down.

When R starts up, it loads the .Rprofile file in your home directory and executes the .First() function.

11.2.1 Where is the R Profile File?

It is located in the \etc folder of your R folder—the folder where you installed R.

In Windows the folder will be of the format "C:\Program Files\R\R-x.ab.c\etc" where x.ab.c is the R version number (like 2.14.1).

Example: *.First <- function(){ library(rattle) rattle() cat("\nHello World", date(), "\n") }* will automatically start the Rattle GUI for data mining and print "Hello World" with the date in your session.

11.2.2 Modify Settings

You can also modify the Rcmd_environ file in the same \etc folder if you want particular settings.

Default browser R_BROWSER=${R_BROWSER: 'C:\Documents and Settings\ abc\Local Settings\ Application Data\Google\Chrome\Application\ chrome.exe}'

Default editor EDITOR=${EDITOR-${notepad++}} will change the default Web browser to Chrome and the default editor to Notepad++ which is an enhanced code editor.

We will discuss code editors in the next section for the added functionality we can get from them.

11.3 Code Editors

Code editors are specific software that help in writing code for developers who need flexibility and an easy-to-write environment. The following is a list of notable code editors or integrated development environment (IDE) for R.

- Notepad++: Enhanced code editor; can be downloaded from http://notepad-plus-plus.org/. It supports R and also has a plugin called NPP to R that is available at http://sourceforge.net/projects/npptor/. Notepad++ can be used for a wide variety of other languages as well and has all the features mentioned above. There is an auto-completion XML/plugin that can be downloaded from http://yihui.name/en/wp-content/uploads/2010/08/R.xml and installed using multiple instructions for different packages from http://yihui.name/en/2010/08/auto-completion-in-notepad-for-r-script/. For a simple install put the XML file under "plugins/APIs" in the directory of Notepad++ (in Windows this can be C:\Program Files\Notepad++\plugins\APIs), and enable auto-completion in Notepad++ (Settings–>Preferences–>Backup/Auto-completion).

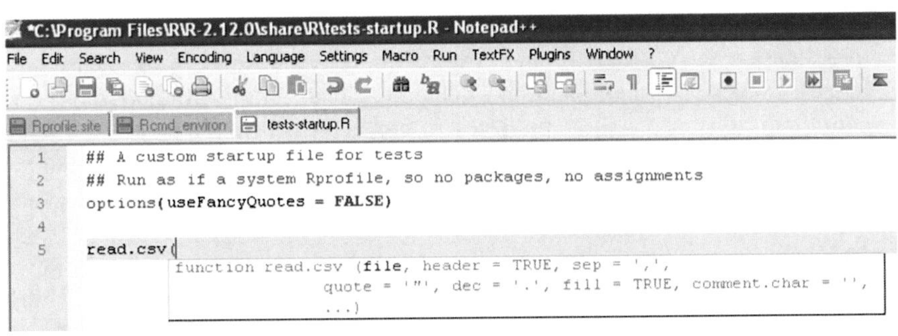

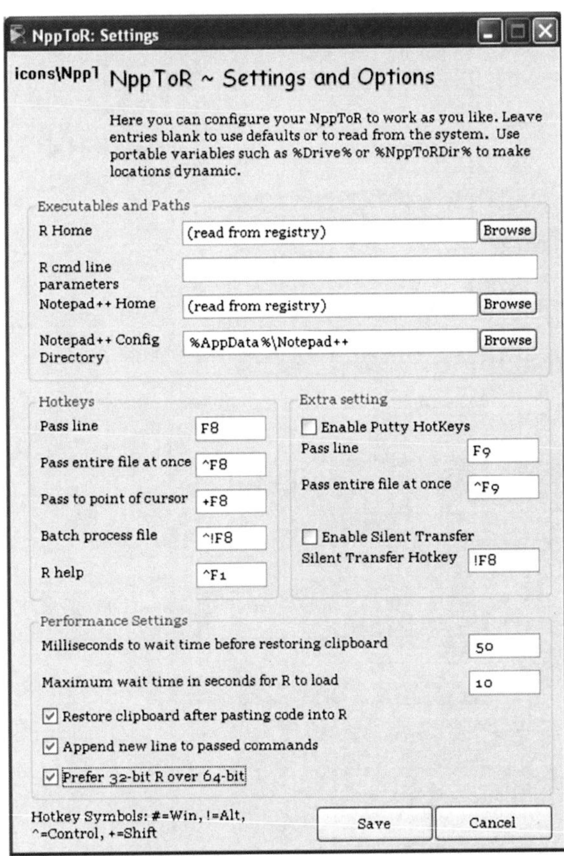

- Rstudio: Latest IDE for R. An IDE is more powerful than a simple code editor and almost functions like a separate GUI. The RStudio IDE is a more attractive toolkit for users than the default command line. However, it has code completion and allows for submitting multiple lines of code, and syntax color highlighting is supported. RStudio is, however, designed to set at ease any migrant user from other statistical software as it divides the screen into familiar zones of console, history, and workspace. It is expected that the next version of RStudio will further enhance the ease of usage of R as an analytical tool. A comparative advantage of RStudio over other environments like Eclipse is the smaller size of the QT-based installer of RStudio and its availability across all operating systems. An additional advantage is the support for $\TeX$ and Sweave (which would be useful for typesetting, but it is not the typical analyst's tool). The facility to use RStudio through a Web browser (by running through a Web server) makes it a key innovation for this software. It is now one of the most popular environments for developers.

- Due to its commercial support and rapid acceptance by the R community, this book recommends RStudio as the code editor of choice for business analytics developers.

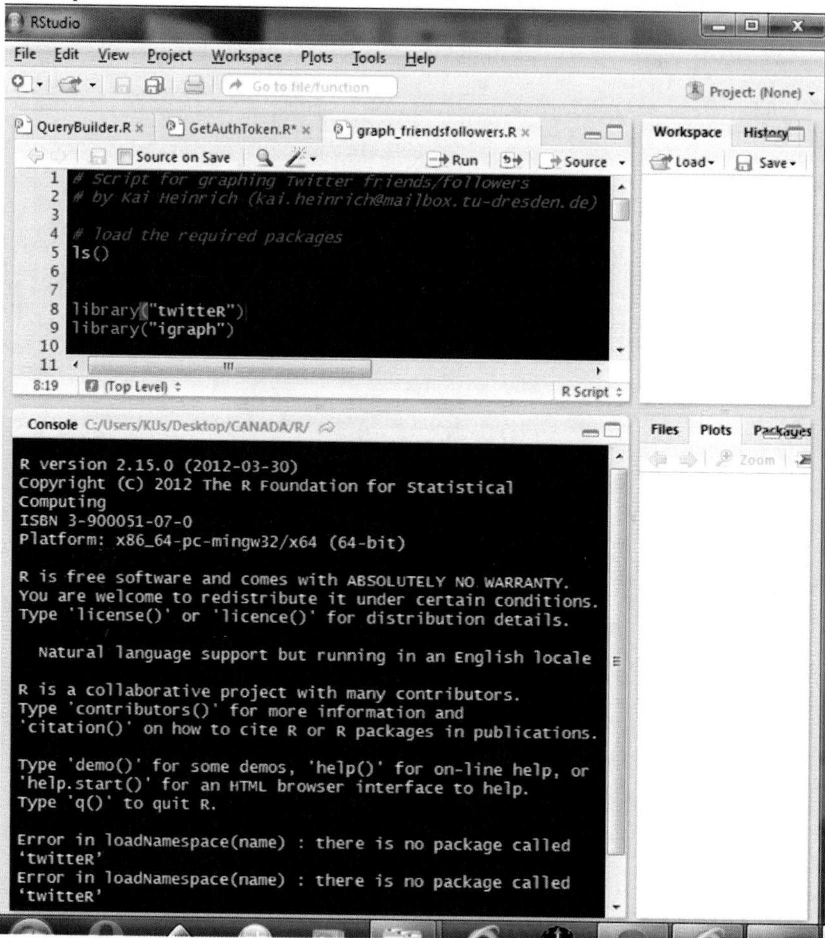

- TinnR is a basic and a easy-to-use code editor. It is available at http://www.sciviews.org/Tinn-R/. Its disadvantage is that it only supports the Windows operating system. It is useful for comparing code using syntax highlight and also passing instructions straight to R.

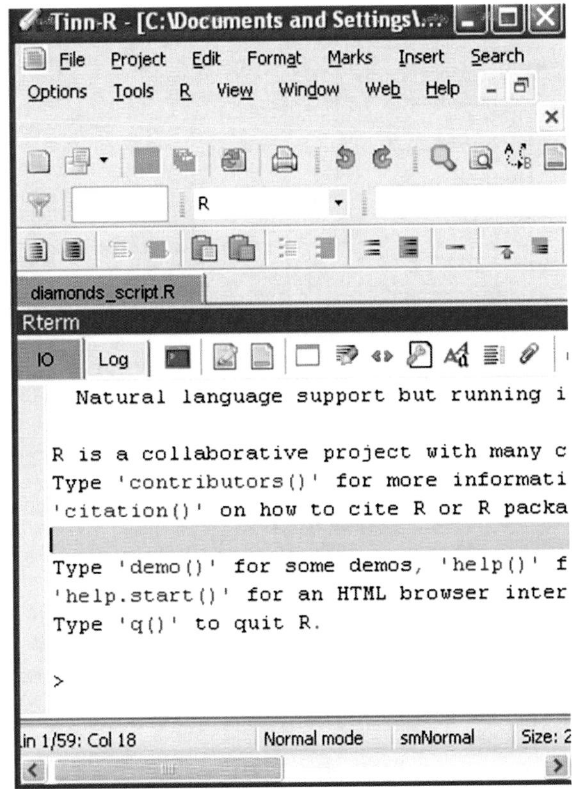

- Eclipse with R plugin: Eclipse can be downloaded from http://www.eclipse.org/downloads/packages/eclipse-classic-361/heliossr1, and the plugin can be configured from within Eclipse from http://www.walware.de/goto/statet. The installation requires the Java Runtime Environment (from http://www.oracle.com/technetwork/java/javase/downloads/index-jdk5-jsp-142662.html). This plugin is recommended especially for people working with Eclipse and for developers. It enables you to do most of the productivity enhancement featured in other text editors including submitting code to an R session. In addition, it has very easy-to-use help (called cheat sheets to set up R, as well as Sweave for Latex/R). An additional document for help is available at http://www.splusbook.com/RIntro/R_Eclipse_StatET.pdf.

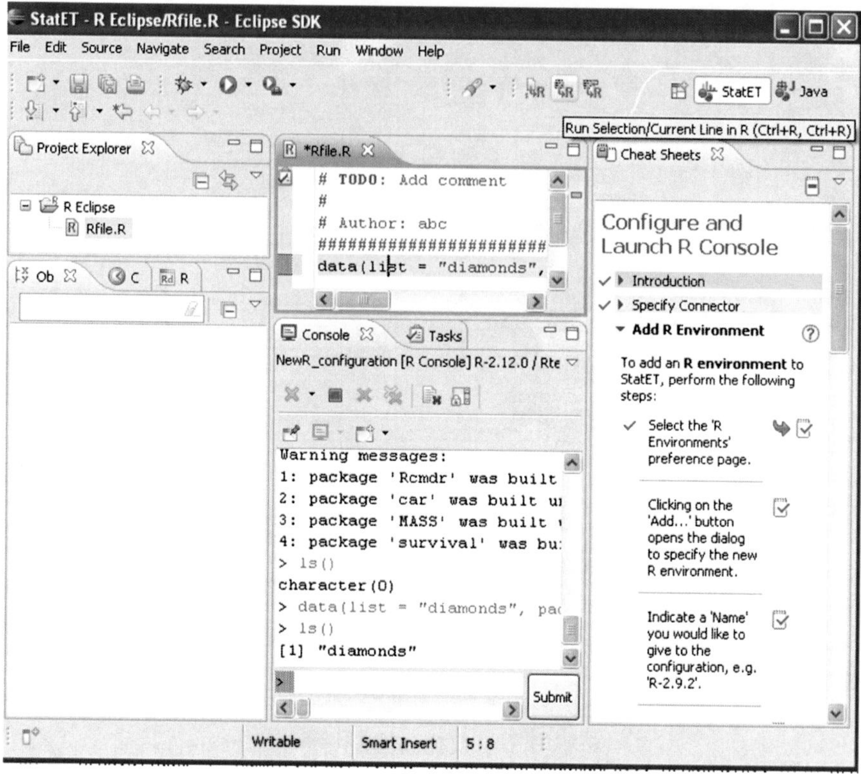

- Gvim is a popular code editor—http://www.vim.org/download.php – along with Vim-R-plugin2—(http://www.vim.org/scripts/script.php?script_id=2628). The Vim-R-plugin developer recently added Windows support to a lean cross-platform package that works well.

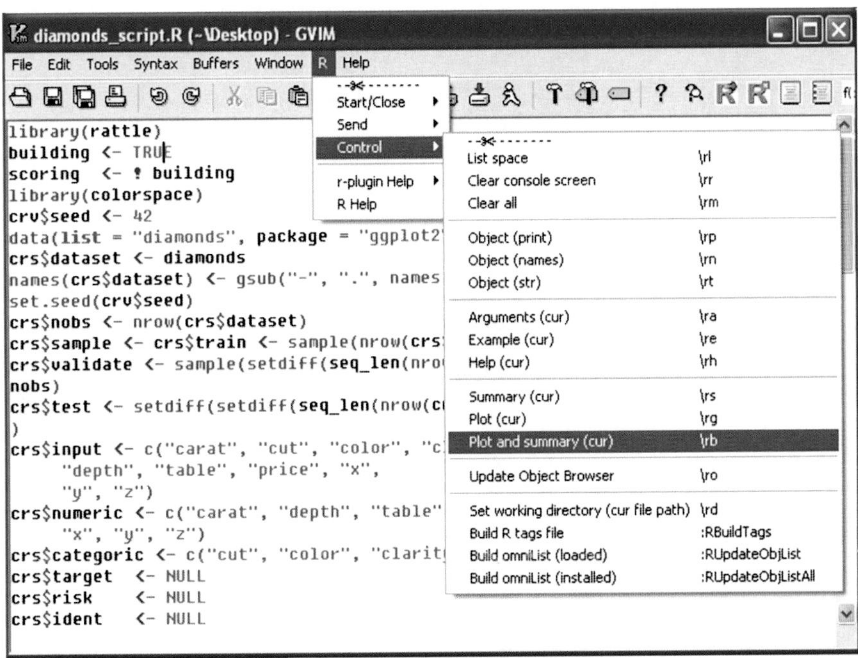

- Highlight: This code editor converts source code in many formats with color syntax highlighting and is customizable for languages. It is suitable for presenting R code. It can be downloaded from http://www.andre-simon.de/zip/download.html.

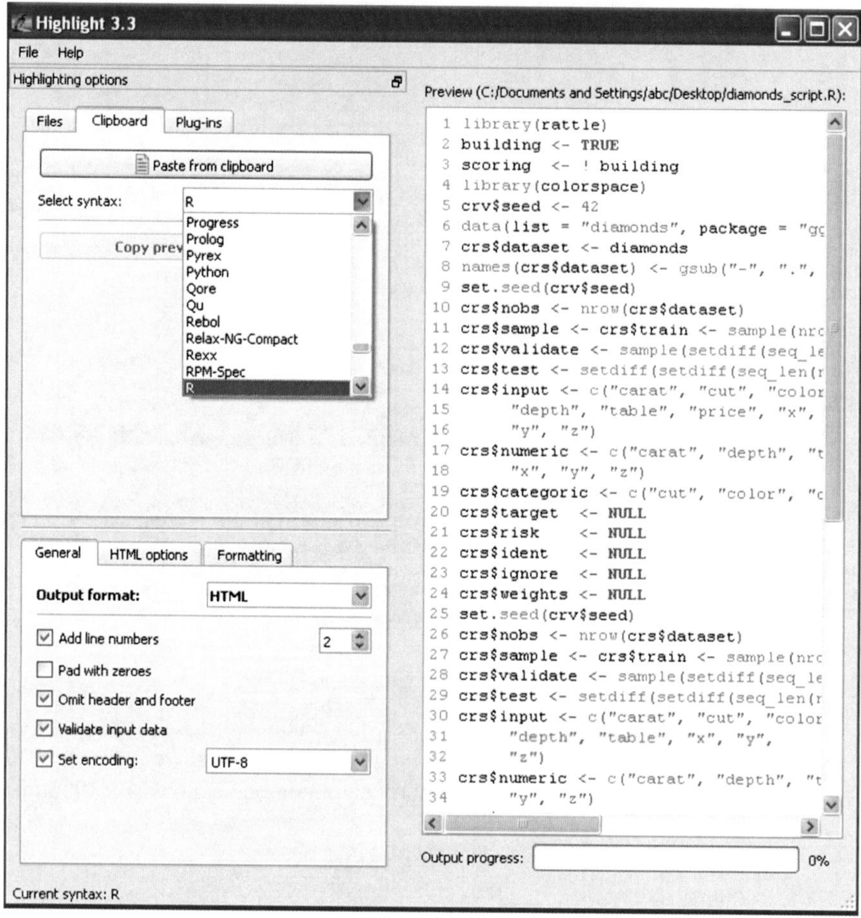

11.4 Advantages of Enhanced Code Editors

Some of the advantages of using enhanced code editors are as follows:

1. **Readability:** Features like syntax coloring help make code more readable for documentation as well as debugging and improvement. Example functions may be colored in blue, input parameters in green, and simple default code syntax in black. Especially for lengthy programs or tweaking auto-generated code by GUI, this readability comes in handy.
2. **Automatic syntax error checking:** Enhanced editors can prompt you if certain errors in syntax (like brackets not closed, commas misplaced) occur, and errors may be highlighted in color (mostly red). This helps a lot in correcting code,

especially if you are either new to R programming or your main focus is business insights and not coding. Syntax debugging is thus simplified.

3. **Speed of writing code:** Most programmers report an increase in code-writing speed when using an enhanced editor.
4. **Point breaks:** You can insert breaks at certain parts of code to run some lines of code together, or to debug a program. This is a big help given that the default code editor makes it very cumbersome and you have to copy and paste lines of code again and again to run selectively. On an enhanced editor you can submit lines as well as paragraphs of code.
5. **Auto-completion:** Auto-completion enables or suggests options for you to complete the syntax even when you have typed a small part of the function name.

11.5 Interview: J.J. Allaire, Creator of R Studio

What follows is a January 2012 interview with J.J. Allaire, founder of RStudio. RStudio is an IDE that has overtaken other IDEs within the R community in popularity.

Ajay: So what is new in the latest version of RStudio and how exactly is it useful for people?

JJ: The initial release of RStudio as well as the two follow-up releases we did last year were focused on the core elements of using R: editing and running code, getting help, and managing files, history, workspaces, plots, and packages. In the meantime users have also been asking for some bigger features that would improve the overall workflow of doing analysis with R. In this release (v0.95), we focused on three of these features:

Projects: R developers tend to have several (and often dozens) of working contexts associated with different clients, analyses, datasets, etc. RStudio projects make it easy to keep these contexts well separated (with distinct R sessions, working directories, environments, command histories, and active source documents), switch quickly between project contexts, and even work with multiple projects at once (using multiple running versions of RStudio).

Version control: The benefits of using version control for collaboration are well known, but we also believe that solo data analysis can achieve significant productivity gains by using version control (this discussion on Stack Overflow talks about why: http://stackoverflow.com/questions/2712421/r-and-version-control-for-the-solo-data-analyst). In this release we introduced integrated support for the two most popular open-source version control systems: Git and Subversion. This includes changelist management, file diffing, and browsing of project history, all right from within RStudio.

Code navigation: When you look at how programmers work, a surprisingly large amount of time is spent simply navigating from one context to another. Modern programming environments for general-purpose languages like C++ and Java solve this problem using various forms of code navigation, and in this release we've brought these capabilities to R. The two main features here are the ability to type the name of any file or function in your project and go immediately to it and the ability to navigate to the definition of any function under your cursor (including the definition of functions within packages) using a keystroke (F2) or mouse gesture (Ctrl+Click).

Ajay: What's the product road map for RStudio? When can we expect the IDE to turn into a full-fledged GUI?

JJ: Linus Torvalds has said that "Linux is evolution, not intelligent design." RStudio tries to operate on a similar principle—the world of statistical computing is too deep, diverse, and ever-changing for any one person or vendor to map out in advance what is most important. So, our internal process is to ship a new release every few months, listen to what people are doing with the product (and hope to do with it), and then start from scratch again making the improvements that are considered most important. Right now some of the things which seem to be top of mind for users are improved support for authoring and reproducible research, various editor enhancements including code folding, and debugging tools. What you'll see us do in a given release is to work on a combination of frequently requested features, smaller improvements to usability and workflow, bug fixes, and finally architectural changes required to support current or future feature requirements. While we do try to base what we work on as closely as possible on direct user feedback, we also adhere to some core principles concerning the overall philosophy and direction of the product. So, for example, the answer to the question about the IDE turning into a full-fledged GUI is: never. We believe that textual representations of computations provide fundamental advantages in transparency, reproducibility, collaboration, and reusability. We believe that writing code is simply the right way to do complex technical work, so we'll always look for ways to make coding better, faster, and easier rather than try to eliminate coding altogether.

11.6 Revolution R Productivity Environment

The existing enhancements to Revolution R include a code editor called RPE. Syntax color highlighting is already included. The most innovative feature is Code Snippets.

Code Snippets work in a fairly simply way. Right click on Insert Code Snippet. You can get a dropdown of tasks to do (like Analysis). Selecting Analysis we get another list of subtasks (like Clustering). When you click on Clustering, you get various options. Clicking clara will auto-insert the code for clara clustering. Now even if you are averse to using a GUI or GUI creators don't have your particular analysis, you can basically type in code at an extremely fast pace. It is useful to even experienced people who do not have to type in the entire code, but it is a boon to beginners as the parameters in the function inserted by Code Snippet are automatically selected in multiple colors. And it can help you modify the auto-generated code by your R GUI at a much faster pace.

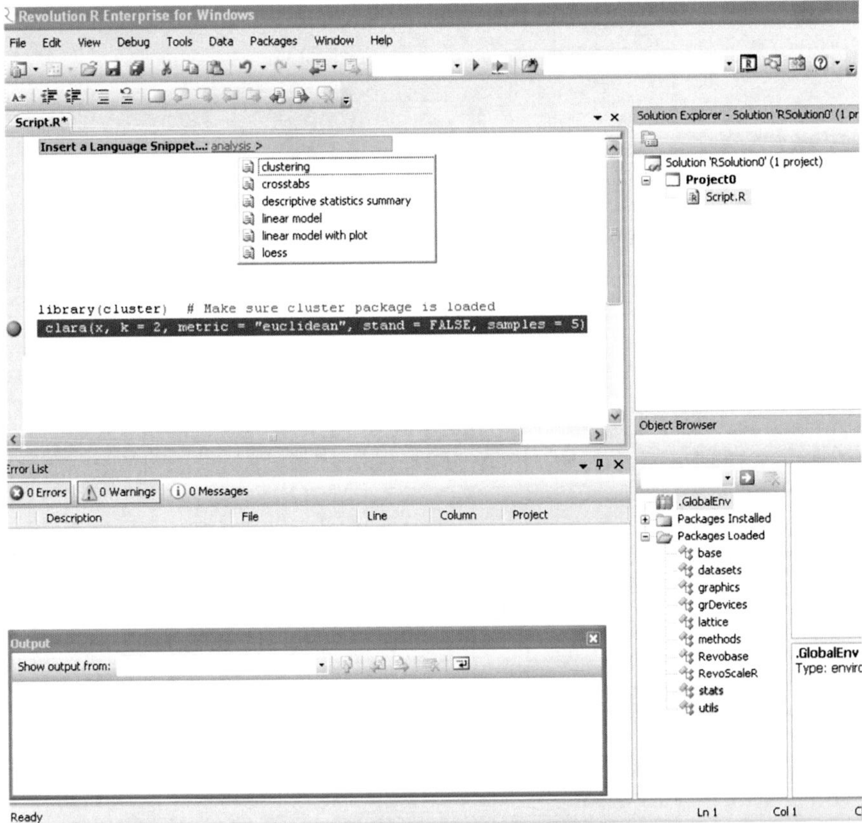

11.7 Evaluating Code Efficiency

For advanced users, it makes sense to track memory usage and measure the time for each small step in running code. Measuring memory use in R code is useful either when the code takes more memory than is conveniently available or when memory allocation and copying of objects is responsible for slow code.

For more details, see http://cran.r-project.org/doc/manuals/R-exts.html#Tidying-and-profiling-R-code.

Profiling works by first recording at fixed intervals (by default every 20 ms) which R function is being used and then recording the results in a file (default Rprof.out in the working directory). Then the function summaryRprof can be used to summarize the activity.

Measuring your coding efficiency consists of three primary steps:

- The system.time function for simple measurements
- The Rprof function for profiling R code
- The Rprofmem function for profiling R memory usage

options(memory.profiling=TRUE)
Rprof("profile.out")
#Rprofmem("profmem.out")
x=rnorm(1e7,10,5)
par(mfrow=c(3,3))
plot(iris$Sepal.Length, main="Scatter Plot with Rug ")
rug(iris$Sepal.Length,side=2)
barplot(table(iris$Sepal.Length), main="Bar Plot")
plot(iris$Sepal.Length,type="l", main="Line Plot")
plot(iris$Sepal.Length, main="Scatter Plot")
boxplot(iris$Sepal.Length, main="Box Plot")
stripchart(iris$Sepal.Length, main="Strip Chart")
sunflowerplot(iris$Sepal.Length, main="Sunflower Plot")
hist(iris$Sepal.Length, main="Histogram")
plot(density(iris$Sepal.Length), main="Density Plot")
Rprof(NULL)
#Rprofmem(NULL)
summaryRprof("profile.out")
#summaryRprof("profmem.out")
#Now you can see how much time it takes to run each part of the code.
You can also see the package for visual rendering of the profiling information generated by Rprof.
profr http://cran.r-project.org/web/packages/profr/index.html

```
> summaryRprof("profile.out")
$by.self
            self.time self.pct total.time total.pct
rnorm            1.08    93.10       1.08     93.10
<Anonymous>      0.04     3.45       0.04      3.45
axis             0.02     1.72       0.02      1.72
plot.xy          0.02     1.72       0.02      1.72

$by.total
                   total.time total.pct self.time self.pct
rnorm                    1.08     93.10      1.08    93.10
<Anonymous>              0.04      3.45      0.04     3.45
axis                     0.02      1.72      0.02     1.72
plot.xy                  0.02      1.72      0.02     1.72
Axis                     0.02      1.72      0.00     0.00
Axis.default             0.02      1.72      0.00     0.00
boxplot                  0.02      1.72      0.00     0.00
boxplot.default          0.02      1.72      0.00     0.00
boxplot.stats            0.02      1.72      0.00     0.00
par                      0.02      1.72      0.00     0.00
points                   0.02      1.72      0.00     0.00
points.default           0.02      1.72      0.00     0.00
rug                      0.02      1.72      0.00     0.00
stripchart               0.02      1.72      0.00     0.00
stripchart.default       0.02      1.72      0.00     0.00

$sample.interval
[1] 0.02

$sampling.time
[1] 1.16
```

Exercise: Check the time for ggplot and the various steps it executes and compare it with a normal plot.

Solution:

options(memory.profiling=TRUE)
Rprof("profileD.out")
x=rnorm(1e5,10,10)
y=rnorm(1e5,10,10)
library(ggplot2)
x1=as.data.frame(x)
ggplot(x1, aes(x,y))+ geom_point()
qplot(x,y)
plot(x,y)
hist(x)
Rprof(NULL)
summaryRprof("profileD.out")

In addition, the profr and proftools packages on CRAN can be used to visualize Rprof data.

Also see garbage collection.

A call of gc causes a garbage collection to take place. This will also take place automatically without user intervention, and the primary purpose of calling gc is for the report on memory usage. However, it can be useful to call gc after a large object has been removed as this may prompt R to return memory to the operating system.

http://stat.ethz.ch/R-manual/R-devel/library/base/html/gc.html.

```
> gc()
          used (Mb) gc trigger (Mb) max used (Mb)
Ncells  405925 21.7     818163 43.7   599648 32.1
Vcells 1757420 13.5    7503606 57.3  9379227 71.6
> gcinfo(TRUE)
[1] FALSE
> gc(TRUE)
Garbage collection 172 = 76+41+55 (level 2) ...
21.7 Mbytes of cons cells used (50%)
13.4 Mbytes of vectors used (29%)
          used (Mb) gc trigger (Mb) max used (Mb)
Ncells  405907 21.7     818163 43.7   599648 32.1
Vcells 1755781 13.4    6002884 45.8  9379227 71.6
> |
```

Beginner or intermediate users of R for business analytics should just stick to checking system.time for how long your code is running.

11.8 Using system.time to Evaluate Coding Efficiency

You can use the system.time function to evaluate how long the particular function takes to do a task. Based on this, you can tweak your custom-built function or use parallel processing to run loops and shorten time. It can also help you eliminate any inefficiencies or redundant tasks in your R code. The basic syntax of system.time is

system.time(expr, gcFirst = TRUE)

expr Valid R expression to be timed.

gcFirst Logical: should a garbage collection be performed immediately before the timing? The default is TRUE.

system.time calls the function proc.time, evaluates expr, and then calls proc.time once more, returning the difference between the two proc.time calls.

proc.time determines how much real and CPU time (in seconds) the currently running R process has already taken.

The result of using system.time will be an output as follows:

user system elapsed

0.28 0.00 0.28

system.time(rnorm(1e8,10,10))

user system elapsed

9.86 0.17 10.13

11.9 Using GUIs to Learn and Code R Faster

Using the right GUI can dramatically improve your learning and coding speed in R. For more on R GUIs, see Chap. 3.

11.10 Parallel Programming

With version 2.14 R supports parallel programming with the help of a package called parallel.

> library(parallel)

> detectCores() #logical cores

[1] 4 >

detectCores(logical = FALSE) #physical cores

[1] 2

Writing parallel code to make use of multiprocessor systems is surprisingly easy in R. There are multiple packages like doMC and doSNOW across operating systems, and foreach for iterating loops. An example with system.time is given below.

```
require(doSNOW)
cl<-makeCluster(2)
# I have two physical processor cores
registerDoSNOW(cl) #connects SNOW
# create a function to run in each iteration of the loop
check <-function(n) {
+ for(i in 1:1000)
+ {
+ sme <- matrix(rnorm(100), 10,10)
+ solve(sme)
+ }
+ } times <- 100 # times to run the loop # Using parallel processing
system.time(x <- foreach(j=1:times ) %dopar% check(j))
user system elapsed
0.16 0.02 19.17
# Using serial code (default) system.time(for(j in 1:times )
x <- check(j))
user system elapsed
39.66 0.00 40.46
stopCluster(cl)
```

The example shows that by running a loop parallel over the two processors, speed was boosted by up to two times. This can especially aid you in running complex calculations on larger datasets, and with the help of Amazon EC2 you can rent a high-end computer with as many as eight cores.

Additional reference: http://blog.revolutionanalytics.com/2009/08/block processing-a-data-frame-with-isplit.html.

11.11 Using Hardware Solutions

The use of blade servers, of GPU-based computers, multiple-processor systems, and cloud computing can help make the execution of your analytical tasks faster and much better. Please note that, despite claims that a given hardware solution is the best, it is advisable to run benchmark tests to finally settle on a hybrid and optimized custom hardware solution for your analytical output.

11.12 Summary of Commands Used in This Chapter

11.12.1 Packages

- parallel
- doSNOW
- profr

- proftools
- foreach

11.12.2 Functions

- system.time(): shows time taken to run a particular operation
- gc()
- sample()
- %dopar

Citations and Sources

- http://www.decisionstats.com/interview-jj-allaire-founder-rstudio/
- The 'parallel' package is part of R. To cite R in publications use:

R Development Core Team (2012). R: A language and environment for statistical computing. R Foundation for Statistical Computing, Vienna, Austria. ISBN 3-900051-07-0, URL http://www.R-project.org/.

- Notepad++ John Ho and http://notepad-plus-plus.org/contributors/
- RStudio is a trademark of RStudio, Inc.: http://www.rstudio.org/docs/about
- Tinn R copyright ï£¡ Philippe Grosjean: http://www.sciviews.org/Tinn-R/
- Revolution Analytics (2012) foreach: Foreach looping construct for R. R package version 1.3.5. http://CRAN.R-project.org/package=foreach
- Revolution Analytics (2011) doSNOW: Foreach parallel adaptor for the snow package. R package version 1.0.5. http://CRAN.R-project.org/package=doSNOW

Chapter 12
Additional Training Literature

Blogs, email help groups, and Web sites are important sources of training literature as well as tutorials. While choosing the mix of books, journal articles, blog posts, and online content is often a matter a personal choice, the reader should choose based on his or her own business or analytical needs.

12.1 Cran Views

For help regarding a particular kind of analytics (like graphics, finance, time series, or cluster analysis) the best resource aggregators are views on CRAN at http://cran.r-project.org/web/views/. Views will give you a list of all packages in that particular class of analytics with brief explanations.

The complete list of views is

Bayesian Bayesian Inference
ChemPhys Chemometrics and Computational Physics
ClinicalTrials Clinical Trial Design, Monitoring, and Analysis
Cluster Cluster Analysis & Finite Mixture Models
Distributions Probability Distributions
Econometrics Computational Econometrics
Environmetrics Analysis of Ecological and Environmental Data
ExperimentalDesign Design of Experiments (DoE) & Analysis of Experimental Data
Finance Empirical Finance
Genetics Statistical Genetics
Graphics Graphic Displays & Dynamic Graphics & Graphic Devices & Visualization
gR gRaphical Models in R
HighPerformanceComputing High-Performance and Parallel Computing with R
MachineLearning Machine Learning & Statistical Learning

A. Ohri, *R for Business Analytics*, DOI 10.1007/978-1-4614-4343-8_12,

MedicalImaging Medical Image Analysis
Multivariate Multivariate Statistics
NaturalLanguageProcessing Natural Language Processing
OfficialStatistics Official Statistics & Survey Methodology
Optimization Optimization and Mathematical Programming
Pharmacokinetics Analysis of Pharmacokinetic Data
Phylogenetics Phylogenetics, Especially Comparative Methods, Psychometrics, and Psychometric Models and Methods
ReproducibleResearch Reproducible Research
Robust Robust Statistical Methods
SocialSciences Statistics for the Social Sciences
Spatial Analysis of Spatial Data
Survival Survival Analysis
TimeSeries Time Series Analysis

To automatically install views (note this will install ALL packages of that particular view), install the ctv package
install.packages("ctv")
library("ctv")
and then the views can be installed
e.g., *install.views("TimeSeries")* or *update.views("TimeSeries")*

12.2 Reading Material

- *Journals*:
  - Journal of Statistical Software http://www.jstatsoft.org/
  - R Journal (site of official R Journal): http://journal.r-project.org/
- *Blogs*:
  - R Bloggers: list of more than 300 blogs aggregated on one site http://www.r-bloggers.com/
- *Tutorials*:
  - Quick R: http://www.statmethods.net/
  - R for Stats: http://r4stats.com/
- *Graph Tutorials*:
  - Producing Simple Graphs with R: https://www.harding.edu/fmccown/r/
  - R Graph Gallery: list of Graphs: http://addictedtor.free.fr/graphiques/allgraph.php
  - Graphics by examples, UCLA: Academic Technology Services, Statistical Consulting Group: https://www.ats.ucla.edu/stat/R/gbe/default.htm (accessed 10 February 2011)

- *Videos*: There are some very good video tutorials for R as well. You can see a relevant list at http://rmc.ncr.vt.edu/forum/index.php?topic=40.0
- *Books*:
  - R for SAS and SPSS Users: http://www.springer.com/statistics/computanional+statistics/book/978-0-387-09417-5

12.3 Other GUIs Used in R

Here is a list of some other GUIs also available in R. If the business need for the reader or analytics practitioner is not satisfied by R Commander, GrapheR, Deducer/JGR, or Rattle, have a look at these options for using R for business analytics.

12.3.1 Red-R: A Dataflow User Interface for R

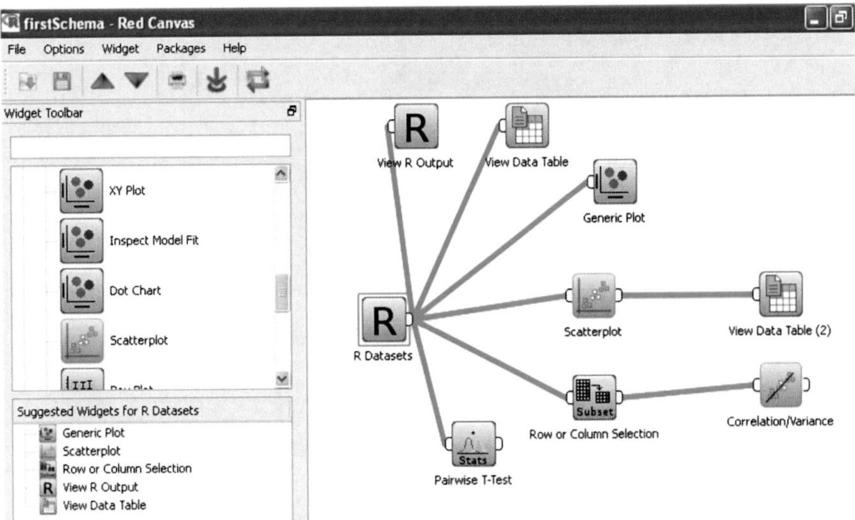

Red-R uses dataflow concepts as a user interface rather than menus and tabs. Thus it is more similar to Enterprise Miner or RapidMiner in design. For repeatable analysis dataflow programming is preferred by some analysts. Red-R is written in Python.

12.3.1.1 Advantages of Red-R

1. Dataflow style makes it very convenient to use. It is the only dataflow GUI for R.

2. You can save the data as well as analysis in the same file.
3. The user interface makes it easy to read generated R code and commit code.
4. For repeatable analysis like reports or creating models it is very useful as you can replace just one widget, and other widget operations remain the same.
5. It is very easy to zoom into data points by double-clicking on graphs as well as to change colors and other options in graphs.
6. One minor feature is that it asks you to set the CRAN location just once and stores it even for the next session.
7. It has automated bug report submission.

12.3.1.2 Disadvantages of Red-R

1. The current version is 1.8 and it requires considerable improvement for building more modeling types as well as debugging errors.
2. It has limited features.
3. It lacks installer packages for MacOS and has limited installation help for Linux.

12.3.2 RKWard

RKWard is primarily a KDE GUI for R, so it can be used on Ubuntu Linux. A Windows version is also available.

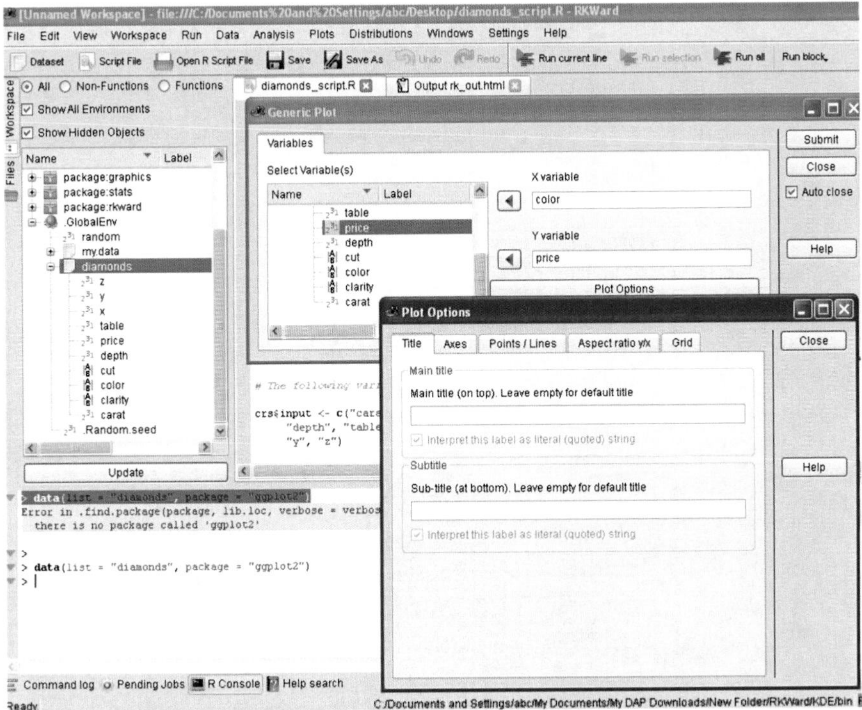

12.3.2.1 Advantages of RKWard

1. It seems like the only R GUI explicitly for Item Response Theory (which includes credit response models, logistic models) and plots contain Pareto charts.
2. It offers a lot of detail in analysis, especially in plots (13 types of plots), analysis, and distribution analysis (eight tests of normality, 14 continuous and six discrete distributions). This detail makes it more suitable for advanced statisticians than business analytics users.
3. Output can be easily copied to Microsoft Office documents.

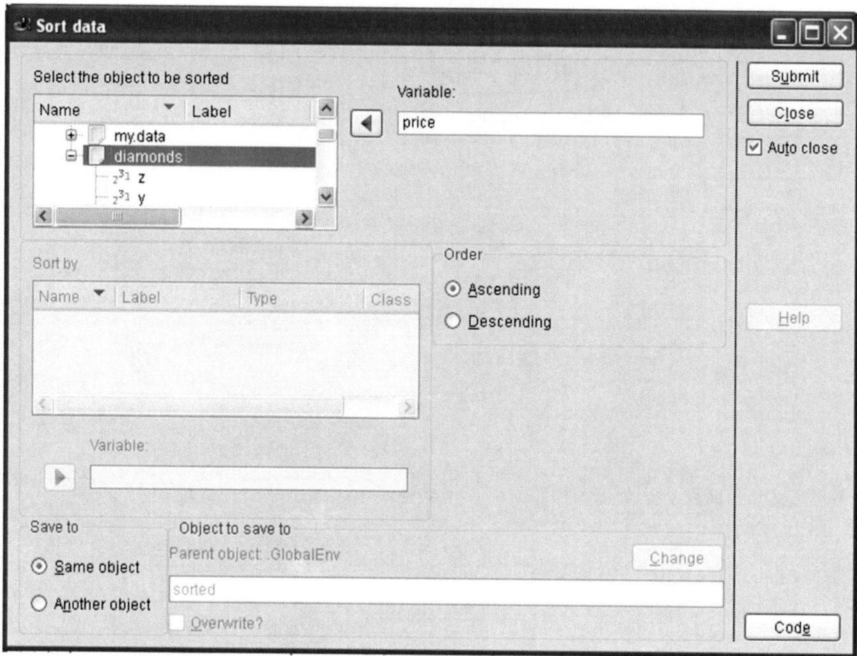
Sort data
Select the object to be sorted
Name
Label
my.data
diamonds
z
y
Variable:
price
Submit
Close
Auto close
Sort by
Name
Label
Type
Class
Variable:
Order
Ascending
Descending
Help
Save to
Same object
Another object
Object to save to
Parent object: .GlobalEnv
Change
sorted
Overwrite?
Code

12.3.2.2 Disadvantages of RKWard

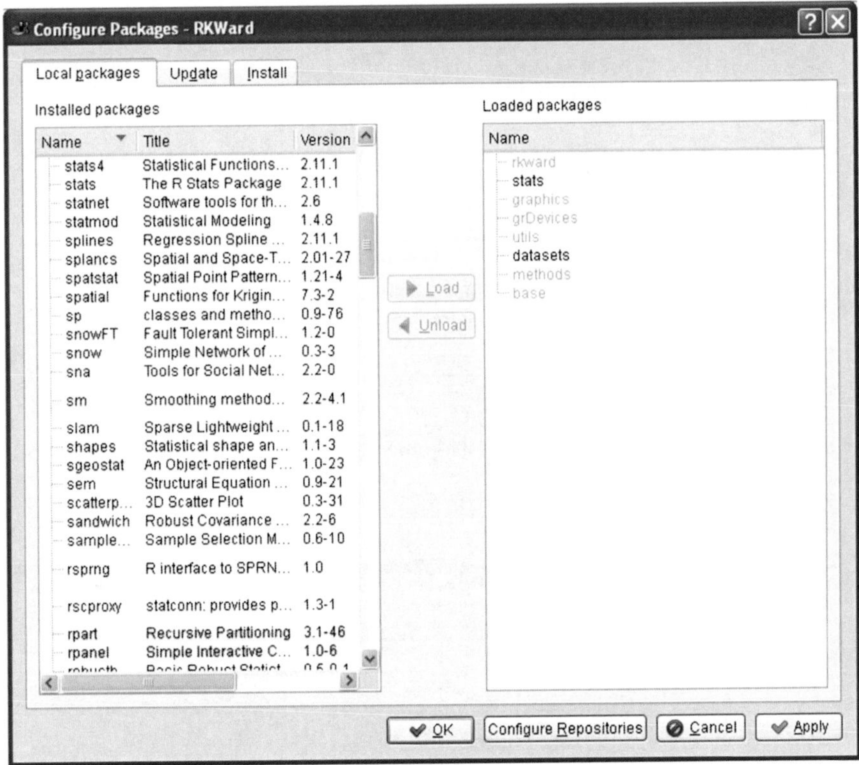

- It has a lot of dependencies and so may have some issues in installing on Windows.
- The design categorization of analysis, plots, and distributions seems a bit unbalanced considering other tabs are File, Edit, View, Workspace, Run, Settings, Windows, and Help. Some of the other tabs can be collapsed, while the three main tabs of Analysis, Plots, and Distributions can be better categorized (especially into modeling and nonmodeling analysis).
- There are few options for data manipulation (like subset or transpose) by the GUI.

Components: Analysis, Plots, and Distributions are the main components, and they are very extensive, covering perhaps the biggest range of plots, analysis, and distribution analysis that can be done. Thus RKWard is best combined with some other GUI when doing advanced statistical analysis.

12.3.3 *Komodo Sciviews-K*

This combines Komodo Edit with Sciviews. There is a bundle of packages in Sciviews-R that need to be downloaded to configure this, apart from downloading Komodo Edit.

12.3.3.1 Advantages of Komodo Sciviews-K

It gives a lot more control to the programmer than the default GUI including syntax auto-completion, color highlighting, and passing line-by-line code.

12.3.3.2 Disadvantages of Komodo Sciviews-K

It is relatively more complex for a beginner than other GUIs and has a more elaborate installation procedure.

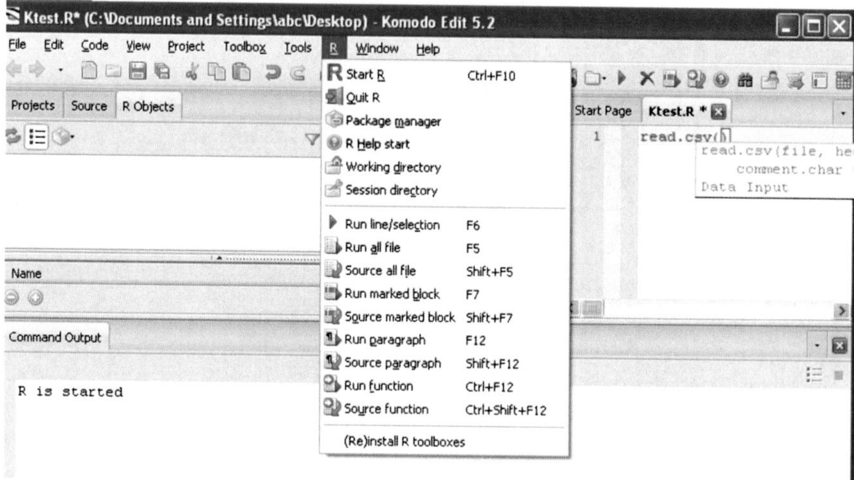

12.3.4 PMG (or Poor Man's GUI)

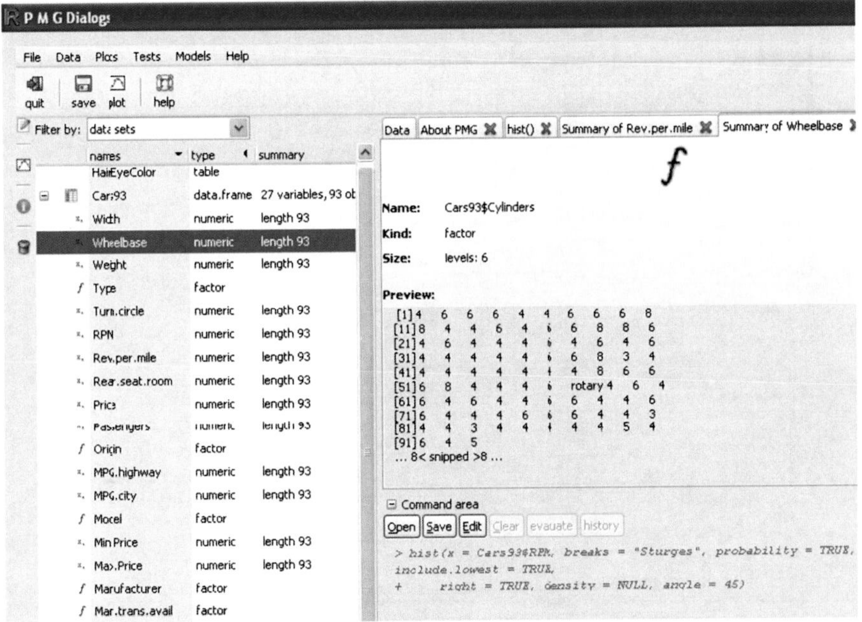

12.3.5 R Analytic Flow

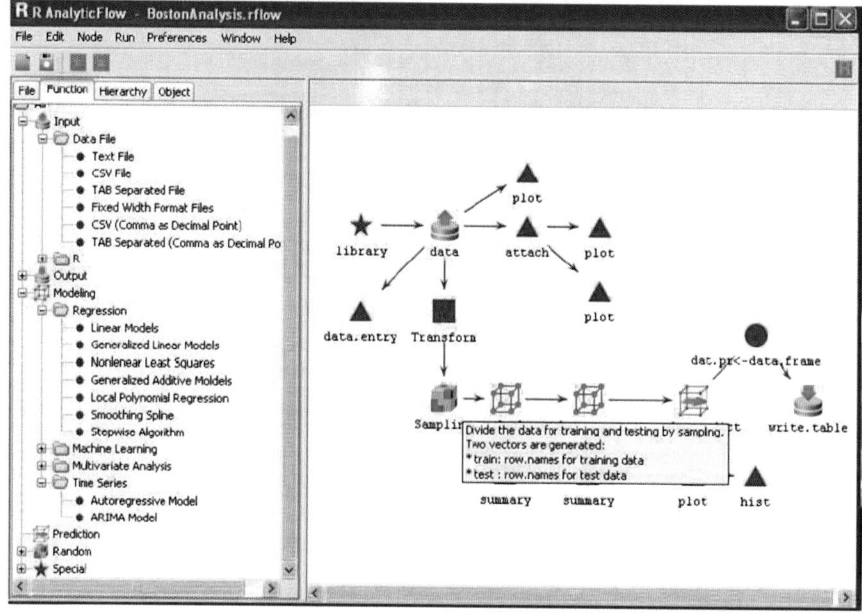

12.4 Summary of Commands Used in This Chapter

12.4.1 Packages

- ctv
- R Analytic Flow
- Red-R
- PMG
- RKWard
- Komodo
- Sciviews-K

12.4.2 Functions

- *install.views*
- *update.views*

Citations and Sources

- Achim Zeileis (2005). CRAN Task Views. R News 5(1), 39-40. URL http://CRAN.R-project.org/doc/Rnews/
- John Verzani with contributions by Yvonnick Noel (2010). pmg: Poor Man's GUI. R package version 0.9-42. http://CRAN.R-project.org/package=pmg
- Anup Parikh with Kyle Covington Red-R *visual programming for R* http://www.red-r.org/contacts
- Thomas Friedrichsmeier, Pierre Ecochard and others RKWard http://sourceforge.net/apps/mediawiki/rkward/index.php?title=About
- Ef-prime, Inc. R Analytic Flow http://www.ef-prime.com/products/ranalyticflow_en/
- Philippe Grosjean http://www.sciviews.org/SciViews-K/

Chapter 13
Appendix

13.1 Web Analytics Using R

Google Analytics is the most widely used Web analytics software on the Internet, and using R we can do advanced analytics or build a custom Web analytics solution with it.

13.1.1 Google Analytics with R

The Google Analytics R package is hosted at http://code.google.com/p/r-google-analytics/.

This project provides access to Google Analytics data natively from the R Statistical Computing programming language. You can use this library to retrieve an R data.frame with Google Analytics data. Then perform advanced statistical analysis, like time series analysis and regressions.

Supported features

1. Access to v2 of the Google Analytics Data Export API Data Feed
2. A QueryBuilder class to simplify creating API queries
3. API response is converted directly into R as a data.frame
4. Library returns the aggregates, and confidence intervals of the metrics, dynamically if they exist
5. Auto-pagination to return more than 10,000 rows of information by combining multiple data requests (upper limit 1M rows)
6. Authorization through the ClientLogin routine
7. Access to all the profile IDs for authorized users
8. Full documentation and unit tests

You can use the following code to read and plot your Google Analytics data, and maybe build a time series forecast for Web site views.

A. Ohri, *R for Business Analytics*, DOI 10.1007/978-1-4614-4343-8_13,

I am writing this code to get Web site views for my blog Decisionstats.com. You can modify the source folder packages and the username and password for your Google Analytics account and run the code to get your Google Analytics data.

```
library(XML)
library(RCurl)
Loading required package: bitops
#Change path name in the following to the folder you downloaded:
# the Google Analytics Package from http://code.google.com/p/r-google-
analytics/
source("C:/Users/R/RGoogleAnalytics/R/RGoogleAnalytics.R")
source("C:/Users/R/RGoogleAnalytics/R/QueryBuilder.R")
# download the file needed for authentication
download.file(url="http://curl.haxx.se/ca/cacert.pem", destfile="cacert.pem")
# set the curl options
curl <- getCurlHandle()
options(RCurlOptions = list(capath = system.file("CurlSSL", "cacert.pem",
package = "RCurl"), ssl.verifypeer = FALSE))
curlSetOpt(.opts = list(proxy = 'proxyserver:port'), curl = curl)
# 1. Create a new Google Analytics API object
ga <- RGoogleAnalytics()
# 2. Authorize the object with your Google Analytics Account Credentials
ga$SetCredentials("USERNAME", "PASSWORD")
# 3. Get the list of different profiles to help build the query
profiles <- ga$GetProfileData()
profiles #Error check to see if we get the right Web site
# 4. Build the Data Export API query
#Modify the start.date and end.date parameters based on data requirements
#Modify the table.id at table.id = paste(profiles$profile[X,3]) to get the Xth
Web site in your profile
# 5. Build the Data Export API query
query <- QueryBuilder()
query$Init(start.date = "2012-01-09",
+end.date = "2012-03-20",
+ dimensions = "ga:date",
+ metrics = "ga:visitors",
+ sort = "ga:date",
+ table.id = paste(profiles$profile[3,3]))
#6. Make a request to get the data from the API
ga.data <- ga$GetReportData(query)
#7. Look at the returned data
str(ga.data)
head(ga.data$data)
```

#8. Plotting the traffic
plot(ga.data$data[,2],type="l")

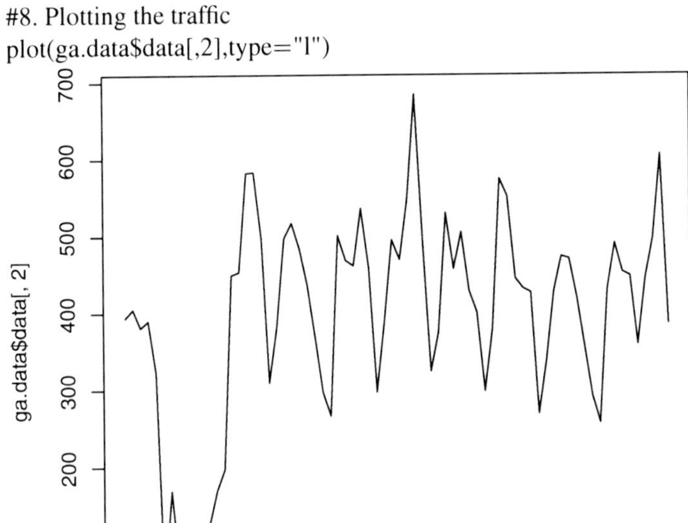

You can get many more dimensions and metrics from Google Analytics. A modified query would be as follows:

```
query <- QueryBuilder() query$Init(start.date = "2011-08-20", end.date =
"2012-08-25",
dimensions = c("ga:date","ga:hour","ga:dayOfWeek"),
metrics = c("ga:visitors","ga:visits","ga:pageviews","ga:timeOnSite"),
sort = c("ga:date","ga:hour","ga:dayOfWeek"),
table.id = paste(profiles$profile[3,3]))
#5. Make a request to get the data from the API
ga.data <- ga$GetReportData(query)
#6. Look at the returned data
str(ga.data)
head(ga.data$data)
```

Note we can even forecast future visits using a time series analysis.

13.2 Social Media Analytics Using R

13.2.1 Using Facebook Data with R

Here is an example of using R for plotting Facebook Network

```
#access token from https://developers.facebook.com/tools/explorer Please generate your own access token
access_token="AAuFgaOcVaUZAss.your own access tokenhO1DcJgSSahd67LgZDZD"
require(RCurl)
require(rjson)
# download the file needed for authentication http://www.brocktibert.com/blog/2012/01/19/358/
download.file(url="http://curl.haxx.se/ca/cacert.pem",
destfile="cacert.pem")
# creating a Facebook function
# http://romainfrancois.blog.free.fr/index.php?post/2012/01/15/Crawling-facebook-with-R
facebook <- function( path = "me", access_token = token, options){ if( !missing(options) ){ options <- sprintf( "?%s", paste( names(options), "=", unlist(options), collapse = "&", sep = "" ) ) } else { options <- "" } data <- getURL( sprintf( "https://graph.facebook.com/%s%s&access_token=%s", path, options, access_token ), cainfo="cacert.pem" ) fromJSON( data ) }
# see http://applyr.blogspot.in/2012/01/mining-facebook-data-most-liked-status.html
# scrape the list of friends
friends <- facebook( path="me/friends" , access_token=access_token)
# extract Facebook IDs
friends.id <- sapply(friends$data, function(x) x$id)
# extract names
friends.name <- sapply(friends$data, function(x) iconv(x$name,"UTF-8","ASCII//TRANSLIT"))
# short names to initials
initials <- function(x) paste(substr(x,1,1), collapse="")
friends.initial <- sapply(strsplit(friends.name," "), initials)
# friendship relation matrix
#N <- length(friends.id) we use N=500 to limit size and time of query
N <- 500
friendship.matrix <- matrix(0,N,N) for (i in 1:N) { tmp <- facebook( path=paste("me/mutualfriends", friends.id[i], sep="/") , access_token=access_token)
mutualfriends <- sapply(tmp$data, function(x) x$id) friendship.matrix[i,friends.id %in% mutualfriends] <- 1 }
require(network)
net1<- as.network(friendship.matrix)
plot(net1, label=friends.initial, arrowhead.cex=0)
```

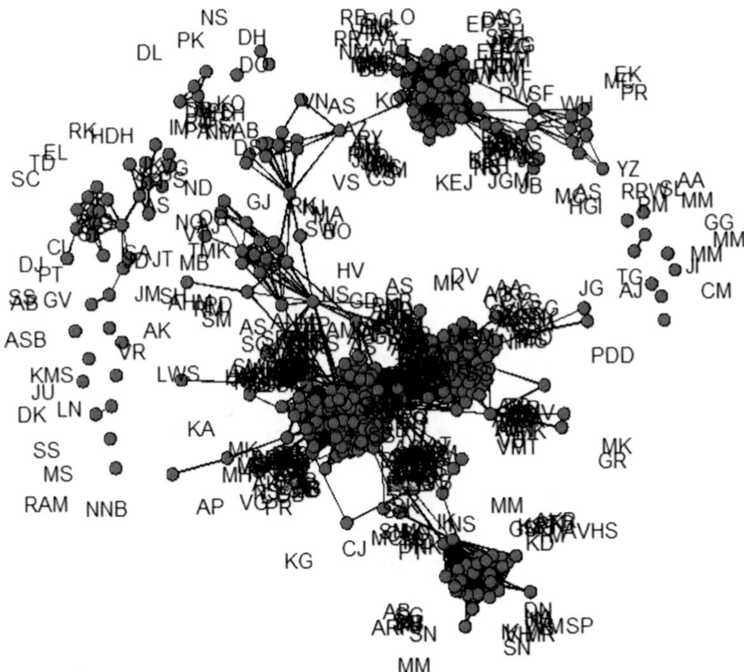

13.2.2 Using Twitter Data with R

TwitteR R Package: http://cran.r-project.org/web/packages/twitteR/vignettes/twitteR.pdf

13.2.2.1 Sentiment Analysis Using Twitter

http://www.slideshare.net/jeffreybreen/r-by-example-mining-twitter-for

```
>#loadthepackage
>library(twitteR)>
#get the 500 most recent tweets mentioning"#rstats":
>rstats.tweets=searchTwitter("#rstats", n=500)
```

13.2.2.2 Graphing Twitter Social Network

```
# Script for graphing Twitter friends/followers
# by Kai Heinrich (kai.heinrich@mailbox.tu-dresden.de)
# load the required packages
library("twitteR")
library("igraph")
```

```
# HINT: In order for the tkplot() function to work on a Mac you need to install
# the TCL/TK build for X11
# (get it here: http://cran.us.r-project.org/bin/macosx/tools/)
# # Get user information with twitteR function getUSer(),
# instead of using your name you can do this with any other username as well
start<-getUser("YOUR_USERNAME")
# Get friends and follower names by first fetching IDs (getFollow-
erIDs(),getFriendIDs()) and then looking up the names (lookupUsers())
friends.object<-lookupUsers(start$getFriendIDs())
follower.object<-lookupUsers(start$getFollowerIDs())
# Retrieve the names of your friends and followers from the friend # and follower
objects. You can limit the number of friends and followers by adjusting the # size
of the selected data with [1:n], where n is the number of followers/friends
# that you want to visualize. If you do not put in the expression the maximum
number of
# friends and/or followers will be visualized.
n<-20
friends <- sapply(friends.object[1:n],name)
followers <- sapply(followers.object[1:n],name)
# Create a data frame that relates friends and followers to you for expression in
the graph
relations <- merge(data.frame(User='YOUR_NAME', Follower=friends),
data.frame(User=followers, Follower='YOUR_NAME'), all=T)
# Create graph from relations
g <- graph.data.frame(relations, directed = T)
# Assign labels to the graph (=people's names)
V(g)$label <- V(g)$name
# Plot the graph using plot() or tkplot(). Remember the HINT at the # beginning
if you are using MAC OS/X
tkplot(g)
```

Source: http://blog.ynada.com/864

Mining Twitter for Sentiment Analysis

```
> library(twitteR)
> library(tm)
#We download 100 tweets only
#My twitter user name is 0_h_r_1 but you can replace this below
> rdmTweets <- userTimeline("0_h_r_1", n=100)
#Data transformation
> df <- do.call("rbind", lapply(rdmTweets, as.data.frame))
#Note we are using Vector Source as the parameter to Corpus here
> b=Corpus(VectorSource(df$text))
#Cleaning up for text mining
> b<- tm_map(b, stripWhitespace) #Strips white space
> b<- tm_map(b, tolower) #Changes case to lowercase
> b <- tm_map(b, removeWords, stopwords("english")) #Removes stop words
```

```
> b <- tm_map(b, removePunctuation) #Removes punctuation
#Creating the frequency of words document or Term Document matrix
> tdm <- TermDocumentMatrix(b)
#Data transformation
> m1 <- as.matrix(tdm)
> v1<- sort(rowSums(m1),decreasing=TRUE)
> d1<- data.frame(word = names(v1),freq=v1)
```

13.3 RFM Analytics Using R

Recency, Frequency, Monetization (RFM) is a technique used for classifying customers, especially in retail. RFM analysis is easy using R.

```
##Creating random sales data of the format CustomerId (unique to each customer), Sales.Date, Purchase.Value
sales=data.frame(sample(1000:1999,replace=T,
size=10000),abs(round(rnorm(10000,28,13))))
names(sales)=c("CustomerId","Sales Value")
sales.dates <- as.Date("2010/1/1") + 700*sort(stats::runif(10000))
#generating random dates
sales=cbind(sales,sales.dates)
str(sales)
sales$recency=round(as.numeric(difftime(Sys.Date(),sales[,3],units="days")) )
library(gregmisc)
##if you have existing sales data, you need to just shape it in this format
rename.vars(sales, from="Sales Value", to="Purchase.Value")#Renaming variable names
## Creating total sales (monetization), frequency, last purchase date for each customer
salesM=aggregate(sales[,2],list(sales$CustomerId),sum)
names(salesM)=c("CustomerId","Monetization")
salesF=aggregate(sales[,2],list(sales$CustomerId),length)
names(salesF)=c("CustomerId","Frequency")
salesR=aggregate(sales[,4],list(sales$CustomerId),min)
names(salesR)=c("CustomerId","Recency")
##Merging R,F,M
test1=merge(salesF,salesR,"CustomerId")
salesRFM=merge(salesM,test1,"CustomerId")
##Creating R,F,M levels
salesRFM$rankR=cut(salesRFM$Recency, 5,labels=F) #rankR 1 is very recent, while rankR 5 is least recent
salesRFM$rankF=cut(salesRFM$Frequency, 5,labels=F)#rankF 1 is least frequent, while rankF 5 is most frequent
```

```
salesRFM$rankM=cut(salesRFM$Monetization, 5,labels=F)#rankM 1 is
lowest sales, while rankM 5 is highest sales
##Looking at RFM table
table(salesRFM[,5:7])
```

13.4 Propensity Models using R

Propensity models in R can be easily built using the lm() function and the Rattle and R Commander GUIs.

13.5 Risk Models in Finance Using R

To use R in finance, one of the most comprehensive R packages is quantmod (available at http://www.quantmod.com/ and ttr package). The quantmod package for R is designed to assist the quantitative trader in the development, testing, and deployment of statistically based trading models. For more advanced modeling (like loss reserve modeling for portfolios, refer to the appropriate package, e.g., http://cran.r-project.org/web/packages/actuar/vignettes/lossdist.pdf.

An excellent case study showing R use within a major bank is at http://files.meetup.com/1685538/R%20ans%20SAS%20in%20Banking.ppt.

13.6 Pharmaceutical Analytics Using R

- A case study on using R in the pharmaceutical industry is available at http://www.r-project.org/conferences/useR-2007/program/presentations/rossini.pdf.
- A software called Decimaker based on R for the pharmaceutical industry is available at http://www.clinbay.com/software/decimaker/.
- Presentation on using R for clinical trials: http://biostat.mc.vanderbilt.edu/Rreport. rreport is an R package that produces statistical reports in LaTeX for clinical trials. It is especially useful for producing interim reports for data monitoring committees (DMCs).
- Using R: Perspectives of a FDA Statistical RevieweR: http://user2007.org/program/presentations/soukup.pdf.
- R: Regulatory Compliance and Validation Issues: A Guidance Document for the Use of R in Regulated Clinical Trial Environments: http://www.r-project.org/doc/R-FDA.pdf.

13.7 Selected Essays on Analytics by the Author

13.7.1 What Is Analytics?

Database mining and analytics are defined as using the power of hidden information locked in databases to reveal consumer and product insights, trends, and patterns for future tactical and operational strategy. A key differentiator between analytics and market research is that analytics relies on data that exist within a database while market research generally involves the collection, collation, and tabulation of the data. Business intelligence is defined as the seamless dissemination of information throughout an organization and is a broader term that involves and includes analytics as well as reporting systems.

The field of data analytics is vast and comprises the following types.

- Reporting or descriptive analytics—Each organization relies on series of management information systems (commonly called MIS) to gather the current state of business as well as any emerging trend. This typically involves sales, finance, and customer and competitor data, which is presented within spreadsheets and presentations. Reports tend to be either regular (like monthly and quarterly) or ad hoc (for special investigative analysis). This is known as descriptive analytics simply because it describes the data that are present. While reporting or descriptive analytics is often the starting point in analytics careers, a proper grounding in this domain is necessary both to build an eye for detail in dealing with large amounts of data and to polish presentation skills for presenting insights from the data.
- Modeling or predictive analytics—Predictive analytics refers to the art and science of using statistical tests, hypotheses, and methods to build up predictive recommendations. These recommendations can range from which type of customer to call by phone for a credit card or insurance, to which type of mobile plan to offer a cell-phone customer with a short message (sms), to what kinds of customer are likely to default on the loans they have taken. Predictive analytics includes techniques like segmentation and regression modeling. It is generally considered to be of high value, and a background in statistics helps in preparing for predictive analytics careers.
- Data-driven strategy—This is also called test–control or champion–challenger testing. This is done by segmenting the data population into test (on which a new strategy, called the challenger strategy, is to be tested) and control (which uses an existing strategy, called the champion strategy). Building association rules that describe which parts of product or customer data are clustered or correlated together is also part of analytics.

13.7.2 *What Are the Basic Business Domains Within Analytics?*

Data-driven analytics by definition thrives in industries that have large amounts of data and high-volume transactions that require systematic and scientific analytics to cut costs and grow sales. The following domains offer employment opportunities to both newcomers and experienced analytics professionals. These can be in domestic firms, captive outsourcing firms, or third-party business process outsourcing companies.

- **Retail sales analytics** deals with the handling of vast amounts of point-of-sales data, inventory data, payment data, and promotional data to help increase sales in retail stores, especially in organized retail. The use of RFIDs, electronic payment, and bar scanning helps capture data better and store them in vast databases. An example of this is the famous Thursday baby diaper–beer sales phenomenon. A big retailer found that on Thursday evening sales of beer and baby diapers were highly correlated. It then found that is was due to young couples preparing for the weekend by buying supplies of diapers and beer. Thus placing diapers and beer closer together boosted sales. This is an example of market basket analytics in which a large amount of data is scrutinized to see which products sell well together. Wal-Mart, the American retail giant, established a competitive edge over its rivals by proactively using data-driven analytics to cut costs and thus offer goods cheaper than others. Another example of a big retailer is Target, which has its own captive backend analytics in India. In India, since Reliance Retail, Future Group, and Walmart-Bharti have set up shop, this is a sector that is bound to grow even within the domestic sector, as these high-volume retailers need data-driven decisions to squeeze the most from their retail stores.
- **Financial services analytics** is widely used. This is because financial services is a very competitive field, and firms have millions of customers and many transactions. It is extremely important for them to store data for billing purposes and to recover the money they lend out as well the deposits they collect. An incremental gain of a few basis points (one-hundredth of 1 % is a basis point) in profitability can lead to millions of dollars in aggregate profits. Within financial services analytics the broad subcategories are risk and credit analytics; Risk and credit functions measure the ability of a customer to pay back loans or debt owed by them. Delinquent customers are those that have fallen behind in paying back debt as per an agreed schedule. Debt can be paid back in fixed installments, like EMIs for a personal loan, or it can be revolving as in variable amounts that can be paid against outstanding credit card balances including a minimum payment. Debt can also be secured, such as against a house, consumer durables, or motor vehicles as collateral, or it can be unsecured as in personal loans or credit card debt that has no collateral or backing. A risk analyst develops scorecards that help measure the credit worthiness of both new and existing customers. As financial service instruments are priced against risk, riskier customers are charged more in

terms of interest rate. But this must be balanced by the customer's total repayment ability, which is reflected by sources of income and current leverage. In addition, the income of customers, especially in India, is changing rapidly, and there is also undeclared income as black money. The ability to conduct an analysis on millions of customers is what makes risk and credit analytics one of the hottest sectors to be in as credit analysts are in demand with all banks and outsourcing and finance companies. ICICI has a big analytics unit (called Business Intelligence Unit), and Citigroup has both domestic analytics (in Chennai) and international analytics centers (in Bangalore).

- **Marketing analytics** helps in customer acquisition and retention. It does so by identifying more responsive customers and selling through a wide variety of channels like call centers, direct mail, sms through mobile phones, and e-mail. It is marketing analytics that helps to bring in new customers by giving inputs to the marketing team and feedback to sales and distribution channels.
- **Collections analytics** focuses on recovery from delinquent customers using optimized efforts like phone calls, direct mail, e-mail, or personal visits. Its aim is to maximize recovery at minimum cost.
- **Fraud analytics** seeks to build in triggers or automated alarms if there is any unusual trend or behavior in spending by a customer, especially involving credit cards.
- **Pricing analytics** tries to determine the most optimized price, adequately compensating for risk as well as the competition. Pricing analytics is a vast field and is also a part of financial services analytics, especially in products like insurance.
- **Telecommunication** or **telecom analytics** encompasses the fields of marketing analytics defined above, but an important part of it is also attrition modeling or churn analytics. It also analyzes the wide variety of pricing schemes and options and customers' response to them. It includes delinquency analytics as well.
- **Pharmaceutical or clinical analytics** relates to clinical trials, which depend on the testing and control of thousands of patients on new drugs. Clinical trial analytics focuses on large numbers of variables that may or may not affect drug response.
- **Supply chain analytics** comprises inventory optimization, tracking turnaround time, multiple reports, and minimizing distribution costs.
- **Transportation analytics**, while covered more extensively in the field of operations research, seeks to minimize route length, fuel costs, or pricing of fares.
- **Online or Web site analytics** focuses on analyzing traffic to a Web site from given sources and keeping shoppers on the Web site for a longer time to purchase more goods. It also involves a bit of search engine optimization to make sure the Web site is relevant in searches by search engines. This branch of analytics now has a subbranch, social media analytics, that deals with numbers regarding interactions of consumers of social media.

13.8 Reasons a Business Analyst Should Learn R

Having worked with SAS and SPSS language products for over 10 years and the R language for over 4 years now, I now believe that learning at least two platforms is essential to staying relevant in the field of business analytics. You can call it risk minimization by portfolio diversification of technical skills. These are some of the principal reasons why any business analyst, young or experienced, should learn R.

- **Great Job Security with R:** R is open source and free. It saves on software as a capital cost, thus freeing up financial resources for hiring more and more analysts with advanced degrees. With the paradigm of cloud computing, hardware costs are minimized as well, thus reducing the overall cost of statistical computing and making it much more affordable and accessible even to companies that are not used to investing thousands of dollars in annual software licenses or companies that have analytical needs but a small software budget. R brings back the focus of analysis to analysts rather than software. Having R language skills can also make analysts heroes within their analytical department since they can help trim the annual budget of their department. With an uncertain economic environment and widespread job layoffs, it makes sense to learn a language that is free to use; if your organization faces a crunch, it can simply switch to R rather than or laying off trained and experienced analysts.
  - With open source R you can build in customized applications, especially if you wish to build proprietary algorithms or software. This is particularly true for use in banking and financial services.
- **Better Business Acceptability with R**
  - Businesses are moving rapidly toward R. According to Rexer Analytics, the foremost surveyor of data miners, R became the dominant platform in 2010: http://www.rexeranalytics.com/Data-Miner-Survey-Results-2010.html.
  - Companies as diverse as Google, Pfizer, Merck, Bank of America, the InterContinental Hotels Group, Shell, Microsoft (Bing), Facebook, Llyod's Insurance, ANZ Bank, New York Times, and Thomas Cook use the R language for their analytical tasks.
- **Software Vendors Support or Plan to Support R**
  - There is broad consensus on using the R platform within the fields of statistical computing and analytics. The range of companies that support R include the SAS Institute, which enables working with R through SAS/IML and JMP software; Oracle, which is building its own version of R Enterprise; Microsoft, which has both invested in Revolution Computing (Analytics) and is building solutions for high-performance computing with R (); IBM, which uses R though its acquired companies (Netezza and SPSS); and SAP, which has proposed integration of HANA with R, Teradata's support for R.

- **No longer so difficult to learn:** R was considered difficult to learn. Now GUIs within R have evolved and made using R much easier. These include packages like Rattle (which is specialized for data mining), Deducer (specialized for data visualization), R Commander (which allows extensions of other statistical packages through e-plugins), and interfaces like RStudio and RExcel that allow very easy usage and adaptability even to newer users and learners in R.
- **Graphs are better in R:** R is a good platform to learn data visualization and data exploration through the creation of graphs and diagrams. This is because R's graphical support is better than any class of analytics software and includes interactive, 3D, and a wide array of publication-ready templates for customizing graphical output. Since analytical results are mostly presented graphically, the use of R can help elucidate a statistical solution, especially to a business audience.
- **R has a fast rising pool of students and future analysts:** Object-oriented programming and statistical thinking are becoming more widespread; hence R has become the de facto language to learn on college campuses and in statistics departments. It thus easily represents the platform with the greatest potential availability of analysts in the future.
- **R can handle big datasets now:** Thanks to advances made with packages like Rcpp, big data packages in open source R, and the RevoScaleR package by Revolution Analytics, working with big datasets is just as easy for a trained analyst as it is for any other analytical platform.

13.9 Careers in Analytics

When I was starting my career, I thought it would just be a matter of time before I learned some analytics software, picked up some domain expertise, and mastered the way to a growing career in analytics. After 10 years along this path, I am afraid that what I do not know about analytics still exceeds what I do know about analytics. My goal of preventing my ignorance from manifesting itself too much and still contributing some value to clients, partners, and myself has been pursued broadly along the lines of the following ten activities:

1. **Learn continuously**. Boost existing skills. Analytics is a continuously changing field. Even analytics software and platforms go through annual upgrades. Thanks to the Internet, the cost of learning continuously is very low, and analytics professionals simply need to remain aware of what is going on in their field.
2. **Network tirelessly**. It is not what you know but who you know that gets you fresh projects. What you know can only get you repeat projects or burnish your credibility. The field of analytics is relatively small in terms of the number of professionals involved in it, and your actions and manners will garner you either positive or negative attention. Again thanks to the Internet, networking is much easier today than it was a decade ago, and you can seamlessly network with

professionals in your field and current, past, and future employers and clientele over social networks like LinkedIn or Twitter.

3. **Market yourself**. No one sees the Invisible Man. To market yourself, realize that marketing is in no way less honorable than learning new coding skills. In fact, the more you learn, the more you should share your learning, and marketing is just another way to share and spread learning and knowledge.
4. **Code relentlessly**. Learning how to code is a skill that should be acquired as early as possible in your analytics career. After that, the more you code, the better you become at it. There is no better way to learn coding than to execute it on projects. The trickier the data quality and the tougher the deadline, the more you will need coding skills that are both robust and creative.
5. **When stuck, search for better options**. Careers, like tides, ebb and flow. It is better to cut your losses than be stuck in dead-end areas where you lack passion. Similarly, when writing code, find alternative ways to do the same thing. Working smarter, and not just harder, is the way to be flexible and agile in your analytics career.
6. **Hedge your risk against technological obsolescense**. Learn cross-platform skills and be paranoid about technology shifts. Computing paradigms shift every 15 years, and analytics paradigms shift even faster. Learn more skills than just a single platform, and broaden your career growth beyond the boom or bust in demand for a single technical skill. Similarly, learn at least two business domains, have a foot in both open source and enterprise software, and learn something about Web analytics, data mining, data visualization, and text mining as secondary and tertiary skill sets.
7. **Soft skills matter**. Put yourself through periodic soft skill trainings and tutorials. Nice people make lasting impressions. Leadership is a science to be learned, not an art to be inherited from divine providers. My personal motto is this: I can get more business over lunch with people than in a series of dull meetings. Sometimes the best soft skill is to be an empathetic listener with impeccably groomed manners.
8. **Certify yourself**. If you think you are an expert, take the time to obtain certification. If you have a certification, take the time to build a stack of certifications. The more certifications you have, the more statistical credibility you can lend to your skills. Beware of expensive and shallow certifications, though, as they can be a drain on your resources. It is best to search for enhancements in employability (based on frequency of keywords on job sites) as a benchmark for determining the right certification in which to invest your time, money, and effort.
9. **Read regularly**. Analytics is a learning domain. The more blogs, papers, Web sites, and tutorials you read, the lesser are your chances of being blindsided with new developments. I personally find half an hour of daily reading is the best mental exercise or workout and provides the optimum balance of effort versus time cost.
10. **Love your work**. It is very difficult for someone to be truly world class in something they do not really love with a passion. Sure, you can love the big

salary, perks, and acclaim that often accompany a successful analytics career. But to truly be the best in any professional environment, you must be passionate about learning, problem solving, cleaning data, writing code, building models, and delivering business insights. Passion for your work often provides the extra push to your analytics career when times are slow and can make the difference between an above-average and a truly superb analytics career.

These are some insights I have gained from my experience. What else could help build your career in analytics?

13.10 Summary of Commands Used in This Chapter

13.10.1 Packages

- *gregmisc*
- *tm*
- *twitteR*

13.10.2 Functions

- *cut*
- *aggregate*
- *rename.vars*

Citations and Sources

Ingo Feinerer (2012). tm: Text Mining Package. R package version 0.5–7.1.

- Ingo Feinerer, Kurt Hornik, and David Meyer (2008). Text Mining Infrastructure in R. Journal of Statistical Software 25/5. URL: http://www.jstatsoft.org/v25/i05/.
- Gregory R. Warnes. (2011). gregmisc: Greg's Miscellaneous Functions. R package version 2.1.2. http://CRAN.R-project.org/package=gregmisc
- Jeff Gentry (2012). twitteR: R based Twitter client. R package version 0.99.19. http://CRAN.R-project.org/package=twitteR
- Google Analytics R Package Google Inc http://code.google.com/p/r-google-analytics/

Index

A
Allaire, J.J., 273
Amazon, 12, 41–46, 53, 189–190, 279
Analytics, 1, 9, 27, 57, 104, 171, 197, 225, 241, 259, 263, 281, 291
API. *See* Application programming interface (API)
Application programming interface (API), 12, 39, 50, 51, 61, 78, 220–222, 293–295
ARIMA. *See* Auto regressive integrated moving average (ARIMA)
Association, 4, 49, 57, 77, 101, 106, 123, 189, 199, 207–209, 229, 235, 273, 301
Auto regressive integrated moving average (ARIMA), 53, 241, 242, 245–248, 253, 255, 256

B
Bleicher, P., 148
Business, 3, 10, 34, 35, 47, 52, 58–59, 62, 103, 197, 213, 222, 229, 273, 301, 306
Business analytics, 3–4, 9–13, 21, 27, 30, 34, 36, 57, 58–59, 64, 87, 90, 104, 151, 160, 171, 172, 188, 193, 201, 225, 226, 241, 263, 268, 277, 281, 283, 285, 304–305

C
Categorical, 64, 65, 119, 122–124, 128, 153, 203, 207
Cloud, 12, 41–43, 47, 53, 129–131, 166, 190, 220
Cloudnumbers.com, 47, 53–54
Clustering, 1, 12, 41, 53, 54, 100, 160, 189, 193, 198–201, 206, 207, 225–239, 274, 281, 301
Code, 2, 4, 6, 10, 13, 18, 19, 28–30, 41, 47–53, 59, 73, 75, 90–92, 94, 104, 106, 124, 127, 132, 139, 152, 156, 165, 168, 199–202, 208, 210, 212–214, 221, 259, 263–280, 284, 293, 294, 306
Colors, 30, 104–107, 117, 121–125, 143–145, 148, 157, 161–164, 166, 209, 217, 220, 254, 267, 271, 272, 274, 284, 288
Computing, 1, 3–5, 7, 10–13, 37–38, 41–43, 48–50, 52, 53, 65, 90–92, 95–97, 101, 165, 168, 262, 274, 280, 281, 283, 293, 304, 306
Concerto, 38, 40, 41, 54
Consulting, 21, 89, 90, 142–144, 190, 213, 282
CRAN, 13–15, 25, 28, 53, 90, 221, 265, 281–282, 284
Croker, S., 4

D
Data, 1, 10, 25, 57, 103, 173, 193, 225, 241, 259, 263, 281, 293
  exploration, 64, 168, 174, 205, 305
  frame, 20, 21, 26, 50–52, 57, 62–67, 70, 76–78, 86, 89–91, 93, 95, 97, 98, 100, 106, 112, 127, 131, 137–139, 142, 201, 214, 216, 217, 219, 263–265, 277, 293, 298, 299
  manipulation, 1, 30, 57–101, 161, 287
  mining, 2, 4, 27, 30, 34, 35, 42, 93, 160, 189, 193–223, 226, 239, 240, 266, 305, 306
  visualization, 1–4, 27, 34, 50, 59, 61, 64, 103, 104, 136, 155–161, 164–165, 168, 227, 305, 306

A. Ohri, *R for Business Analytics*, DOI 10.1007/978-1-4614-4343-8,

Databases, 2, 5, 7, 12, 19, 35, 38, 41, 49, 52, 55, 57, 77–89, 91–98, 100, 146, 147, 193, 194, 202, 301, 302
Datasets, 10–12, 21, 30, 36, 42, 48, 52, 54, 57–68, 70, 73, 74, 76, 77, 86, 90, 108, 110, 114, 119, 127, 156, 160, 161, 168, 173, 174, 200–203, 208–211, 213, 218, 226–227, 230, 231, 247, 255, 256, 263–265, 273, 279, 305
Date, 14, 64–67, 148, 247, 248, 257, 258, 266, 294, 295, 299
Deducer, 4, 12, 27, 34, 42, 45, 49–50, 54, 69, 70, 104, 147, 151, 156–161, 166, 283, 305
Dixon, J., 87

E
EC2. *See* Elastic compute cloud (EC2)
Elastic compute cloud (EC2), 12, 41–46, 53, 189–190, 279
Enterprise, 4, 9–11, 35, 42, 45, 86, 90–95, 165, 190, 304
Enterprise software, 42, 87, 306
Excel, 1, 4, 7, 9, 13, 30, 35–37, 60–61, 151, 305

F
Facebook, 4, 92, 104, 295–297, 304
Few, S., 136
Forecast, 3, 241, 242, 245–247, 254, 256, 257, 293, 295
Forecasting, 27, 53, 241–257
Fox, J., 21, 28
Frequency, 12, 48, 64, 104, 137, 161, 208–210, 214, 215, 217, 299, 306
Fun, 20, 23, 65, 96
Function, 50, 53, 59, 61, 62, 65–68, 70, 72, 74, 95–101, 105, 111–113, 116, 117, 127, 131, 146, 156, 189, 209, 211, 216–218, 227, 241, 242, 244, 246, 247, 249, 250, 252, 259, 260, 264–266, 273–275, 277–279, 296, 298, 300
Functionality, 6, 36, 87, 91–94, 100, 160, 211, 255, 266

G
Ggplot, 4, 36, 104, 118, 136, 137, 142, 143, 151, 156, 160, 167, 277
Ggplot2, 49, 52, 103, 118, 136, 142, 151–152, 164–167, 208, 277
Google, 4, 12, 47–53, 61, 78, 92, 104, 156, 220–222, 304
Google Analytics, 50, 293–295
Graphical user interfaces (GUIs), 1, 4, 10, 12–13, 19, 21–22, 25, 27–34, 41, 42, 44, 45, 49, 53, 54, 57, 59, 64, 69, 70, 73, 76–77, 93, 104, 119–121, 127, 147, 149, 151–161, 164, 165, 173, 182, 199–213, 222, 226–230, 236, 241, 248–256, 260, 261, 266, 267, 272, 274, 278, 283–290, 300, 305
Graphs, 29, 30, 34, 39, 98, 99, 104–152, 154–155, 158, 161, 173, 174, 177, 184, 185, 197, 200, 201, 206, 259–260, 283, 284, 305
Graphs, Analytics, 29, 34, 42, 104, 151, 164–165, 174, 197, 200, 201, 259, 281, 305
Grossman, R., 41
GTK, 19, 28, 120, 121, 152, 166, 167, 201, 236
GUIs. *See* Graphical user interfaces (GUIs)

H
Hadoop, 5, 78, 88, 90, 92, 93, 100, 101
Hadoop Distributed File System (HDFS), 92, 100, 101
Hardware, 11–13, 41, 42, 90, 93, 100, 279, 304
Harell, F.E. Jr., 189
HDFS. *See* Hadoop Distributed File System (HDFS)
Heavlin, W., 50
Hierarchical, 160, 206, 225, 226, 229, 230, 234, 235
Holland, P., 4
Hornick, M., 91

I
Interactive, 6, 28, 36, 37, 50, 52, 129, 152–154, 161, 205, 227, 305
Interview, 5–7, 21–23, 40–41, 51–54, 87, 91, 164–165, 189, 212–213, 273–274

J
Java, 2, 38, 39, 86, 100, 159, 273
Java GUI for R (JGR), 12, 27, 34, 54, 77, 156, 157, 159, 160, 166, 283
JGR. *See* Java GUI for R (JGR)
JMP, 1, 4–7, 13, 35, 304
Journals, 2, 4, 226, 281, 282

K

Kinds, 9, 11, 22, 30, 41, 60, 116, 161, 230, 234, 281, 301
Kosinski, M., 40

L

Lapsley, M., 78
Layers, 43, 50, 91–100, 136, 151, 152, 165
Linux, 4, 10, 11, 14, 18, 19, 23, 34, 36, 42–46, 59, 93, 123, 213, 220, 236, 274, 284
Literature, 2, 193, 225, 281–290
Loading, 23, 25, 30, 31, 41, 49, 58, 61, 70, 75, 85, 94, 146, 156, 159, 174, 201, 208, 248, 253, 265, 294, 297
Logistic, 161, 171, 172, 188, 211, 287
Longworth, N., 194

M

Manipulation, 1, 6, 30, 57–101, 161, 287
Market, 87, 104, 190, 225, 301, 302, 306
Marketing, 3, 10, 58, 87, 104, 172, 173, 190, 225, 301–303, 306
Matrix, 6, 21, 30, 63, 110, 124, 125, 136, 138, 139, 172, 174, 177, 179, 211, 214, 216, 219, 223, 259, 265, 279, 296, 299
Mean, 62, 63, 74, 119
Miclaus, K., 5
Microsoft, 1, 4, 10, 13, 35–37, 45, 60, 285, 304
Miner, 1, 4, 32, 34, 35, 201, 210, 212, 213, 222, 283, 304
Mining, 2–4, 27, 30, 34, 35, 42, 90, 93, 160, 189, 193–223, 226, 266, 298, 301, 305, 306
Models, 4, 9, 27, 59, 161, 171, 193, 225, 242, 260, 263, 281, 300
MySQL, 1, 5, 37, 38, 41, 78–85

N

Networks, 4, 13, 41, 52, 173, 189, 295–299, 305–306
Numeric, 20, 49, 64, 67, 71, 74, 106, 110, 119, 124, 177, 183, 201, 203, 209, 299

O

ODBC. *See* Open database connectivity (ODBC)
Open database connectivity (ODBC), 1, 19, 57, 78, 79, 84, 85, 93, 202
Oracle, 1, 5, 34, 35, 78, 91–101, 304

P

Packages, 1, 10, 25, 57, 103, 171, 200, 225, 241, 259, 263, 281, 293
Parallel, 12, 45, 48, 90–94, 97, 201, 227–279, 281
Pentaho, 1, 78, 87, 89
Playwith, 152, 153, 166, 227–229, 239
PMML. *See* Predictive modeling markup language (PMML)
Prediction, 3, 7, 12, 35, 51, 61, 78, 87, 186, 188, 189, 193, 220–222
Predictive modeling markup language (PMML), 189–190, 200
Python, 2, 7, 52, 53, 283

R

R, 1, 9, 25, 57, 104, 173, 199, 225, 242, 259, 263, 281, 293
Rather, R.J., 260
Rattle, 4, 12, 19, 27, 34, 42, 45, 54, 64, 76, 77, 153, 157, 160, 166, 173, 199–213, 222, 223, 226, 229–239, 265, 266, 283, 300, 305
R Commander, 12, 18, 21–23, 27–34, 36, 44, 45, 54, 64, 69, 70, 73, 76, 77, 127, 149, 157, 160, 166, 177, 182, 186, 190, 199, 200, 241, 248–256, 265, 283, 300, 305
Regression, 4, 50, 53, 112, 113, 160, 171–191, 193, 198, 203, 211, 226, 246, 293, 301
Reporting, 4–5, 36, 52, 59, 86, 87, 263, 273, 277, 284, 301, 303
Revolution, 10, 42, 45, 86, 94, 274–275
Revolution Analytics, 2, 4, 9–12, 39, 50, 86, 87, 90, 93, 94, 148, 226–227, 279, 305
RevoScaleR, 4, 10, 42, 90, 94, 226–227, 239, 305
Ripley, B., 78
Rstats, 47, 51, 213, 297
RStudio, 46, 53, 267–268, 273–274, 305
Rules, 4, 47, 183, 189, 193, 207–210, 213, 301

S

SAP. *See* System application program (SAP)
SAS, 1, 4–7, 13, 21, 34, 35, 52, 91, 212, 213, 283, 304
Sashikanth, V., 91
SAS Institute, 5, 34, 304
Scatterplots, 29, 30, 74, 108–111, 115, 116, 127, 174
Schmidberger, M., 53
Sentiment analysis, 297, 298
Shaddick, G., 241

Smith, D., 148
Social media analytics, 295–299, 303
Software, 1–2, 4, 6, 9–14, 19, 22, 23, 25, 29, 30, 34–37, 39, 42, 49, 51, 52, 57, 62, 63, 78, 86, 87, 89–90, 92, 93, 100, 106, 147, 164–165, 171, 173, 193, 205, 213, 226, 229, 265–268, 282, 293, 300, 304–307
Spatial, 70, 145–147, 160, 282
SPSS, 1, 4, 5, 21, 22, 30, 34, 35, 41, 52, 55, 91, 156, 159, 213, 283, 304
SQL. *See* Structured Query Language (SQL)
SQLite, 85–86, 88
Statistical, 2, 3, 7, 30, 36, 40, 41, 49, 52, 58, 87, 91, 188, 193, 197, 199, 229, 259, 300, 301, 305, 306
  analysis, 3, 42, 50, 193, 287
  computing, 1, 3, 5, 10, 48, 50, 52, 91, 274, 293, 304
Structured Query Language (SQL), 44, 51, 78, 85, 86, 91–93, 95–98
Support, 4–7, 10, 11, 32, 39, 49, 52, 55, 90, 92–94, 96, 105, 119, 152, 165, 171, 189, 209, 210, 212, 213, 225–226, 234, 266–268, 270, 273, 274, 278, 293, 304–305
System application program (SAP), 35, 304
System.time, 62, 63, 66, 67, 239, 275, 277–280

T

Tables, 30, 51, 59, 60, 64, 65, 73, 74, 76, 78, 83, 85, 86, 88, 93, 95, 97–100, 114, 118–123, 131, 148, 161, 166, 167, 202, 204, 209, 218–220, 259, 264, 276, 294, 295, 300
Time series (TS), 4, 5, 20, 27, 32, 34, 50, 53, 62, 76, 131, 148, 241–257, 281, 282, 293, 295
TS. *See* Time series (TS)
Tutorials, 20–21, 38, 41–47, 61, 63, 64, 78, 84, 90, 160, 165, 226, 241, 281–283, 306
Twitter, 297–299, 305–306

U

Ubuntu, 10, 11, 14, 42–44, 123, 236, 284
University, 28, 38–41

V

Variables, 6, 21, 26, 29, 30, 47, 57–60, 64–68, 70–71, 73–75, 90, 97, 103, 104, 106–110, 112–114, 119, 121, 127, 128, 142, 143, 145, 151, 152, 156, 157, 159–161, 172, 174, 177, 178, 181, 183, 186, 201–204, 209, 226–227, 229, 241, 246, 247, 263, 299, 302, 303
Vendors, 1, 2, 4, 9, 12, 86, 87, 92, 94, 190, 212, 213, 274, 304

W

Web, 1, 4, 13, 14, 19, 25, 30, 37–41, 44, 46, 49–54, 60–62, 70, 78, 86, 87, 89, 90, 94, 99, 106, 116, 120, 123, 145, 150–152, 154, 160, 164, 171, 173, 190, 215–220, 223, 226, 246, 266, 268, 276, 281, 293, 294, 297, 300, 303, 306
Web analytics, 173, 293–295, 306
Wickham, H., 164
Williams, G., 4, 212, 226